Propositional and Predicate Calculus: A Model of Argument

Derek Goldrei

Propositional and Predicate Calculus

A Model of Argument

 Springer

Derek Goldrei, MA, MSc
Open University
Milton Keynes
UK

Mathematics Subject Classification (2000): 03B05, 03B10, 03C07

British Library Cataloguing in Publication Data
Goldrei, Derek
    Propositional and predicate calculus: a model of argument
    1. Propositional calculus 2. Predicate calculus
    I. Title
    511.3
ISBN 1852339217

Library of Congress Cataloging-in-Publication Data
Goldrei, Derek.
    Propositional and predicate calculus: a model of argument/Derek Goldrei.
       p. cm.
    Includes bibliographical references and index.
    ISBN 1-85233-921-7 (acid-free paper)
    1. Propositional calculus. 2. Predicate calculus. I. Title.
QA9.3.G65 2005
511.3--dc22   2005040219

ISBN-10: 1-85233-921-7
ISBN-13: 978-1-85233-921-0
Springer Science+Business Media
springeronline.com

Typesetting: Camera-ready by author
Printed in the United States of America
12/3830-543210 Printed on acid-free paper SPIN 11345978

PREFACE

How to Use This Book

This book is intended to be used by you for independent study, with no other reading or lectures etc., much along the lines of standard Open University materials. There are plenty of exercises within the text which we would recommend you to attempt at that stage of your work. Almost all are intended to be reasonably straightforward on the basis of what's come before and many are accompanied by solutions – it's worth reading these solutions as they often contain further teaching, but do try the exercises first without peeking, to help you to engage with the material. Those exercises without solutions might well be very suitable for any tutor to whom you have access to use as the basis for any continuous assessment of this material, to help you check that you are making reasonable progress. But beware! Some of the exercises pose questions for which there is not always a clear-cut answer: these are intended to provoke debate! In addition there are further exercises located at the end of most sections. These vary from further routine practice to rather hard problems: it's well worth reading through these exercises, even if you don't attempt them, as they often give an idea of some important ideas or results not in the earlier text. Again your tutor, if you have one, can guide you through these.

The book is also peppered with notes in the margins, like this! They consist of comments meant to be on the fringe of the main text, rather than the core of the teaching, for instance reminders about ideas from earlier in the book or particularly subjective opinions of the author.

If you would like any further reading in logic textbooks, there are plenty of good books available which use essentially the same system, for instance those by Enderton [12], Hamilton [18], Mendelson [25] and Cori and Lascar [7].

Acknowledgments

I would like to thank all those who have in some way helped me to write this book. My enthusiasm for the subject was fuelled by Robin Gandy, Paul Bacsich, Jane Bridge, Angus Macintyre and Harold Simmons, when I studied at the Universities of Oxford and Aberdeen. Anything worthwhile I have successfully learnt about teaching stems from my colleagues at the Open University and the network of mathematicians throughout the UK who support the Open University by working as Associate Lecturers, external assessors and examiners. They have taught me so much. It has been particularly stimulating writing this book alongside producing the Open University's course on Mathematical Logic (with a very different angle on the subject) with Alan Pears, Alan Slomson, Alex Wilkie, Mary Jones, Roger Lowry, Jeff Paris and Frances Williams. And it is a privilege to be part of a university which puts so much care and effort into its teaching and the support of its students. The practicalities of producing this book owe much to my publishers, Stephanie Harding and Karen Borthwick at Springer; and to my colleagues at the Open University who have done so much to provide me with a robust and attractive LaTeX system: Alison Cadle, David Clover, Jonathan Fine, Bob Margolis and Chris Rowley. And thanks to Springer, I have received much invaluable advice on content from their copy-editor Stuart Gale and their anonymous,

Plainly the blame for any errors and inadequacies of this book lies entirely with me. But perhaps at some deep and subtle level, the fault lies with everyone else!

very collegial, reviewers. I would also like to thank Michael Goldrei for his work on the cover design.

Perhaps the main inspiration for writing the book is the enthusiasm and talent for mathematical logic displayed by my old students at the Open University and at the University of Oxford, especially those of Somerville, St. Hugh's and Mansfield Colleges. In particular I'd like to thank the following for their comments on parts of the book: Dimitris Azanias, David Blower, Duncan Blythe, Rosa Clements, Rhodri Davies, David Elston, Michael Hopley, Gerrard Jones, Eleni Kanellopoulou, Jakob Macke, Zelin Ozturk, Nicholas Thapen, Matt Towers, Chris Wall, Garth Wilkinson, Rufus Willett and especially Margaret Thomas.

This book is dedicated to all those whose arguments win me over, especially Jennie, Michael, Judith and Irena.

CONTENTS

1 INTRODUCTION

1.1 Outline of the book

Mathematics abounds with theoretical results! But on what basis do we trust any of them? Normally we rely on seeing some sort of justification for results which we, or someone we feel we can trust, can scrutinize and then verify. The justification will normally involve some sort of argument showing that a result holds. This book is about a mathematical model of such arguments, rather than the mathematics being argued about. We shall not attempt to cope with the full range of mathematical arguments, but only look at a fragment covered by what is called the *predicate calculus*. The predicate calculus is an important part of *logic*, the science of reasoning and the laws of thought.

The sort of argument we shall try to model is one which starts from given assumptions and moves by steps to a conclusion. For instance, in everyday maths we might start from the assumption that f is a differentiable function from the set $\mathbb{R}$ of real numbers to itself and conclude after various steps that f is continuous. The book is written for readers familiar with this sort of argument. So you will know that the steps involve things like: use of the assumption that f is differentiable; the definition of the words 'differentiable', 'function' and 'continuous'; facts about inequalities and arithmetic involving real numbers; and forms of reasoning enabling us to infer each step from previous assertions in the argument. An essential feature of such an argument is our ability to follow and agree on each step, almost in a mechanical way, like a computer recognizing whether input data conforms to agreed rules.

If a step seems too large to be followed, we usually seek an explanation using more elementary steps.

Let's expand on what we call facts about the set $\mathbb{R}$ of real numbers. The modern approach to $\mathbb{R}$ is to describe it not in terms of its members, but in terms of the properties its members have. We are usually given some of their very basic properties, called *axioms*, from which one can infer more complicated properties.

We shall give such axioms and illustrate some of these inferences soon.

We have two different but connected sorts of expectation of the axioms for $\mathbb{R}$:

(i) that the axioms and any statements we infer from them are true statements about $\mathbb{R}$;

(ii) that any statement that we feel is true about $\mathbb{R}$ can be inferred from these axioms.

Investigation of the connection between these expectations is a major theme of the book.

We have phrased our expectations above in terms of what is true about $\mathbb{R}$ and what can be inferred from axioms. In this book, we shall attempt to explain on the one hand what is meant by a statement being true in a structure like $\mathbb{R}$ and, on the other, what constitutes inference using acceptable forms of reasoning. Establishing a connection between these very different concepts, called the *completeness theorem*, is a major goal of the book.

We must reiterate that what the book covers is a mathematical model. Just as with a mathematical model of, say, the motion of the planets in the solar system, we shall need to make compromises and simplifications within our model to produce something that is both mathematically tractable and provides useful insights into the actual and arguably much more complex world. Here our model will be of truth and proof, and we shall investigate the connection between them.

As we shall be modelling proofs from axioms, let us look at a set of axioms for $\mathbb{R}$ that could be found in a standard textbook on real analysis.

Axioms for $\mathbb{R}$

The real numbers system consists of a set S, usually written as $\mathbb{R}$, with binary operations $+$ and $\cdot$, unary operations $-$ and $^{-1}$, a binary relation $\leq$ (besides $=$, equality), and special elements written as 0 and 1, such that $0 \neq 1$, which satisfy the following properties.

1. For all $x, y, z \in S$, $x + (y + z) = (x + y) + z$.

2. For all $x \in S$, $x + 0 = 0 + x = x$.

3. For each $x \in S$, $x + (-x) = (-x) + x = 0$.

4. For all $x, y \in S$, $x + y = y + x$.

5. For all $x, y, z \in S$, $x \cdot (y \cdot z) = (x \cdot y) \cdot z$.

6. For all $x \in S$, $x \cdot 1 = 1 \cdot x = x$.

7. For all $x \in S$ with $x \neq 0$, $x \cdot x^{-1} = x^{-1} \cdot x = 1$.

8. For all $x, y \in S$, $x \cdot y = y \cdot x$.

9. For all $x, y, z \in S$, $x \cdot (y + z) = (x \cdot y) + (x \cdot z)$.

10. For all $x \in S$, $x \leq x$.

11. For all $x, y \in S$,   if $x \leq y$ and $y \leq x$, then $x = y$.

12. For all $x, y, z \in S$,   if $x \leq y$ and $y \leq z$, then $x \leq z$.

13. For all $x, y \in S$, $x \leq y$ or $y \leq x$.

14. For all $x, y, z \in S$,   if $x \leq y$, then $x + z \leq y + z$.

15. For all $x, y, z \in S$,   if $x \leq y$ and $0 \leq z$, then $x \cdot z \leq y \cdot z$.

16. *(Completeness axiom)* Any non-empty subset A of S which is bounded above has a least upper bound in S.

We are deliberately writing the set as S rather than $\mathbb{R}$ to reinforce the point that the axioms involve no knowledge of what the objects in S are. The axioms are often said to axiomatize a *complete ordered field*.

'Bounded above' and so on are defined in terms of more basic terminology.

Within the context of a first course on real analysis, one would build up further properties of the set S, starting with elementary properties, such as

for all $x, y \in S$, if $x + y = x + z$, then $y = z$.

This is sometimes grandly called the *left cancellation law* for $+$.

An argument proving this might run as follows. Suppose that $x + y = x + z$. Then

$$(-x) + (x + y) = (-x) + (x + z).$$

By axiom 1, $(-x) + (x + y)$ equals $((-x) + x) + y$, which by axiom 3 equals $0 + y$, which by axiom 2 equals y. Similarly $(-x) + (x + z) = z$, so the above gives $y = z$, as required.

It might be good for you to try proving a few elementary properties of S for yourself, just to get a feeling of what sort of features of proofs our model might have to take into account.

Exercise 1.1

Give proofs purely from these properties of each of the following. (You may of course prove and then exploit subsidiary results, or lemmas like the left cancellation law for $+$ above.)

(a) For all $x \in S$, $(-(-x)) = x$.

(b) If $x, y \in S$ satisfy $x \cdot y = 1$, then $y = x^{-1}$.

(c) For all $x \in S$, $x \cdot 0 = 0$.

(d) For all $x \in S$, $(-1) \cdot x = -x$.

Solution

We shall give a solution only for part (a).

Take any $x \in S$. Then by axiom 3,

$$(-x) + x = 0,$$

and by the same axiom used with the element $-x$ instead of x,

$$(-x) + (-(-x)) = 0,$$

so that

$$(-x) + x = (-x) + (-(-x)).$$

By the left cancellation law for $+$, we can conclude that $x = (-(-x))$, or equivalently $(-(-x)) = x$, as required.

Here are just a few of the features in arguments of this sort which we shall try to incorporate into our mathematical model of proof.

- Proofs consist of statements, like '$(-x) + (x + y)$ equals $((-x) + x) + y$', connected by justification for these statements.

- The statements in a proof use a fairly limited technical language, including symbols like $+$, $\cdot$ and $=$.

- Proofs should be presented in a way that allows others to follow each step.

- Proofs should involve no properties of the set S not ultimately traceable back to the axioms.

- Proofs should make no assumptions about the nature of the elements x, y, 0, 1 and so on, or the operations $+$, $\cdot$ and $-$, other than what we are told about the symbols by the axioms.

- To prove a statement of the form 'for all $x \in S$, x has some property', take a typical $x \in S$, show that it has the property, and conclude the property holds for all x.

- To prove a statement of the form 'if something then something else', assume the first 'something' and prove the 'something else'.

Our list barely scrapes the surface of the sort of reasoning that is employed, even in simple arguments like our solution to Exercise 1.1(a) or our proof of the left cancellation law for $+$. For instance, there are a host of ways in which we use $=$, the equality symbol, like the step from

$$x + y = x + z$$

to

$$(-x) + (x + y) = (-x) + (x + z)$$

in our proof of the left cancellation law for $+$, and at least some of these should be covered within our model.

Our model of proof will ultimately cope with all the features needed in these proofs. We shall look at a formal symbolic language from which statements like 'for all x, $(-1) \cdot x = -x$' can be constructed and at a formal system for proofs which can handle such statements and, indeed, for this particular example, derive it from some of the axioms given above. But a limitation of this model is that the formal system will not be able to derive all the statements one can derive in everyday maths about $\mathbb{R}$ from the axioms we have given, as for important technical reasons it will not be able to handle the completeness axiom.

The axioms for $\mathbb{R}$ are not entirely typical of the way axioms are used in modern maths, in that they have the following special property: any two sets, equipped with suitable operations and relations, which both satisfy the axioms, are essentially the same, so that the axioms tie down the one system. An example of a set of axioms for which there are many essentially different sets satisfying the axioms is obtained by taking just the first three axioms of those given for $\mathbb{R}$, which axiomatize the theory of *groups*, a group being the name for a set and suitable operation matching the $+$ symbol for which all three axioms are true. Our formal proof system will be able to cope with these axioms. Just as for the axioms for $\mathbb{R}$, we shall expect a connection between the statements we can derive from these axioms and the statements which are true in all groups. Importantly, we shall explain what we mean by a statement being true in a group as something more than that it can be proved from the axioms, so that the major result, the completeness theorem, connecting the notions of truth and proof, genuinely connects different notions.

In representing mathematical statements within a formal symbolic language, it will be important to have strict construction rules to delineate the strings of symbols which are to be considered as statements, for instance to exclude the equivalent of expressions like $8) - \times 5))7+$ which signify nothing meaningful in everyday maths. Likewise the rules have to provide expressions for which, when meanings are given to the symbols, there is an unambiguous interpretation. For instance, we want to avoid the analogue of expressions like $1 + 2 \times 3$ in normal maths, for which it is unclear whether is meant $(1 + 2) \times 3$ or $1 + (2 \times 3)$. A key feature of our formal language, and indeed the formal proof system for handling statements in it, is what we shall describe as its mechanical nature. The rules governing which strings of symbols are formal statements and which combinations of statements constitute a formal proof will all be ones that a computer could be programmed to check, with no understanding of any intended meanings of the symbols or rules. This aspect of our undertaking not only predates the development of modern computers, but has had a strong influence on both their development and how they are

The proper technical description is that the two sets with their operations are *isomorphic*. Another important axiom system with this property is Peano's axioms for the natural numbers, which we shall meet in Chapter 6.

The first nine axioms are axioms for the theory of *fields*. While we hope that you have encountered the theories of groups and fields, we are not relying on this experience in this book, but will give you a brief background in Chapter 4.

See Davis [10] for a very readable account interweaving the history of logic with the development of the modern computer.

used – at a certain level, what computers do is the manipulation of strings of symbols, e.g. 0s and 1s, according to rules, and the distinction and interplay between the formal language and its interpretation in the real world, first made and investigated within the context outlined in this book, has proved to be vital in computer science.

We've not said what we mean by words like 'computer' and 'program' above, and the subject of this book can be developed without ever settling their meanings or referring to them at all. But we think that it helps to know that there is an undercurrent of mechanical, or algorithmic, processes. Roughly speaking, an *algorithm* is a finite set of instructions which, given input data, can be followed by a person or a machine in a deterministic way, ideally so that the process arrives at a conclusion after a finite number of steps. Each instruction should require some humanly or mechanically feasible action, for instance adding 1 to an integer and comparing two words to see whether one is an anagram of the other. You can see that it would be desirable to have some sort of *algorithmic procedure* (that is, a procedure based on an algorithm) for checking the correctness of a proof. The issue of whether there is an algorithm for doing something will crop up occasionally in the book, and we shall give (and expect from you) only very informal descriptions of algorithmic procedures.

By being deterministic, if several people follow the instructions with the same inputs, all will obtain the same results at each step of their computations.

Bound up with this mechanical vision are ideas about finiteness. Somehow mathematical (or any other) statements should be expressed, and their truth or falsity decided upon, in a finite way, using a finite amount of time and paper, with only finitely many symbols on each page. The controversy and paradoxes surrounding the development of the theory of infinite sets in the latter part of the 19th century resulted in a sharpening of ideas about how finiteness was built into mathematics. The German mathematician David Hilbert (1862–1943), who played a key role in developing the material covered in this book, wanted to base proofs on what he called finitary methods of reasoning. Roughly speaking, this meant producing finitely long proofs involving finitely expressed statements, starting from finitely described axioms and using a finitely many allowed rules of inference. Infinity creeps in by putting no finite upper bounds on the number or length of any of the constituent parts of the system, e.g. the alphabet of the symbols or the length of expressions formed from this. The modern theory is very interested in the outcome of stretching some of the features into the infinite, e.g. mathematical theories described by infinitely many axioms, albeit that each axiom is finitely expressed. You will see what this means as the book goes on.

You might like to suggest ways in which an infinite amount of information could be written on a single A4 sheet of paper, recoverable by someone with appropriately acute eyesight!

The use of infinite sets in the definition of the real numbers was one of the reasons behind the renewal of interest in the axiomatic approach to mathematics.

The structure of the book is as follows. In Chapters 2 and 3, we shall look at a mathematical model for dealing with very limited sorts of statements within the framework of what is called *propositional calculus*. The formal statements within this fall far short of the mathematical statements, like those earlier about $\mathbb{R}$, which are our ultimate goal. But the propositional calculus gives us valuable experience of some of the issues and methods that will be of importance for these more mathematical statements. In Chapter 2 we look at the formal language for propositional calculus and the way it can be interpreted to talk about statements as being true or false. In Chapter 3 we look at a formal proof system for propositional calculus and the completeness theorem connecting truth and formal proof. In Chapters 4 and 5 we look at our promised mathematical model of reasoning called the *predicate calculus*, which can handle at least some interesting fragments of everyday mathematics. In Chapter 4 we look at the formal language and how to interpret it and talk about the truth or falsity of statements within interpretations. In Chapter 5 we look at a formal proof system for the predicate calculus and prove the completeness theorem for it. Although the completeness theorem is an end in itself, it has very interesting mathematical consequences. We investigate some of these in Chapter 6.

The first published system of predicate logic was devised by the German mathematician and philosopher Gottlob Frege (1848–1925). This seminal work of mathematical logic, entitled *Begriffsschrift* (meaning 'conceptual notation'), can be found in Heijenoort [19], which also contains much further source material for material covered in this book.

1.2 Assumed knowledge

The book is written on the basis that you have already had some experience of using sets and functions, and that you are familiar with a variety of mathematical words and notations. Perhaps most surprisingly for a book about logic, we assume that you already know something about logic and reasoning!

What you already know about logic

An underlying feature of statements about mathematics and arguments involving them is the use of words and phrases like 'for all', 'if ... then', 'not' and 'and'. These will play a major role in our model of argument and we assume that you know something about how they are used in everyday maths. We don't expect that you will have thought about the use of these words in quite the formal way that will be adopted later in the book, but you will know something about what we are trying to model.

For instance, we assume that you are happy to infer from the statements

'for all functions $f\colon \mathbb{R} \longrightarrow \mathbb{R}$, if f is differentiable then f is continuous'

and

'the sine function is differentiable'

the statement 'the sine function is continuous'.

Let us break down this inference into the small steps which we are going to model. From the statement 'if f is differentiable then f is continuous' holding in general for all functions $f\colon \mathbb{R} \longrightarrow \mathbb{R}$, it will hold for the sine function in particular, so we can infer from it the statement

'if the sine function is differentiable then the sine function is continuous'.

We assume that you are happy that this is just how we naturally use the words 'for all'.

From the additional statement 'the sine function is differentiable', the way we use the words 'if' and 'then' enable us to infer that 'the sine function is continuous'. Again we assume that you are happy with this use of 'if ... then'.

Suppose that you are assured that

'there is some infinite ordinal which is not a limit'

and you asked whether it follows from this that 'all infinite ordinals are limits'. Even if you don't know anything about infinite ordinals, we hope that you

You can find out more about ordinals in e.g. Goldrei [16].

would give the answer no, on the basis of how we use the words 'there is some', 'not' and 'for all'.

In the formal treatment of argument in this book, we shall generalize the way these words are used in the following sort of way. Looking at the example about ordinals, we shall regard the statement 'there is some infinite ordinal which is not a limit' as a particular case of a statement of the form

'there is some x which does not have the property $\phi(x)$',

where x stands for an object, here an infinite ordinal, and $\phi(x)$ stands for a property that objects might or might not have, here that x is a limit. With this notation, the question of whether it follows that 'all infinite ordinals are limits' becomes one of whether

'all x have the property $\phi(x)$'.

The reason why the answer to this question is no is because from

'there is some x which does not have the property $\phi(x)$'

we can infer that

'it is not the case that all x have the property $\phi(x)$'.

We hope that, even though you might not be familiar with the abstract way of phrasing this using $\phi(x)$ to stand for a property of x, you are comfortable with the roles played by 'there is some', 'not' and 'for all' in this inference.

Likewise, from our first example about the sine function, we can abstract a rule that from a statement of the form '$\phi(x)$ holds for all x', we can conclude that '$\phi(t)$ holds' for a particular example t of the xs in question. This is indeed one of the ways we use 'for all' in general. Similarly, letting θ and ψ stand for statements, from the statements 'if θ holds then ψ holds' and 'θ holds', we can always infer 'ψ holds', this being the way we use 'if ... then'.

From now on we shall usually rephrase 'θ holds', where θ is a statement, more simply as 'θ', dropping the 'holds', and often even more simply as θ without the quotation marks. The context will normally make it clear whether we are talking about some aspect of the statement like its grammatical structure or we are *asserting* it, that is, claiming that it holds or is true. So we shall rephrase our immediately preceding observation about 'if ... then' as saying that from 'if θ then ψ' and θ we can infer ψ. We shall abbreviate 'it is not the case that ψ holds' as 'not ψ' and statements like 'it is not the case that all x have the property $\phi(x)$' as 'not for all x, $\phi(x)$'. With conventions of these sorts, try the following exercises.

Quotation marks ' and ' will often be useful for clarity when we are discussing a complicated statement.

Exercise 1.2 ____

Let θ and ψ stand for statements. Which, if any, of the following statements follows from one of the others?

(a) If θ then ψ.

(b) ψ implies θ.

(c) If 'not θ' then 'not ψ'.

(d) 'Not ψ' implies 'not θ'.

We hope that you take 'ψ implies θ' to mean the same as 'if ψ then θ'.

Solution

We shall show that (d) follows from (a). Suppose that (a) is true, that is, that if θ then ψ. Now suppose that 'not ψ' is true. If θ is also true, then as 'θ then ψ' is true, ψ must also be true. But this contradicts that 'not ψ' is true. Thus our supposition that θ is true leads to a contradiction, so that it must instead be the case that θ is false, or equivalently, that 'not θ' is true. We conclude that from the original supposition that (a) is true, it follows that 'not ψ' implies 'not θ'.

We hope that you can see and argue persuasively that (d) follows from (a), so that (a) and (d) are essentially equivalent. Likewise (b) and (c) are equivalent.

In our argument above that (d) follows from (a), we made use of what is known as *proof by contradiction*: by showing that from various assumptions and the assumption θ one can derive a *contradiction*, namely that some statement is both true and false, we infer that 'not θ' follows from the other assumptions. We hope that you have seen this sort of proof before. Famous and very antique examples of its use are in the proofs that $\sqrt{2}$ is irrational and that there are infinitely many prime numbers.

Likewise to prove from (a) that 'not ψ' implies 'not θ', we assumed 'not ψ' and derived from it that 'not θ', a style of proof with which we hope you are familiar.

Note that there are no other pairs of statements in Exercise 1.2 which follow from each other, even though their shapes are related. For instance, 'ψ implies θ' (statement (b)) is called the *converse* of 'if θ then ψ' (statement (a)). But in general the one does not follow from the other. For instance, taking θ to be the statement $n > 5$ about an integer n and ψ to be the statement $n > 1$, the statement 'if $n > 5$ then $n > 1$' is true, while the statement '$n > 1$ implies $n > 5$' need not be true, for instance when $n = 4$. This provision of a *counterexample* is another feature of everyday mathematical argument which we expect you to know and understand.

> The classic proof that $\sqrt{2}$ is irrational assumes that it can be written as a fraction a/b for integers a, b with highest common factor 1 and then proves that 2 is a factor of both a and b, giving a contradiction. The classic proof that there are infinitely many primes assumes that there are only finitely many primes, all listed as $p_1, p_2, \ldots, p_n$, and shows (by looking at the number $p_1 p_2 \ldots p_n + 1$) that there is a prime number not in the list, giving a contradiction.

Exercise 1.3

Let $\phi(x)$ stand for a statement that x has a particular property. Which, if any, of the following statements follows from one of the others?

(a) For all x, $\phi(x)$.

(b) For some x, $\phi(x)$.

(c) For no x, $\phi(x)$.

(d) For all x, not $\phi(x)$.

(e) For some x, not $\phi(x)$.

(f) For no x, not $\phi(x)$.

Solution

We shall show that (d) follows from (c). Suppose that for no x, $\phi(x)$. Then for any x it cannot be the case that $\phi(x)$ holds, so that 'not $\phi(x)$' holds. Thus it is the case that for all x, not $\phi(x)$. We hope that you can see that (c) follows from (d), so that (c) and (d) are equivalent. Likewise (a) and (f) are equivalent, each one following from the other.

Assuming that there are actually some xs being talked about, it follows from (a), namely for all x, $\phi(x)$, and taking any one of these xs, that there is some x for which $\phi(x)$ holds, so that (b) holds. As (a) and (f) are equivalent, (b) also follows from (f). In the same way (e) follows from each of (c) and (d).

The point about whether there are actually any numbers x being talked about when we state that for all x, $\phi(x)$, is rather subtle. We could in principle state that for any property $\phi(x)$ which takes our fancy

'for all integers x which are simultaneously even and odd, $\phi(x)$ holds',

where of course there are no such integers x! We shall avoid this problem by insisting on the convention that we only use 'for all x' when there are some xs of the sort we are talking about. With this convention we can rephrase our curious statement above as

'for all integers x, if x is both even and odd, then $\phi(x)$ holds',

abiding by our convention.

We have of course just given you a small push down the road of formalizing arguments, which will occupy much of the rest of the book, and will give you one further push, following the theme of building on the assumption that you already know how to argue within everyday mathematics. Consider the following argument.

All square numbers are non-negative.	(1)
All non-negative numbers have a fourth root.	(2)
Therefore all square numbers have a fourth root.	(3)

We hope that from your everyday experience of argument that you accept that, as the 'therefore' suggests, statement (3) can be inferred from statements (1) and (2). (We are not asking you at this stage whether you think that statement (3) is true in its own right.) The correctness of the argument stems from its shape, which very crudely is

All B are C.
All A are B.
Therefore all A are C.

We hope that you agree that this argument, where the third statement follows from the first two, is correct in general, not just for our particular example. This general argument is an example of a *syllogism*, as first defined and discussed by the Greek philosopher Aristotle (384 BC–322 BC), who is widely regarded as the founder of the study of logic. Aristotle introduced the idea of looking at the form of statements and seeing how the correctness of many arguments stemmed from their shape, rather than the particular statements involved. Accepting that this form of argument is correct means that if, in a given set of circumstances, statements (1) and (2) are true, then statement (3) must also be true.

Although not all arguments are syllogisms, they often have the same very general shape, starting from some initial statements, called *premises* (here statements (1) and (2)), leading to a statement which is the *conclusion* (here statement (3)). A desired property of an argument is that in a given set of circumstances in which all the premises are true, the conclusion must also be true.

Exercise 1.4

Which of the following arguments are correct, so acceptable in general?

(a) No B are C.
  All A are C.
  Therefore no A are B.

(b) Some A are B.
  All A are C.
  Therefore some B are C.

(c) Some A are B.
  All C are B.
  Therefore some A are C.

Solution

The arguments in (a) and (b) are both correct (and are further examples of syllogisms). For instance, for argument (a), you might argue in your head as follows. Assume that no B are C and that all A are C. If some A is B, then as no B is C, that A cannot be C. But all A are C, so there can after all be no A that is B.

Our apologies if our analysis simply made your head hurt! In general we shall try to formalize arguments using even smaller steps of reasoning than this 2000 year old example.

The argument in (c) doesn't always hold. Consider the following example.

  Some odd integers are perfect squares.
  All positive powers of 4 are perfect squares.
  Therefore some odd integers are positive powers of 4.

The first statement is true, e.g. the odd number 9 is a perfect square. The second statement is true, as for any positive integer n, $4^n = (2^2)^n = 2^{2n}$, which is a perfect square. But the third statement is false, as a positive power of 4 is even, so there cannot be an odd integer of this form.

Exercise 1.5

Consider the argument:

  All square numbers are non-negative.   (1)
  All non-negative numbers have a fourth root.   (2)
  Therefore all square numbers have a fourth root.   (3)

Do you think that statement (3) is true?

Solution

Note that this argument is correct: statement (3) does follow from statements (1) and (2) simply because of the shape of the statements. But whether statement (3) is true will depend on what sort of numbers we are talking about. For instance it is true for real numbers. But it would usually be regarded as false for integers, as implicit within everyday maths discussion of this sort about integers is that the roots in question are also integers, rather than non-integer real numbers.

It will also depend on what is meant by 'square' and 'fourth root'; but these have well established meanings, whereas 'number' is not at all specified here.

A point arising from the last exercise is that, while the argument is correct, its applicability to conclude that the third statement is true depends on the interpretation of the terms being used, most importantly here what sort of numbers we mean. If we mean the real numbers, then statements (1) and (2) are true, so that as the argument form is acceptable, it follows that statement (3) is true. If we mean the integers, statement (1) is true, but statement (2) is not, so that the argument, while correct, cannot be exploited to give any information about the truth or falsity of statement (3).

Other assumed knowledge

Set Notation

A *set* X is a collection of objects called the *elements*, or *members*, of X. We write $x \in X$ to express that the object x is an element of the set X and $y \notin X$ to say that y is *not* an element of X.

When $x \in X$, we also say 'x is in X' or 'x belongs to X'.

We use curly brackets, { and }, around a list of objects to signify the set of all those objects. For instance $\{3, 8, 9\}$ is the set with elements 3, 8, 9.

The order in which the elements are listed inside the curly brackets doesn't change the set, nor does listing some element more than once. Thus $\{9, 3, 7\}$ and $\{3, 3, 7, 9\}$ both represent the same set as $\{3, 7, 9\}$. In general, two sets X and Y are equal if and only if they contain the same elements or, equivalently, if and only if every element of X is an element of Y and vice versa.

We use standard notation for the most common sets of numbers: $\mathbb{N}$ for the set of natural numbers, $\mathbb{Z}$ for the set of all integers (positive, negative and zero), $\mathbb{Q}$ for the set of rational numbers, $\mathbb{R}$ for the set of real numbers and $\mathbb{C}$ for the set of complex numbers.

In this book we take $\mathbb{N}$ to include all the positive integers *and* the number 0.

We use the notation $\varnothing$ for the *empty set*, the set which contains no elements.

We can also describe a set using curly brackets in terms of a property possessed by all its elements, as with $\{n : n \text{ is an even integer}\}$ or, equivalently, $\{n \in \mathbb{Z} : n \text{ is even}\}$ for the set of all even integers $\{\ldots, -4, -2, 0, 2, 4, 6, \ldots\}$. In general we write

$$\{x : \phi(x)\}$$

The colon ':' is read as 'such that'.

for the set of all x such that $\phi(x)$ holds, where $\phi(x)$ stands for a property which may or may not be possessed by a given object x.

We shall occasionally use standard notation for intervals of the real line:

(a, b) for the *open interval* $\{x \in \mathbb{R} : a < x < b\}$;

$[a, b]$ for the *closed interval* $\{x \in \mathbb{R} : a \leq x \leq b\}$;

$(a, b]$ and $[a, b)$ for the *half open and closed* intervals $\{x \in \mathbb{R} : a < x \leq b\}$ and $\{x \in \mathbb{R} : a \leq x < b\}$ respectively;

$(-\infty, b)$ and (a, ∞) for the open intervals $\{x \in \mathbb{R} : x < b\}$ and $\{x \in \mathbb{R} : x > a\}$ respectively;

$(-\infty, b]$ and $[a, \infty)$ for the closed intervals $\{x \in \mathbb{R} : x \leq b\}$ and $\{x \in \mathbb{R} : x \geq a\}$ respectively.

Given two sets X and Y, we write

$X \cup Y$ for the *union* of X and Y, that is, the set of elements belonging to X or Y (or both);

$X \cap Y$ for the *intersection* of X and Y, that is, the set of elements belonging to both X and Y;

$X \setminus Y$ for the *complement* of Y in X, that is, the set of elements of X *not* in Y.

> We shall adopt the standard mathematical use of the word 'or' as allowing the 'or both' case – what's called the *inclusive* use of 'or'.

Given a set $\mathcal{F}$ whose elements are sets, we write $\bigcup\{X : X \in \mathcal{F}\}$ for the union of all the sets in the set $\mathcal{F}$, that is, the set $\{x : x \in X \text{ for some } X \in \mathcal{F}\}$. Such a set $\mathcal{F}$ of sets might sometimes be indexed by another set, for instance the family of all open intervals of $\mathbb{R}$ of the form $(\frac{1}{n+1}, \infty)$ for $n \in \mathbb{N}$ is effectively indexed by the set $\mathbb{N}$: in such a case we would write the family as $\{(\frac{1}{n+1}, \infty) : n \in \mathbb{N}\}$ and the union of the family as $\bigcup\{(\frac{1}{n+1}, \infty) : n \in \mathbb{N}\}$ (which happens to equal the set $(0, \infty)$).

X is a *subset* of the set Y means that X is a set of which every element is also an element of Y (so that, for all x, if $x \in X$ then $x \in Y$). We write $X \subseteq Y$ for 'X is a subset of Y'. A subset X of Y is said to be *proper* if $X \neq Y$. The *power set* of Y, written as $\mathscr{P}(Y)$, is the set of all subsets of Y.

We write $X \times Y$ for the *Cartesian product* of X and Y, that is, the set of all ordered pairs (x, y) with $x \in X$ and $y \in Y$. We use X^2 as shorthand for $X \times X$, X^3 for $(X \times X) \times X$ and so on. An element $(x_1, x_2, \ldots, x_n)$ of X^n is often described as an n-tuple.

Function Notation

A *function* f from a set X to a set Y associates an element, $f(x)$, of Y with each element x of X. The *rule* of f, written as $x \longmapsto f(x)$, describes this process of association. The element $f(x)$ of Y is called the *image of x under f*. The *domain* of f is the set X and the *codomain* of f is the set Y. We use the standard arrow notation for such a function, combining the information of its domain, codomain and rule:

$$f : X \longrightarrow Y$$
$$x \longmapsto f(x).$$

If A is a subset of the domain of this function f, the *restriction* of f to A, written as $f|_A$, is the function

$$f|_A : A \longrightarrow Y$$
$$x \longmapsto f(x),$$

that is, $f|_A$ has the same rule and codomain as f, but has its domain restricted to A.

The *image set* or *range* of $f : X \longrightarrow Y$, written as $\text{Range}(f)$, is the set of images of f, namely $\{f(x) : x \in X\}$. For any subset A of the domain X the set $\{f(x) : x \in A\}$ is called the *image set of A under f*.

A function is said to be *onto* if for each $y \in Y$ there is an $x \in X$ with $f(x) = y$.

> So that f is onto exactly when $\text{Range}(f) = Y$.

The function f is said to be *one–one* if for all $x, x' \in X$, if $f(x) = f(x')$ then $x = x'$ (or, equivalently, if $x \neq x'$ then $f(x) \neq f(x')$).

If f is a one–one function, then its *inverse function* f^{-1} is defined as the function

$$f^{-1}\colon \operatorname{Range}(f) \longrightarrow X$$
$$y \longmapsto \text{the unique } x \text{ such that } f(x) = y.$$

For any subset B of the codomain Y, its *inverse image set* under f, written as $f^{-1}(B)$, is the set $\{x \in X : f(x) \in B\}$.

If $f\colon X \longrightarrow Y$ and $g\colon Y \longrightarrow Z$ are functions then the *composite* function (or *composition of f and g*) $g \circ f$ is the function

$$g \circ f\colon X \longrightarrow Z$$
$$x \longmapsto g(f(x)).$$

If f is both one–one and onto, then f is a *bijection*. If f, g are both bijections with the codomain of f equal to the domain of g, then the composition $g \circ f$ is also a bijection.

Countable sets

A set X is *finite* if it is empty or there is a bijection to it from the n-element set $\{i \in \mathbb{N} : i < n\}$ for some $n \in \mathbb{N}$, where $n = 0$ corresponds to X being the empty set $\varnothing$. A set is *infinite* if it is not finite. It is *countably infinite* if there is a bijection to it from the set $\mathbb{N}$ of natural numbers. It is *countable* if it is finite or countably infinite; otherwise it is *uncountable*.

Examples of countably infinite sets are $\mathbb{N}$, $\mathbb{Z}$ and $\mathbb{Q}$. Uncountable sets include $\mathbb{R}$, $\mathbb{C}$, $\mathscr{P}(\mathbb{N})$ and the set of functions from $\mathbb{N}$ to itself.

Results about finite and countable sets include the following.

- For any finite set X, any one–one function from X to itself must be onto. This is a version of the *pigeon-hole principle*.

- Any subset X of a countable set Y is also countable.

- The union $X \cup Y$ of countable sets X, Y is countable.

- The Cartesian product $X \times Y$ of countable sets X, Y is countable, so in particular $\mathbb{N} \times \mathbb{N}$ is countable.

- The set of finite subsets of $\mathbb{N}$ is countable.

- The set of finitely long sequences $(n_1, n_2, \ldots, n_k)$ of natural numbers is countable.

- Given a countable set of symbols, the set of all finite sequences of these symbols is countable.

- The countable union of countable sets is countable, that is, if $\mathcal{F}$ is a set whose elements X are countable sets, then $\bigcup\{X : X \in \mathcal{F}\}$ is a countable set.

Well-order property of $\mathbb{N}$

Every non-empty subset B of $\mathbb{N}$ contains a least element b_0; that is, there is $b_0 \in B$ such that $b_0 \leq b$, for all $b \in B$.

Mathematical induction

The *principle of mathematical induction* can be stated as follows: if A is a subset of $\mathbb{N}$ such that $0 \in A$ and whenever $n \in A$, then $n + 1 \in A$, then $A = \mathbb{N}$.

The use of f^{-1} in this context does *not* mean that the inverse function f^{-1} exists – a well-known source of confusion!

We define $g \circ f$ in exactly the same way when the domain of g contains the range of f as a subset, rather than requiring that the domain of g coincides exactly with the codomain of f.

For background, see e.g. Halmos [17], Goldrei [16] or Enderton [13].

This result is needed to prove our major result, the completeness theorem in Chapter 5. It depends on a principle called the axiom of choice. We do not assume that you know about this principle, but discuss it along with further facts about infinite sets in Section 6.4 of Chapter 6.

The method of *proof by mathematical induction* used to show that a set A of natural numbers with a given property is all of $\mathbb{N}$ is as follows. Prove that

$$0 \in A$$
Called the basis of induction.

and that for all $n \in A$,

$$\text{if } n \in A, \text{ then } n + 1 \in A,$$
Called the inductive step.

and conclude from the principle of mathematical induction that $A = \mathbb{N}$.

An important variant of this method of proof used often in this book is as follows. As before, first prove that

$$0 \in A$$

and then prove that for all $n \in A$,

$$\text{if } k \in A \text{ for all } k \leq n, \text{ then } k \in A \text{ for all } k \leq n + 1,$$

to conclude that $A = \mathbb{N}$.

Exercise 1.6 ⎯⎯⎯⎯⎯⎯⎯⎯⎯⎯⎯⎯⎯⎯⎯⎯⎯⎯⎯⎯⎯⎯

Explain how the above variant of the method of proof by mathematical induction follows from the principle of mathematical induction. [*Hint*: You might wish to exploit the well-order property of $\mathbb{N}$.]

Further exercises

Exercise 1.7 ⎯⎯⎯⎯⎯⎯⎯⎯⎯⎯⎯⎯⎯⎯⎯⎯⎯⎯⎯⎯⎯⎯

What, if anything, is wrong with the following argument about real numbers?

> Let x be a real number.
> Suppose that $x = 0$.
> Then $x^2 = 0^2 = 0 = x$,
> so that for all $x \in \mathbb{R}$, $x^2 = x$.

Exercise 1.8 ⎯⎯⎯⎯⎯⎯⎯⎯⎯⎯⎯⎯⎯⎯⎯⎯⎯⎯⎯⎯⎯⎯

What, if anything, is wrong with the following solution of the inequality $\sqrt{x^2 - 1} < x$ involving real numbers?

> Let x be a real number.
> If $\sqrt{x^2 - 1} < x$
> then $(\sqrt{x^2 - 1})^2 < x^2$,
> i.e. $x^2 - 1 < x^2$,

which is true for all x. Therefore all real numbers x satisfy the inequality.

Exercise 1.9 ⎯⎯⎯⎯⎯⎯⎯⎯⎯⎯⎯⎯⎯⎯⎯⎯⎯⎯⎯⎯⎯⎯

Explain why the existence of a function $f : X \longrightarrow X$ which is one–one but not onto means that the set X is infinite.

Exercise 1.10

Explain why it follows from the set of finitely long sequences $(n_1, n_2, \ldots, n_k)$ of natural numbers being countable that for a given countable set of symbols, the set of all finite sequences of these symbols is countable.

Exercise 1.11

Use mathematical induction to prove each of the following. In both cases you should use the variant of the method where for the inductive step you assume that the relevant property holds for all $k \le n$.

(a) The Fibonacci numbers F_n, $n = 0, 1, 2, \ldots$, are defined by

$$F_0 = 0, \ F_1 = 1 \text{ and } F_{n+2} = F_{n+1} + F_n, \text{ for all } n \ge 0.$$

Show that

$$F_n = \frac{1}{\sqrt{5}} \left(\frac{1 + \sqrt{5}}{2} \right)^n - \frac{1}{\sqrt{5}} \left(\frac{1 - \sqrt{5}}{2} \right)^n,$$

for all $n \ge 0$.

(b) A sequence $\{x_n\}$ of integers is defined as follows:

$$x_0 = 1; \quad x_n = x_0 + x_1 + x_2 + \ldots + x_{n-1}, \text{ for all integers } n \ge 1.$$

Show that $x_n = 2^{n-1}$, for all integers $n \ge 1$.

Exercise 1.12

For each of the following sets of statements, what can you conclude from them? In each case, give a conclusion which depends on all the statements in the set.

(a) (1) Every one who is sane can do Logic;

(2) No lunatics are fit to serve on a jury;

(3) None of *your* sons can do Logic.

(b) (1) Every idea of mine, that cannot be expressed as a Syllogism, is really ridiculous;

(2) None of my ideas about Bath-buns are worth writing down;

(3) No idea of mine, that fails to come true, can be expressed as a Syllogism;

(4) I never have any really ridiculous idea, that I do not at once refer to my solicitor;

(5) My dreams are all about Bath-buns;

(6) I never refer any idea of mine to my solicitor, unless it is worth writing down.

The examples come from Carroll [5], one of Lewis Carroll's serious attempts as a mathematician, as well as whimsical author, to teach classical logic exploiting symbolic reasoning. There are many more modern books of challenging logic puzzles, for instance Smullyan [28] and [29], often drawing inspiration from Carroll's wit.

Exercise 1.13

Let X be a set with n elements, where $n \in \mathbb{N}$. How many elements are there in each of the following sets?

(a) The set $\mathscr{P}(X)$ of all subsets of X.

(b) The set of all functions from $\mathscr{P}(X)$ to the 2-element set $\{0, 1\}$.

Exercise 1.14 —————————————————————————————

Let A, B, C be sets. Prove the following set identities.

(a) $A \cap (B \cup C) = (A \cap B) \cup (A \cap C)$

(b) $A \cup (B \cap C) = (A \cup B) \cap (A \cup C)$

(c) $A \cap (B \cup A) = A$

(d) $A \cup (B \cap A) = A$

Exercise 1.15 —————————————————————————————

Let A and B be subsets of a set X. Prove the following set identities.

(a) $X \setminus (A \cap B) = (X \setminus A) \cup (X \setminus B)$

(b) $X \setminus (A \cup B) = (X \setminus A) \cap (X \setminus B)$

2 PROPOSITIONS AND TRUTH ASSIGNMENTS

2.1 Introduction

In this chapter we shall look at statements with a very simple form and arguments about them which rely only on how we use words like 'and', 'or', 'not' and 'implies'. We shall also look at the truth or falsity of the statements and the validity of arguments built up from them. It is best to give one or two examples. For instance, suppose that we are told that

'the temperature outside is at most 20°C or the drains smell'

and believe this statement to be true. Suppose that the weather forecast for tomorrow predicts a temperature of over 20°C. Then we would predict that the drains will smell.

As another example, suppose that we are told about some function $f \colon \mathbb{R} \longrightarrow \mathbb{R}$ that

'f is not differentiable or f is continuous'

and that we have the further piece of information that f is differentiable. Then from this information we can infer that f is continuous.

These arguments cover entirely different areas of experience, but at a certain level they have a common shape. The statement 'the temperature outside is at most 20°C or the drains smell' in the first argument is built up from two shorter statements

'the temperature outside is at most 20°C'

and

'the drains smell'

connected by the word 'or'. The extra information that the statement 'the temperature outside is over 20°C' is true tells us that the statement 'the temperature outside is at most 20°C' is false, from which we can infer that 'the drains smell' will be true, using our understanding of the word 'or'.

Indeed we call the word 'or' a *connective* as it connects shorter statements to produce a longer one.

Similarly the statement 'f is not differentiable or f is continuous' is built up from the statements

'f is not differentiable'

and

'f is continuous'

by connecting them with 'or'. And given the extra information that 'f is differentiable', so that the statement 'f is not differentiable' is false, our understanding of the word 'or' helps us infer that 'f is continuous'.

The 'not' converts 'f is differentiable' into the longer statement 'f is not differentiable'. We shall also describe 'not' as a connective and consider it in this chapter.

We shall summarize the common feature of these two arguments that will particularly interest us in this chapter as follows. Using the letters p and q to stand for statements like 'the drains smell' and 'f is not differentiable', if we believe the more complicated statement 'p or q' to be true and the statement p to be false, we can infer that q is true.

There are other features of the arguments which are of interest. For instance, one might reasonably question whether the statement 'the temperature outside is at most 20°C or the drains smell' is actually true – it might be true for one person's drains but not for somebody else's – whereas anyone who has studied real analysis would know that for any function $f\colon \mathbb{R} \longrightarrow \mathbb{R}$, the statement '$f$ is not differentiable or f is continuous' is always true. We shall refine our description of the common features of the argument to account for these factors by saying that

> under any set of circumstances for which the statement 'p or q' is true and p is false, then q is true.

This is then something to do first with how a statement is built up from its component parts, here using 'or', and second how the truth of the statement depends on the truth of these component parts. It is nothing to do with the content of the statements for which p and q stand.

In this chapter we shall discuss a formal language within which we can build up more complicated statements from basic component propositions using symbols like $\vee, \wedge, \rightarrow$ to stand for connecting words, here respectively 'or', 'and', 'implies'. The formal language will have construction rules to ensure that any such complicated expression is capable of being judged to be either true or false, given the truth or falsity of the component parts. For instance, we want to avoid the formal equivalent of expressions like 'or the drains smell': without some statement before the 'or', we would be reluctant to describe this as in a fit state to be pronounced true or false. These construction rules are described as the *syntax* of the language. The framework within which we give some sort of meaning to the formal statements and interpret them as true or false in a given set of circumstances is called the *semantics* of the language. After we have established the basic rules of the language and its interpretation, we shall move on to issues like when one statement is a consequence of others. This will lay the ground for the discussion of a formal proof system for such statements. We shall build on this later in the book when we discuss this idea of consequence for much richer languages involving predicates and quantifiers within which we can express some serious mathematical statements.

'Proposition' is often used to mean a statement about which it is sensible to ask whether it is true or false.

There might appear to be potential for confusion between the formal language we study and the language we use to discuss it. The language we use for this discussion is that of everyday mathematical discourse and is often described as the *metalanguage*. We hope that we won't confuse the two sorts of language – usually the context will make it clear when we are talking about the formal language. However there will be strong links between the two levels of language. For instance, the formal rules for the use of the symbol $\wedge$ intended to represent the word 'and' will inevitably be based on how we use the word 'and' in everyday discourse. Also the desire to represent some part of everyday language in a formal way can force us to tie down how we use everyday language correctly.

What we are about to describe in this chapter, and indeed in the book as a whole, is a mathematical model of a fragment of natural language and arguments using it, not capturing fully their richness and variety. The importance of the model resides in the richness of the resulting theory, its applicability to large tracts of mathematics and, historically, in giving a paradigm for more refined modern analyses of language and argument – indeed, it is *the* model used by virtually all mathematicians and users of logic. Our model will make some hard and fast decisions about how to use terms like 'true', 'false' and 'or' which could legitimately be challenged in terms of how well they model natural language. Your attitude as a reader and student should be to run with our decisions for the purposes of this book, and probably for all the mathematics you will ever do, but to have an open mind to well-reasoned objections to them!

2.2 The construction of propositional formulas

In this section we shall describe the formal language which we shall use to represent statements. The language will consist of some basic symbols and we shall give rules for combining these into more complicated expressions, giving what is called the syntax of the language. We shall describe the way in which we shall give meaning to the formal language, that is, give its semantics, in the following section. However, the syntax and semantics are, perhaps not surprisingly, intertwined, so that considerations of the semantics will influence the specification of the syntax.

We have already indicated that we shall use letters like p and q to stand for basic component propositions, like 'the drains smell' and 'f is continuous', and that we shall use symbols to stand for connectives like 'or' and 'and' to build more complicated propositions from these basic ones. The building process involves stringing together these symbols and letters. We need to be clear what is meant by a string of symbols. A *string* is a finite sequence of symbols. Furthermore, we normally specify the set of symbols which can be used to form a string and we shall do this soon. Just as in everyday language, we have to distinguish which strings of symbols represent anything to which we can usefully give a meaning. In most normal uses a string like)9X7a)) would normally mean nothing and signify that some error has occurred, e.g. a cat has danced on a computer keyboard. For our purposes in this chapter, we will have particular requirements of a statement and this will have a knock-on effect on the strings of symbols in which we shall be interested. For instance, we want to represent statements for which it is meaningful to talk in terms of their truth or falsity. So we would want to exclude from our set of formal statements a string representing

> 'I'll go down to the shops and'

on the grounds that there's something missing after the 'and', preventing us from deciding on its truth on the basis of the truth or falsity of its component parts. Just because we can *utter* the words in this string, we are not necessarily *stating* any idea. More subtly, we want to avoid ambiguity in our formal statements, as for instance with

> 'it is snowing or the bus doesn't come and I'll be late for work'.

The finiteness of the sequence is of considerable importance in this book. There are other contexts where it makes sense to allow strings to be infinite.

These are examples of how the intended meaning, the semantics, will influence the formal rules of the syntax.

The truth of this statement depends on whether its component parts are bracketed together as

'it is snowing or the bus doesn't come' and 'I'll be late for work'

or

'it is snowing' or 'the bus doesn't come and I'll be late for work'.

In the case when 'it is snowing' is true, but both 'the bus doesn't come' and 'I'll be late for work' are false, the first way of bracketing the components gives false while the second gives true. So without some form of bracketing (perhaps done by pausing or emphasis when speaking, or extra punctuation in writing) the original 'statement' is ambiguous, that is, it admits more than one interpretation. For the precision in mathematical argument which the framework of logic helps to achieve, such ambiguity has to be avoided and this is one of the features we shall build into our syntax.

Exercise 2.1

Which, if any, of the following statements is ambiguous?

(a) If it is snowing and the bus doesn't come, then I'll be late for work.

(b) If it is snowing then the bus doesn't come and I'll be late for work.

Solution

We think that (a) is unambiguous, but that (b) is possibly ambiguous. In the context of everyday life, we would normally interpret (b) as saying that 'the bus doesn't come' and 'I'll be late for work' are both consequences of 'it is snowing'. But it is also possible as interpreting it as saying 'if it is snowing then the bus doesn't come' and 'I'll be late for work', so that I'll be late for work regardless of whether it is snowing!

The meaning of (b) could be made clearer by the insertion of some punctuation, like a comma, in an appropriate place.

With considerations like these in mind, we shall define our formal statements as follows. First we shall specify the *formal language*, that is, the symbols from which strings can be formed. We shall always allow brackets – these will be needed to avoid ambiguity. We shall specify a set P of basic statements, called *propositional variables*. From these we can build more complex statements by joining statements together using brackets and symbols in a set S of *connectives*, which are going to represent ways of connecting statements to each other, like $\vee$ for 'or' and other symbols mentioned earlier. We can take any symbols we like for the propositional variables, so long as these symbols don't clash with those used for the brackets and the connectives. To make life easier, we shall adopt the following convention for the symbols we'll use.

Later in this chapter, we shall allow an extra sort of symbol called a *propositional constant*.

Convention for variables

We shall normally use individual lower case letters like $p, q, r, s, \ldots$ and subscripted letters like $p_0, p_1, p_2, \ldots, p_n, \ldots$ for our propositional variables. Distinct letters or subscripts give us distinct symbols. When we don't specify the set P of propositional variables in a precise way, we shall use p, q, r and so on to represent different members of the set.

Use of symbols like these to represent variable quantities is, of course, very standard in normal mathematics.

Our formal version of statements, which we'll call *formulas*, is given by the following definition.

Definition Formula

Let P be a set of propositional variables and let S be the set of connectives $\{\wedge, \vee, \neg, \rightarrow, \leftrightarrow\}$. A *formula* is a member of the set $Form(P, S)$ of strings of symbols involving elements of P, S and brackets (and) formed according to the following rules.

(i)   Each propositional variable is a formula.

(ii) If θ and ψ are formulas, then so are

$$\neg\theta \qquad (\theta \wedge \psi) \qquad (\theta \vee \psi) \qquad (\theta \rightarrow \psi) \qquad (\theta \leftrightarrow \psi)$$

(iii) All formulas arise from finitely many applications of (i) and (ii).

If we use a different set S of connectives, for instance just $\{\vee, \rightarrow\}$, then clause (ii) is amended accordingly to cover just these symbols.

The intended meanings of the connectives are as follows: $\wedge$ will be interpreted by 'and', $\vee$ by 'or', $\neg$ by 'not', $\rightarrow$ by 'implies' and $\leftrightarrow$ by 'if and only if'. With these intended meanings, you can see why clause (ii) of the definition uses $\wedge$, $\vee$, $\rightarrow$ and $\leftrightarrow$ to connect together two formulas, while $\neg$ connects with only one. The brackets used in clause (ii) are a very important part of the definition, playing a crucial part in making it possible to interpret formulas in an unambiguous way.

Taking the set P of propositional variables to be $\{p, q, r\}$, each of the following strings is a formula:

$$q \qquad (p \vee q) \qquad \neg\neg(p \vee q) \qquad (\neg p \wedge (q \rightarrow r)).$$

Let's check that each of these is a formula. The symbol q is a propositional variable, so is a formula by clause (i) of the definition. Each of the symbols p and q are propositional variables and thus formulas by clause (i), so the string $(p \vee q)$ is a formula by clause (ii). A use of the $\neg\theta$ part of clause (ii), taking θ to be $(p \vee q)$, gives that $\neg(p \vee q)$ is a formula; and one more use of the $\neg\theta$ part of clause (ii), this time taking θ to be $\neg(p \vee q)$, gives that $\neg\neg(p \vee q)$ is a formula. Lastly, as p, q, r are formulas, $\neg p$ is a formula by clause (ii), $(q \rightarrow r)$ is a formula by clause (ii), and so $(\neg p \wedge (q \rightarrow r))$ is a formula by clause (ii).

On the other hand, none of the following strings are formulas:

$\qquad\qquad *$ ($*$ is not an allowed symbol)

$\qquad\qquad \rightarrow$ (the only single symbol formulas consist of

$\qquad\qquad\qquad$ a propositional variable)

$\qquad\quad q\neg$ (there's nothing following the $\neg$)

$\qquad\quad p \vee q$ (any formula using $\vee$ must also have some brackets)

$\quad ((\neg r \wedge q))$ (with just one $\wedge$ we can only have one pair of brackets.)

For these last two examples of non-formulas it is tempting to say that we know what they are supposed to represent, so let's call them formulas. But they don't conform to our strict rules – for many purposes in this book, you

In many books the phrase *well-formed formula* is used instead of formula. The 'well-formed' emphasizes that the string has to obey special construction rules.

So all formulas are finitely long.

There are usually several reasons why a string fails to be a formula. In each case, we've just given a single one.

We can't promise a prize to anyone who spots us forgetting brackets in what we say is a formula. But please let the author know about it.

should regard formulas as capable of being recognized and manipulated by a machine, and we shall keep the instructions for such a machine simple by careful use of brackets. These last two examples can be turned into formulas doubtless expressing correctly what was intended by suitable bracketing as

$$(p \vee q) \quad \text{and} \quad (\neg r \wedge q).$$

The use of brackets is a vital part of how we avoid ambiguity. We discussed earlier the 'statement'

'it is snowing or the bus doesn't come and I'll be late for work'.

A corresponding string is

$$(p \vee q \wedge r)$$

and this is not a formula, for instance because any formula containing one $\vee$ and one $\wedge$ would have to have two pairs of brackets (), rather than the one pair in the string. We discussed the two obvious ways of bracketing the statements 'it is snowing', 'the bus doesn't come','I'll be late for work' together, essentially as one of $((p \vee q) \wedge r)$ and $(p \vee (q \wedge r))$. The moral is that brackets matter a lot.

Exercise 2.2

Explain why each of the following are formulas, taking the set P of propositional variables to include p, q, r, s.

(a) $(r \leftrightarrow \neg s)$

(b) $((r \rightarrow q) \wedge (r \vee p))$

(c) $\neg\neg p$

Exercise 2.3

Explain why each of the following is not a formula, taking the set P of propositional variables to include p, q.

(a) $p \leftrightarrow q)$

(b) $(p \,\&\, q)$

(c) $\neg(p)$

(d) $(\neg p)$

(e) $(p \wedge \vee q)$

It's all very well asking you to show that a short string of symbols is a formula – we hope that you had no problem doing this in the last exercise. But for a long string, we really do need something systematic. Likewise it is, we hope, obvious that strings like $(p\neg$ and $(p \wedge q \vee r)$ are *not* formulas. But can we nail down why they are not formulas in a way that will then cope with a long string? We will answer the important question of how one tests whether a string of symbols is a formula first at a fairly informal level and in greater detail later in the section.

Obviously our answer must take account of the definition of a formula.

If a string consists of just a single symbol, then the string is a formula precisely when this symbol is a propositional variable – no string in the sequence can

The issue of checking whether a string of symbols conforms to given construction rules is of considerable practical importance, for instance in many uses of computers. The details of how to do such checks are normally quite complicated and in this book we don't want them to get in the way of our main objective, which is to investigate the properties of strings which *are* formulas.

be empty and if clause (ii) has been used, the resulting string consists of more than one symbol.

When a string ϕ contains more than one symbol, it can only be a formula if it is one of the forms $\neg\theta$, $(\theta \wedge \psi)$, $(\theta \vee \psi)$, $(\theta \rightarrow \psi)$, $(\theta \leftrightarrow \psi)$, for some shorter strings θ, ψ (which of course also have to be formulas). Whichever of the forms ϕ is, the connective that you can see written down in the list we've just given, rather than any connectives that are hidden within the strings θ and ψ, is given a special name, the *principal connective* of ϕ.

If $\neg$ is the principal connective, this would easily be identified by seeing it at the front (i.e. lefthand end) of the string; and that's the only circumstance under which $\neg$ can be the principal connective. If it is one of the other connectives, one way by which the principal connective can be identified is by looking at brackets. Each appearance of one of $\wedge$, $\vee$, $\rightarrow$ and $\leftrightarrow$ brings with it a pair of brackets $(\ldots)$. Brackets give vital information about the way a formula has been constructed and constrain which strings can be formulas. For instance, it is pretty obvious that any formula contains an equal number of left brackets (and right brackets), so that any string for which this fails cannot be a formula. A special property of brackets which identifies the principal connective when it is one of $\wedge$, $\vee$, $\rightarrow$ and $\leftrightarrow$ can be expressed in terms of the number of brackets to its left. For each occurrence of these connectives, we look at

the number of ('s to its left minus the number of)'s to its left.

For instance, in the formula

$$((p \wedge r) \rightarrow (\neg q \vee r)),$$

this 'left minus right bracket count' for the $\wedge$ is 2, for the $\vee$ near the righthand end is also 2, and for the $\rightarrow$, which is the principal connective, is 1. What distinguishes the principal connective when it is one of $\wedge$, $\vee$, $\rightarrow$ and $\leftrightarrow$ is that its 'left minus right bracket count' equals 1, which accords with our example. As another example, in the formula

$$\neg((p \rightarrow \neg(r \vee q)) \rightarrow p)$$

the 'left minus right bracket count' of the leftmost occurrence $\rightarrow$ is 2, for the $\vee$ it's 3, and for the rightmost occurrence of $\rightarrow$, which is the principal connective, it's indeed 1.

Actually, it can be shown that for formulas with more than one symbol whose principal connective isn't $\neg$, there is *exactly one* connective with 'left minus right bracket count' equal to 1, which helps show that strings like

$$p \wedge q \quad \text{and} \quad (p \wedge q \vee r)$$

aren't formulas. For the string $p \wedge q$ there is no connective for which the 'left minus right bracket count' equals 1, while for $(p \wedge q \vee r)$ this equals 1 for more than one connective, both the $\wedge$ and the $\vee$.

A very sketchy description of an algorithm for checking a string of symbols to see whether it is a formula is as follows.

The length of a string will be very important when proving results about those strings which are formulas.

Remember that for this purpose we ignore the $\neg$s.

We shall justify these results about brackets later in this section.

We haven't exhausted here all the ways in which a string can fail to be a formula, but will give an algorithm that detects all of these later in this section.

Look to see if it has a principal connective. If so, split it into the appropriate shorter string(s) and repeat the process: look for the principal connective for each of these shorter string(s) and split them up accordingly, and so on. In this way we analyse successively shorter and simpler strings until we reach strings consisting of just a single propositional variable – the shortest legal sort of string. If we don't trip up at any stage of the process, e.g. by failing to find a principal connective, and every analysis gets down to a propositional variable, our initial string was indeed a formula. In any other case, it was not a formula.

If the principal connective is $\neg$, so the string looks like $\neg\theta$, there's only one shorter string, namely θ. All the other connectives join together two substrings, e.g. the $\wedge$ in $(\theta \wedge \psi)$ joins together the substrings θ and ψ with outer brackets added. In the latter case, if the outer brackets are missing, the string can't be a formula.

Let's illustrate the process for the string $(\neg p \vee ((p \wedge q) \rightarrow \neg r))$, which you can probably see is a formula, by the following diagram.

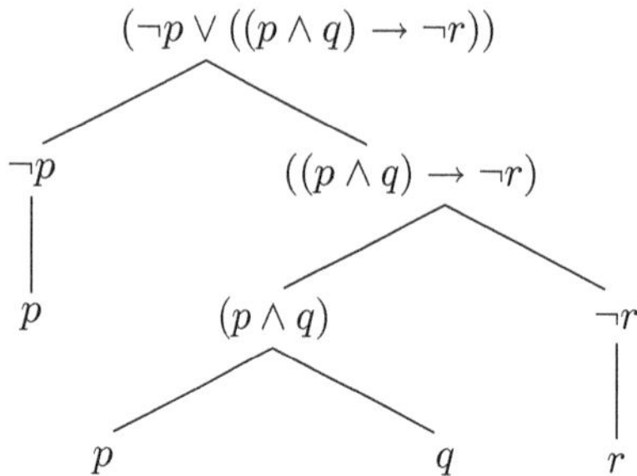

A diagram like this is called a *tree*, even though you might think that it looks like an upside-down tree! The string we are analysing is placed at the top of the diagram. The branches of the tree go downwards and it is no coincidence that, as our string is actually a formula, each branch ends with a propositional variable. How do we construct the tree? We first write down our original string $(\neg p \vee ((p \wedge q) \rightarrow \neg r))$ and attempt to locate its principal connective. If we find a candidate, then what we do next depends on whether it's one of $\{\wedge, \vee, \rightarrow, \leftrightarrow\}$ or it's $\neg$. In this case it's $\vee$, one of the first sort, so under the original string we write the two separate substrings which, when joined together with $\vee$ and a pair of outer brackets added, give the original string – here, these are the strings $\neg p$ and $((p \wedge q) \rightarrow \neg r)$. This gives the first stage of the diagram:

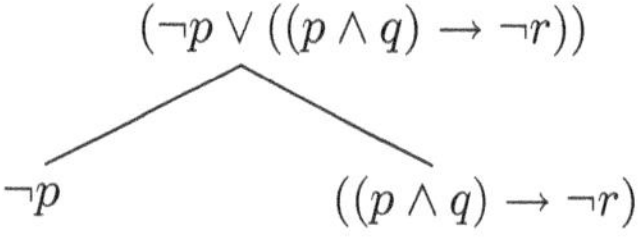

We join the top string to each of these smaller strings with a line to give a sense of them flowing directly from the top string. We repeat the process for each of these shorter strings. One of them begins with a $\neg$, so there's just the one string, namely p to write underneath it; and this is a propositional variable, so this (upside-down!) branch of the tree has successfully stopped at a propositional variable. The diagram now looks like

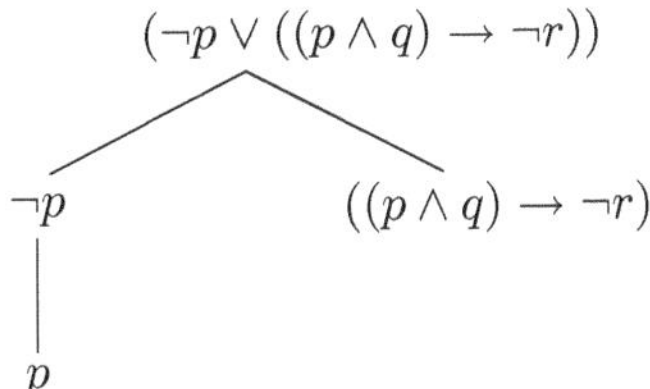

We now do the analysis of the other string $((p \wedge q) \to \neg r)$, finding its principal connective to be the $\to$, and we hope that by now you can see how we obtained the full diagram.

Exercise 2.4

Construct a similar sort of tree for each of the following strings.

(a) $\neg(\neg p \leftrightarrow r)$

(b) $((p \wedge r) \to (\neg p \leftrightarrow q))$

(c) $\neg((\neg r \vee (r \wedge \neg p)) \leftrightarrow \neg\neg\neg q)$

Solution

(a)

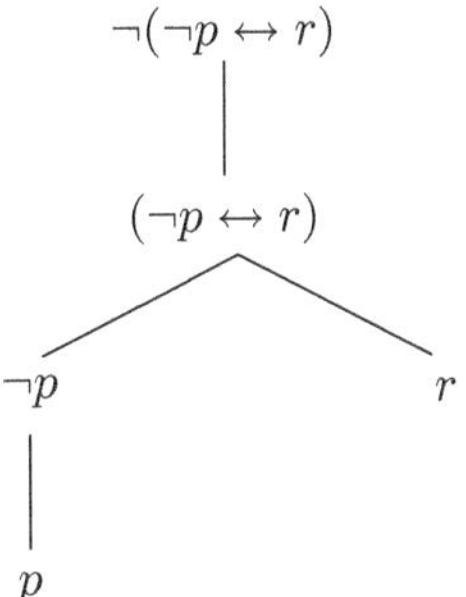

(b) Not given.

(c) Not given.

As we have already observed, in this sort of diagram the constituent parts all are formulas and the branches all end with a propositional variable. It is helpful to have a name for these constituent parts: we call them *subformulas* of the original formula. More formally, we have the following definition.

Definition   Subformula

For all formulas ϕ, their *subformulas* are defined as follows, exploiting the construction rules for formulas.

1.   If ϕ is atomic, then ϕ is the only subformula of itself.

2.   If ϕ is of the form $\neg\psi$, then the subformulas of ϕ are ϕ and all subformulas of ψ.

3.   If ϕ is one of the forms $(\theta \wedge \psi)$, $(\theta \vee \psi)$, $(\theta \rightarrow \psi)$ and $(\theta \leftrightarrow \psi)$, then the subformulas of ϕ are ϕ, all subformulas of θ and all subformulas of ψ.

So the subformulas of $(\neg p \vee ((p \wedge q) \rightarrow \neg r))$ are

$$(\neg p \vee ((p \wedge q) \rightarrow \neg r)), \quad \neg p, \quad p, \quad ((p \wedge q) \rightarrow \neg r),$$
$$(p \wedge q), \quad q, \quad \neg r, \quad r.$$

Note that p occurs as a subformula in more than one place in the original formula, but we need only list it once as a subformula.

We gave the tree diagram for this formula on page 24. Its subformulas are all the formulas involved at some stage of its construction.

Exercise 2.5 ────────────────────────────

Give the subformulas of each of the following formulas (which appeared in Exercise 2.4).

(a) $\neg(\neg p \leftrightarrow r)$

(b) $((p \wedge r) \rightarrow (\neg p \leftrightarrow q))$

(c) $\neg((\neg r \vee (r \wedge \neg p)) \leftrightarrow \neg\neg\neg q)$

We shall now tighten up on some of the details of our sketchy algorithm. First we shall look at an example of how to prove certain sorts of result about formulas, in particular the result that any formula ϕ contains an equal number of left brackets (and right brackets). Although we hope that this is somehow obvious from clause (ii) of the definition of a formula, it is instructive to see how to justify it in a more formal way – we shall need this style of argument several times later in the book to justify much less obvious results about formulas! Our challenge here is to prove something about *all* formulas ϕ, however complex and long they are. The way we shall proceed is by mathematical induction on the *length* of ϕ. There are several sensible measures of the length of a formula ϕ, for instance the total number of symbols in it or the height (that is, length of the longest branch) in its construction tree. Our preferred measure of length is the number of occurrences of connectives in ϕ, so that the length of the formula

$$((q \wedge \neg\neg r) \wedge (p \vee (r \rightarrow \neg q)))$$

We shall say 'the number of connectives' for short.

is 7 (consisting of 3 $\neg$s, 2 $\wedge$s, 1 $\vee$ and 1 $\rightarrow$). The smallest possible length of a formula is then 0, when ϕ is just a propositional variable p, for some p – the definition of formula allows no other possibility.

The structure of this sort of proof is as follows. Show first that the result holds for all formulas of length 0 – the basis of the induction. Then we do the inductive step: assume that the result holds for all formulas of the given type with length $\leq n$ – this is the induction hypothesis for n – and from this show that it holds for all formulas of length $\leq n + 1$. As we are assuming that the hypothesis holds for all formulas of length $\leq n$, this boils down to showing that the hypothesis holds for formulas whose length is exactly $n + 1$. By the principle of mathematical induction the hypothesis then holds for formulas of all lengths, i.e. all formulas.

Let us now use this method of proof to establish the following theorem about brackets.

Theorem 2.1

Any formula ϕ contains an equal number of left brackets (and right brackets).

Proof

Our induction hypothesis is that all formulas of length $\leq n$ contain an equal number of left and right brackets.

If ϕ is a formula of length 0, it can only be a propositional variable, thus containing an equal number, namely zero, of left and right brackets. Thus the hypothesis holds for $n = 0$.

Now suppose that the result holds for all formulas of length $\leq n$. To show from this that the result holds for all formulas of length $\leq n + 1$, all that is needed is to show that it holds for formulas of length $n + 1$, as those of shorter length are already covered by the induction hypothesis for n. So let ϕ be such a formula of length $n + 1$. As ϕ has at least one connective, it cannot be simply a propositional variable, so must be a formula by an application of clause (ii) in the definition of formula, that is, it must be of one of the five forms

$$\neg\theta, \quad (\theta \wedge \psi), \quad (\theta \vee \psi), \quad (\theta \rightarrow \psi), \quad (\theta \leftrightarrow \psi)$$

where θ and ψ are formulas. We must deal with each of these possible forms. In all five cases, as ϕ has $n + 1$ connectives, θ and ψ have at most n connectives, so that the inductive hypothesis will apply to them.

Case: ϕ is of the form $\neg\theta$

As θ has length n, by the induction hypothesis θ contains an equal number, say k, of left and right brackets. The formation of the string $\neg\theta$ from θ doesn't add further brackets, so that this form of ϕ also contains an equal number, namely k, of left and right brackets.

Case: ϕ is of the form $(\theta \wedge \psi)$

As both θ and ψ have length $\leq n$, by the induction hypothesis θ contains an equal number, say k, of left and right brackets, while ψ contains an equal number, say j, of left and right brackets. The formation of the string $(\theta \wedge \psi)$ from θ and ψ adds an extra left bracket and an extra right bracket to those in θ and ψ, so that this form of ϕ contains an equal number, namely $k + j + 1$, of left and right brackets.

You might like to think why our induction hypothesis isn't simply that the result holds for all formulas of length exactly equal to n, rather than $\leq n$. The reason for this will become clear soon!

If ϕ is of the form $\neg\theta$, then θ has exactly n connectives. If ϕ is of the form $(\theta \wedge \psi)$, the $\wedge$ accounts for one of the $n + 1$ connectives in ϕ, leaving the remaining n connectives to be distributed somehow between θ and ψ.

It can be shown that the result holds for ϕ of each of the three remaining forms, completing the inductive step. It follows by mathematical induction that the result holds for all $n \geq 0$, that is for all formulas. ■

Exercise 2.6

Fill in the gaps in inductive step of the proof above for the cases when ϕ is of one of the forms $(\theta \vee \psi)$, $(\theta \rightarrow \psi)$, $(\theta \leftrightarrow \psi)$.

Solution

We hope that this is seen as essentially trivial, simply replacing the $\wedge$ in the argument for the case when ϕ is of the form $(\theta \wedge \psi)$ by, respectively, $\vee$, $\rightarrow$ and $\leftrightarrow$. In all but the most fastidious circles, one would merely complete the proof given above of the theorem by saying that the other cases are similar to that of $(\theta \wedge \psi)$!

Exercise 2.7

(a) Use mathematical induction on the length of a formula to show that the number of occurrences of the symbol $\wedge$ in a formula ϕ is less than or equal to the number of left brackets (in the formula.

(b) Does the result of part (a) hold for the symbol $\neg$? Justify your answer.

Recall that our preferred measure of length of a formula is the number of occurrences of connectives in it.

We shall now describe a more comprehensive algorithm for checking whether a string of symbols is a formula. For simplicity in most of the rest of the section, we shall suppose that the language used involves only two propositional variables p and q, the connective $\wedge$ and brackets. It is very straightforward to extend our algorithm to cope with richer languages. There are several possible algorithms and we shall go for one which treats the brackets as the crucial component. Consider the following string, which is a formula:

$$(((p \wedge q) \wedge p) \wedge (q \wedge p)).$$

By Theorem 2.1, the number of left brackets in a formula equals the number of right brackets. We can check this for the formula above by moving along the string from left to right keeping a count of the difference between the numbers of left and right brackets in the following way. Start at the lefthand end of the string with the count at 0. Whenever we meet a left bracket, we add 1 to the count. When we meet a right bracket, we subtract 1 from the count. In this way we associate a number with each bracket in the string, as follows:

$$(((p \wedge q) \wedge p) \wedge (q \wedge p))$$
$$1\,2\,3 \quad\ 2 \quad\ 1 \quad 2 \quad\ \ 1\,0$$

An undercurrent in the development of the subject is whether there is an algorithm for generating true statements of mathematics. For this to be remotely feasible, we need an algorithm for checking whether a string of symbols is a statement, hence our interest in this algorithm for a very simple language.

We shall call the number associated with each bracket in this way its *bracket count*. The final bracket has bracket count 0, which confirms that the number of left brackets equals the number of right brackets, as we would expect for a formula. Furthermore, the bracket count is greater than zero for any bracket before this final one, where it is 0. These properties apply to all formulas, not just this one, and can be proved using mathematical induction on the length of formulas. Of course, these properties by themselves don't ensure that the string is a formula.

E.g. consider the string
$$(p)$$
$$1 \quad\ 0$$

Another important feature of the bracket count is that it helps identify which occurrence of the connective $\wedge$ is the principal connective of the formula. In our example, observe that the principal connective happens to follow a bracket with a count of 1. This is not a coincidence. In any formula in our restricted language containing at least one occurrence of $\wedge$, there will be *exactly one* occurrence which follows a bracket with a count of 1 and this will be the principal connective. This principle works for the further example

Earlier we described this occurrence as having a 'left minus right bracket count' of 1.

$$(q\wedge((p\wedge p)\wedge q))$$
$$1 \quad 2\,3 \quad\;\; 2 \quad\;\; 1\,0$$

where the relevant bracket is a left bracket (rather than a right bracket) as in our first example. Let's try to explain informally why the principle holds in general.

A formula containing an $\wedge$ will be of the form $(\phi \wedge \psi)$, where the $\wedge$ which is the principal connective is the one that we can see between the ϕ and the ψ – these subformulas might, of course, contain other occurrences of $\wedge$. If the first subformula ϕ contains brackets, then its bracket count is greater than zero for any bracket before its final one, where the count is 0. Thus in the formula $(\phi \wedge \psi)$, which has an extra left bracket at its lefthand end, all the bracket counts for the subformula ϕ increase by 1. That means that the bracket count for $(\phi \wedge \psi)$ looks something like this:

$$\overbrace{}^{\phi} \qquad \overbrace{}^{\psi}$$
$$((\ldots(\ldots)\wedge(\ldots(\ldots)))$$
$$1\,2 \qquad 1 \quad 2 \qquad\;\; 1\,0$$

It starts by going straight from 1 to 2 as ϕ is entered. It then first goes back to 1 at the final bracket of ϕ, just before the principal connective, as required. If the subformula ψ contains brackets, the bracket counts increase – all the counts for ψ on its own also increase by 1, so the count for $(\phi \wedge \psi)$ only gets back to 1 at the final bracket of ψ, which is then followed by the final bracket of $(\phi \wedge \psi)$.

We leave it to you to think about the cases when one or both of ϕ and ψ contain no brackets, meaning that it is simply a propositional variable.

The algorithm will then be as follows, starting with the string which one wishes to test to see whether it is a formula. At any stage when the algorithm declares that the string is a formula, the process halts. Similarly it halts when the algorithm declares that the string is not a formula.

(i) Test the lefthand symbol of the string.

    (a) If it is one of p and q, check if this is the only character in the string. If this is so, the string is a formula; if not, then it isn't a formula.

    (b) If it is $\wedge$ or a right bracket), then the string isn't a formula.

    (c) If it is a left bracket (, then proceed to step (ii).

(ii) Check whether the string ends in a right bracket). If not, the string is not a formula. Otherwise proceed to step (iii).

(iii) Compute the bracket count. Moving along the string from left to right, locate the first occurrence of $\wedge$ following a bracket with count 1. If there is no such occurrence of $\wedge$, the string is not a formula. Otherwise, proceed to step (iv).

(iv) Use this occurrence of $\wedge$ to split the string into two substrings: one consisting of all the symbols to the left of the $\wedge$ except for the initial left bracket (; and the other consisting of all the symbols to the right of the $\wedge$ except for the final right bracket).

(v) Now apply the algorithm starting with (i) to both of these substrings. If both substrings are formulas, then the string is a formula.

Note that at stage (v), the substrings are shorter than the original (finite!) string, so that the algorithm will stop with a result after a finite number of steps.

Exercise 2.8 ⎯⎯⎯⎯⎯⎯⎯⎯⎯⎯⎯⎯⎯⎯⎯⎯⎯⎯⎯⎯⎯⎯⎯⎯⎯

How does our algorithm detect the case when there are two occurrences of $\wedge$ which follow a bracket count of 1?   Of course, in such a case the string is not a formula.

Exercise 2.9 ⎯⎯⎯⎯⎯⎯⎯⎯⎯⎯⎯⎯⎯⎯⎯⎯⎯⎯⎯⎯⎯⎯⎯⎯⎯

Adapt our algorithm for strings built up using the propositional variables p, q and the connective $\wedge$ so that it tests strings which might also include the connective $\neg$.

Now that we have a definition of formula, we can look at how to interpret formulas and discuss their truth or falsity in an interpretation. This is the subject of the next section.

Further exercises

Exercise 2.10 ⎯⎯⎯⎯⎯⎯⎯⎯⎯⎯⎯⎯⎯⎯⎯⎯⎯⎯⎯⎯⎯⎯⎯⎯

Show that in all formulas θ built up using the propositional variables p, q and the connective $\wedge$, the bracket count is greater than zero for any bracket of θ before its final one where the count is 0.

Exercise 2.11 ⎯⎯⎯⎯⎯⎯⎯⎯⎯⎯⎯⎯⎯⎯⎯⎯⎯⎯⎯⎯⎯⎯⎯⎯

Suppose that formula ϕ is built up using only $\wedge$ and $\vee$ and has connective length n. What can you say about the number of subformulas of ϕ? What can be said if ϕ might include the connective $\neg$ as well as $\wedge, \vee$?

Exercise 2.12 ⎯⎯⎯⎯⎯⎯⎯⎯⎯⎯⎯⎯⎯⎯⎯⎯⎯⎯⎯⎯⎯⎯⎯⎯

Show that in all formulas θ built up using the propositional variables p and q using the connective $\wedge$ and containing at least one occurrence of $\wedge$, there is exactly one occurrence of $\wedge$ which follows a bracket with a count of 1.

2.3   The interpretation of propositional formulas

We shall now describe how to give meaning to the formal language, giving
what is called its semantics. Recall that we introduced the simplest (shortest!)
sort of formula, a propositional variable, by saying that it was intended to
stand for a basic component proposition, like 'f is a continuous function'. In
normal mathematics, the truth of this will depend on whether the f we are
given is indeed a continuous function and, for that matter, what we mean by
a continuous function. But for propositional calculus, this level of detail of
how a propositional variable is interpreted is much greater than we shall need
in this chapter. For purposes like deciding whether one statement or formula
is a consequence of others within the propositional calculus, all we shall need
to know about each propositional variable is whether, in a particular set of
circumstances, it is true or false. Once we have specified how to interpret the
connectives, we can then say how the truth or falsity of more complicated
formulas depends on that of the propositional variables, which are the basic
building blocks. This in turn will allow us to say whether one formula is a
consequence of others.

We shall be much more interested in what each basic proposition expresses when we look at the predicate calculus.

Hidden in the preamble above and implicit in earlier discussions in the book
is an important decision. Under a given set of circumstances, a statement is
either true or false – one or the other, and no sort of half-truth in between.
We hope that this seems perfectly reasonable. In everyday mathematics, a
statement like 'f is a continuous function' is just one of true or false, depending
on the f we are given. However, there are circumstances where it might make
sense to describe the truth of a statement in a less black and white way, for
instance giving a probability that the statement is true; and one of the ways
in which you could extend your knowledge beyond this book is by learning
about other ways of analysing what is meant by truth. *For the rest of this
book*, our standard measure of the truth of a statement will be in terms of
the two distinct values 'true' and 'false'. We shall describe each of these as a
truth value and abbreviate them by T for 'true' and F for 'false'. So the set
of truth values is the two element set $\{T, F\}$.

We have talked informally about knowing whether, in a particular set of circumstances, each propositional variable is true or false. More formally and
elegantly, this set of circumstances is a function $v\colon P \longrightarrow \{T, F\}$, where P
is the set of propositional variables in our language. The function v gives
a truth value to each propositional variable in P, thus describing the set of
circumstances. We will explain how to extend such a function v so that it
assigns a truth value to each formula built up from P using connectives in
a set S – the function so obtained will be called a *truth assignment*. A key
step is to specify how to interpret the connectives. For each of these, we shall
explain how the truth of a formula ϕ with it as principal connective depends
on the truth of the subformulas it connects to form ϕ. We shall look at each
of the connectives introduced so far in the book, namely $\neg$, $\wedge$, $\vee$, $\rightarrow$ and $\leftrightarrow$,
the intended meanings of which we have already said are, respectively, 'not',
'and', 'or', 'implies' and 'if and only if'.

The issue of other reasonable connectives used in everyday discourse is delayed until Section 2.5.

¬ (negation)

A formula of the form ¬θ for some formula θ with principal connective ¬ is called the *negation* of θ. We shall specify how its truth value is to be related to the truth value of θ. As our intended way of interpreting ¬ is as 'not', we want ¬θ to have the value F (false) when θ has the value T (true) and the value T when θ has the value F, i.e. ¬θ will have the opposite value to that assigned to θ. We can summarize this by the following table.

We may also sometimes refer to a formula of the form ¬θ as a *negation*.

θ	¬θ
T	F
F	T

∧ (conjunction)

A formula of the form $(\theta \wedge \psi)$ for some formulas θ, ψ with principal connective ∧ is called the *conjunction* of θ and ψ. Each of the formulas θ and ψ is called a *conjunct* of $(\theta \wedge \psi)$. Our intended way of interpreting ∧ is as 'and', so we shall assign $(\theta \wedge \psi)$ the value 'true' exactly when both θ and ψ are assigned the value 'true'. We can summarize this by the following table.

We may also sometimes refer to a formula of the form $(\theta \wedge \psi)$ as a *conjunction*.

So if one or both of θ and ψ are false, then so is $(\theta \wedge \psi)$.

θ	ψ	$(\theta \wedge \psi)$
T	T	T
T	F	F
F	T	F
F	F	F

This sort of table, giving the truth values of a formula constructed from some of its subformulas for all possible combinations of truth values of these subformulas, is called a *truth table*. Here the formula $(\theta \wedge \psi)$ is given in terms of the subformulas θ and ψ. There are four combinations of truth values for θ and ψ, so this truth table has 4 rows. Our earlier table for negation gave the truth values of ¬θ in terms of the values of the subformula θ, so this truth table only required 2 rows.

∨ (disjunction)

A formula of the form $(\theta \vee \psi)$ for some formulas θ, ψ with principal connective ∨ is called the *disjunction* of θ and ψ. Each of the formulas θ and ψ is called a *disjunct* of $(\theta \vee \psi)$. Our intended way of interpreting ∨ is as 'or', but unlike 'not' and 'and' earlier, we run into the problem that there is more than one way of using 'or' in English. One way, called the *exclusive* 'or', makes $(\theta \vee \psi)$ true when exactly one of θ and ψ is true – the truth of one of them excludes the truth of the other. For instance, many restaurants offer a fixed price menu with a choice of dishes for each course. The choice for each course is to be read as a disjunction with the exclusive use of 'or' – you can have any one of the soup, terrine and prawn cocktail, but only one. Another use of 'or' is in what is called an *inclusive* way, where $(\theta \vee \psi)$ is true when one or both of θ and ψ are true. For instance, a common sort of argument in maths is along the lines of 'if x or y are even integers, then xy is even', where 'x or y are even' includes the case that both x and y are even. Because this way of using

We may also sometimes refer to a formula of the form $(\theta \vee \psi)$ as a *disjunction*.

'or' is pretty well standard in mathematics, we shall choose to interpret $\vee$ in the *inclusive* way, as given by the following truth table.

θ	ψ	$(\theta \vee \psi)$
T	T	T
T	F	T
F	T	T
F	F	F

$\rightarrow$ (implication)

A formula of the form $(\theta \rightarrow \psi)$ for some formulas θ, ψ with principal connective $\rightarrow$ is called an *implication*. Our intended way of interpreting $\rightarrow$ is as 'implies', or 'if …then', which suggests some of the rows of the truth table in the following rather backhanded way. In normal use of 'if …then' in English, from being told that 'if θ then ψ' is true and that θ is true, we would expect ψ to be true. Likewise if we are told that θ is true and ψ is false, then 'if θ then ψ' would have to be false. This then settles two of the rows of the truth table, as follows:

The formula θ is called the *antecedent* and ψ the *consequent* of the implication.

θ	ψ	$(\theta \rightarrow \psi)$
T	T	T
T	F	F
F	T	$?$
F	F	$?$

It may not be immediately obvious from normal English how to fill in the remaining two rows of the table, covering the cases where θ is false. It is a constraint of the process of making a simple model of this fragment of natural language and argument that we have to make some sort of decision about the truth value of $(\theta \rightarrow \psi)$ when θ is false, and our decision is to make $(\theta \rightarrow \psi)$ true on these rows, giving the following truth table.

θ	ψ	$(\theta \rightarrow \psi)$
T	T	T
T	F	F
F	T	T
F	F	T

This table might be made more memorable by thinking of it as saying that $(\theta \rightarrow \psi)$ is false only when θ is true and ψ is false – surely circumstances when 'θ implies ψ' has to be false.

The decision we have taken about the bottom two rows in this table is consistent with the way we handle implication in everyday mathematics. We frequently state theorems in the form 'if …then', for instance the following theorem:

'for all $x \in \mathbb{R}$, if $x > 2$, then $x^2 > 4$'.

We hope that this result strikes you not only as correct (which it is!) but as a familiar way of expressing a host of mathematical results using 'if …then'. Given that we regard this result as true, we must surely also regard

'if $x > 2$, then $x^2 > 4$'

as being true for each particular x in $\mathbb{R}$, and we want the truth table for 'if …then' to reflect this. Taking some particular values for x, for instance 3, 1

Giving the value 'true' to the statement 'if $x > 2$, then $x^2 > 4$' even when the particular value of x makes $x > 2$ false is a fair reflection that there is a correct proof of 'if $x > 2$, then $x^2 > 4$'.

and -5, this means we want our truth table for 'if ... then' to give the value T in each of the following circumstances:

when $x = 3$, both $x > 2$ and $x^2 > 4$ have the value T;

when $x = 1$, both $x > 2$ and $x^2 > 4$ have the value F;

when $x = -5$, $x > 2$ has the value F but $x^2 > 4$ has the value T.

These cases correspond to all the rows of our truth table for $\rightarrow$ which result in the value T.

The connective $\rightarrow$ with its intended meaning of 'implies' is perhaps the most important of the connectives we have introduced. This is because a major use of our formal language is to represent and analyze mathematical arguments and theorems, and a salient feature of these is the use of implication both to state and prove results.

Note that the truth of the statement $(\theta \rightarrow \psi)$ does not necessarily entail any special relationship between θ and ψ, for instance that in some sense θ causes ψ. To test your understanding of the truth table of $\rightarrow$, try the following exercise.

Exercise 2.13

The British psychologist Peter Wason (1924–2003) devised a famous experiment involving people's understanding of 'if ... then ...' as follows. The experimenter lays down four cards, bearing on their uppermost faces the symbols A, B, 2 and 3 respectively. The participants are told that each card has a letter on one side and a number on the other side. Their task is to select just those cards that they need to turn over to find out whether the following assertion is true or false: 'If a card has an A on one side, then it has a 2 on the other side.' Which cards should be turned over?

Solution

In the original experiment and in subsequent trials, most people selected the A card and, perhaps, the 2 card. Surprisingly they failed to select the 3 card. According to Wason's obituary in the Guardian (25th April 2003), the experiment 'has launched more investigations than any other cognitive puzzle. To this day – and Wason's delight – its explanation remains controversial.'

The 2 card doesn't need to be turned over to test the assertion!

$\leftrightarrow$ (bi-implication)

A formula of the form $(\theta \leftrightarrow \psi)$ for some formulas θ, ψ with principal connective $\leftrightarrow$ is called a *bi-implication* of θ and ψ. Our intended way of interpreting $\leftrightarrow$ is as 'if and only if', so we shall assign $(\theta \leftrightarrow \psi)$ the value 'true' exactly when the truth value of θ matches that of ψ. We can summarize this by the following truth table.

Why do you think that $\leftrightarrow$ is formally described as 'bi-implication'?

θ	ψ	$(\theta \leftrightarrow \psi)$
T	T	T
T	F	F
F	T	F
F	F	T

Exercise 2.14 ____

We interpret the formula $(\theta \leftrightarrow \psi)$ as 'θ if and only if ψ'. Write down two formulas involving the connective $\rightarrow$, one which represents 'θ if ψ' and the other representing 'θ only if ψ'. Often in everyday maths we interpret $(\theta \leftrightarrow \psi)$ as 'θ is a necessary and sufficient condition for ψ'. Which of your formulas represents 'θ is a necessary condition for ψ' and which represents 'θ is a sufficient condition for ψ'?

And say which is which!

____

An important point to note about the truth tables for the connectives $\neg$, $\wedge$, $\vee$, $\rightarrow$ and $\leftrightarrow$ is that these are not simply conventions for the purposes of this book when working out the truth values of formulas under an interpretation of the formal language. They also reflect how in normal mathematics we determine the truth of statements made involving their standard interpretations as, respectively, 'not', 'and', 'or', 'implies' and 'if and only if'. In particular when we discuss our formalization of statements and arguments in this book, in 'normal' language (what we called earlier the metalanguage), we shall use these standard interpretations.

Now that we have said how to interpret each of the connectives, we can turn to the interpretation of formulas in general. Our aim is to define a *truth assignment*, that is a special sort of function from the set of all formulas to the set $\{T, F\}$ of truth values, which turns particular truth values given to the propositional variables into a truth value for any formula built up from them, exploiting the truth tables of the connectives. We shall approach this by making more precise what we mean by exploiting these truth tables, starting with the truth table for $\wedge$.

Let $Form(P, S)$ be the set of all formulas built up from propositional variables in a set P using connectives in a set S which includes $\wedge$. We shall say that a function $v\colon Form(P, S) \longrightarrow \{T, F\}$ *respects the truth table of* $\wedge$ if

$$v((\theta \wedge \psi)) = \begin{cases} T, & \text{if } v(\theta) = v(\psi) = T, \\ F, & \text{otherwise,} \end{cases}$$

for all formulas $\theta, \psi \in Form(P, S)$. That is, the value of $v((\theta \wedge \psi))$ is related to those of $v(\theta)$ and $v(\psi)$ by the truth table for $\wedge$:

$v(\theta)$	$v(\psi)$	$v((\theta \wedge \psi))$
T	T	T
T	F	F
F	T	F
F	F	F

You can probably guess how we are going to exploit this definition. If we have $v(p) = v(q) = T$ and $v(r) = F$, where p, q, r are propositional variables, and v respects the truth table for $\wedge$, then $v((p \wedge q)) = T$ and $v(((p \wedge q) \wedge r)) = F$.

Likewise we say that v respects the truth tables of $\neg$ and respectively $\vee$ if

$$v(\neg\theta) = \begin{cases} F, & \text{if } v(\theta) = T, \\ T, & \text{if } v(\theta) = F, \end{cases}$$

and

$$v((\theta \vee \psi)) = \begin{cases} F, & \text{if } v(\theta) = v(\psi) = F, \\ T, & \text{otherwise,} \end{cases}$$

for all formulas $\theta, \psi \in Form(P, S)$.

In a similar way we can define that the function v respects the truth tables of other connectives.

In the next section we shall discuss other connectives besides $\neg$, $\wedge$, $\vee$, $\rightarrow$ and $\leftrightarrow$.

Exercise 2.15 ______________________

Suggest definitions for v respects the truth tables of $\rightarrow$ and $\leftrightarrow$.

Solution

$$v((\theta \rightarrow \psi)) = \begin{cases} F, & \text{if } v(\theta) = T \text{ and } v(\psi) = F, \\ T, & \text{otherwise,} \end{cases}$$

and

$$v((\theta \leftrightarrow \psi)) = \begin{cases} T, & \text{if } v(\theta) = v(\psi), \\ F, & \text{otherwise,} \end{cases}$$

for all formulas $\theta, \psi \in Form(P, S)$.

We can now give the key definition giving the truth value of a formula under a given interpretation of the propositional variables it contains.

Definitions Truth assignment

Let P be a set of propositional variables and S a set of connectives. A function $v\colon Form(P, S) \longrightarrow \{T, F\}$ is said to be a *truth assignment* if v respects the truth tables of all the connectives in S. We shall sometimes call $v(\phi)$ the *truth value of ϕ under v*.

We shall sometimes describe this v as a *truth assignment on P*.

If $v(\phi) = T$, we shall often say that '*v makes ϕ true*' or that '*v satisfies ϕ*'; and we adapt this terminology appropriately when $v(\phi) = F$.

Often we shall not be very specific about the sets P and S in this definition, and will rely on the context making it obvious what these sets are being taken to be.

It might appear to be very cumbersome, if not downright impossible, to give an example of a truth assignment v as we would have to give the value of $v(\phi)$ for every formula ϕ, however long and complicated. Fortunately a truth assignment v can essentially be described simply by giving the values of $v(p)$ for all propositional variables p. As v respects all relevant connectives, it can be shown that $v(\phi)$ is completely determined by these values of $v(p)$. Furthermore, for any choice of truth values for the propositional variables, there is a truth assignment v taking these given values on the propositional variables. This is the import of the following vital result.

Theorem 2.2

Let P be a set of propositional variables, let S be the set of connectives $\{\neg, \wedge, \vee, \rightarrow, \leftrightarrow\}$ and let $v\colon P \longrightarrow \{T, F\}$ be a function. Then there is a unique truth assignment $\overline{v}\colon Form(P, S) \longrightarrow \{T, F\}$ such that $\overline{v}(p) = v(p)$ for all $p \in P$.

Proof

Existence can be demonstrated by defining $\overline{v}$ as follows, using what is called *recursion* on the length of a formula, exploiting the construction of a formula from subformulas of shorter lengths, with propositional variables as the basic building blocks with length 0.

For any formula of length 0, such a formula can only be a propositional variable p in P, in which case define $\overline{v}(p)$ to be $v(p)$.

Now suppose that $\overline{v}(\phi)$ has been defined for all formulas of length $\leq n$. Let ϕ be a formula of length $n + 1$. As $n + 1 > 0$, ϕ has a principal connective (which is unique), so is one of the forms $\neg\theta$, $(\theta \wedge \psi)$, $(\theta \vee \psi)$, $(\theta \rightarrow \psi)$ and $(\theta \leftrightarrow \psi)$, where θ and ψ are of length $\leq n$, so that both $\overline{v}(\theta)$ and $\overline{v}(\psi)$ have already been defined. Now define $\overline{v}(\phi)$ by using the appropriate row of the truth table for its principal connective with these values of $\overline{v}(\theta)$ and $\overline{v}(\psi)$.

This process, exploiting what's called the *recursion principle*, defines $\overline{v}(\phi)$ for all formulas ϕ and thus defines a function $\overline{v}\colon Form(P, S) \longrightarrow \{T, F\}$. Plainly the construction guarantees that $\overline{v}$ respects the truth tables of all the connectives in S, that is, $\overline{v}$ is a truth assignment.

We now need to prove that the function $\overline{v}$ is unique. We suppose that $v'\colon Form(P, S) \longrightarrow \{T, F\}$ is another truth assignment with $v'(p) = v(p)$ for all $p \in P$. We shall use mathematical induction to show that for all formulas ϕ of length $\leq n$, $\overline{v}(\phi) = v'(\phi)$, where $n \geq 0$. As every formula has a finite length, this will show that $\overline{v}(\phi) = v'(\phi)$ for all formulas ϕ, so that the functions $\overline{v}$ and v' are equal.

Any formula of length 0 has to be a propositional variable p in P, in which case both $\overline{v}(p)$ and $v'(p)$ are, by definition, $v(p)$, and are thus equal.

Now suppose that for all formulas ϕ of length $\leq n$, $\overline{v}(\phi) = v'(\phi)$, and that ϕ is a formula of length $n + 1$. Then ϕ is one of the forms $\neg\theta$, $(\theta \wedge \psi)$, $(\theta \vee \psi)$, $(\theta \rightarrow \psi)$ and $(\theta \leftrightarrow \psi)$, where θ and ψ are of length $\leq n$, so that

$$\overline{v}(\theta) = v'(\theta) \text{ and } \overline{v}(\psi) = v'(\psi).$$

In all cases, as $\overline{v}$ and v' are truth assignments, and so respect the truth tables of all the connectives in S, we then have $\overline{v}(\phi) = v'(\phi)$. By mathematical induction, we have $\overline{v}(\phi) = v'(\phi)$ for formulas of all lengths $n \geq 0$, i.e. for all formulas ϕ. $\blacksquare$

One way of phrasing this result is that any assignment of truth values to the propositional variables of a language can be extended to a unique truth

If you are not familiar with recursion and the recursion principle, then you can simply take this theorem on trust or you can look at the details in e.g. Enderton [12].

In general, two functions f, g are equal if they have the same domain A and the same effect on each element of the domain, i.e. $f(a) = g(a)$ for all $a \in A$.

assignment. An important consequence of it is that the effect of a truth assignment v is completely determined by the values it gives to the propositional variables, as we claimed earlier. A full explanation of this is as follows.

Given a truth assignment v, look at the restriction w of the function v to the set P of propositional variables (which is a subset of the domain $Form(P, S)$ of v as each propositional variable is a formula). This means that w is the function from P to $\{T, F\}$ defined by $w(p) = v(p)$ for all $p \in P$. Our task is to show that the effect of v on *all* formulas is determined by the effect of this w just on propositional variables. By the result above, w can be extended to a unique truth assignment $\overline{w}$. This means that (a) $\overline{w}$ is a truth assignment, (b) $\overline{w}(p) = w(p) = v(p)$ for all $p \in P$ and (c) $\overline{w}$ is the only truth assignment with property (b). But the v we started with is a truth assignment and satisfies (b). So by (c) $\overline{w}$ is v.

We have used the letter w rather than the standard notation $v|_P$ for this restriction function in the hope that it will make this passage easier to read!

Now we know that a truth assignment v is determined by the values it gives to the propositional variables, we can look at some examples. In practice we are given the value of $v(p)$ for each propositional variable p. How do we then compute the truth value $v(\phi)$ for a formula ϕ? We exploit the fact that a truth assignment respects truth tables and the construction of ϕ from its subformulas. For a simple formula, the process is easy. For instance, suppose that v is a truth assignment with $v(p) = F, v(q) = T$ and that we want the value of $v((p \wedge (q \to \neg p)))$. As v respects truth tables, we have

$$v(\neg p) = T$$
$$v((q \to \neg p)) = T$$
$$v((p \wedge (q \to \neg p))) = F.$$

Exercise 2.16 ____

Suppose that v is a truth assignment with $v(p) = T, v(q) = F$. What is the value of $v((p \wedge (q \to \neg p)))$?

Solution

As v respects truth tables, we have

$$v(\neg p) = F$$
$$v((q \to \neg p)) = T$$
$$v((p \wedge (q \to \neg p))) = T.$$

What we have done for these simple formulas is essentially to build up to the truth value of the whole formula by working out the truth values of its subformulas. This method works just as well for more complicated formulas. Think of a formula in terms of its construction tree. Take for instance the formula $(\neg p \vee ((p \wedge q) \to \neg r))$ and the truth assignment v with $v(p) = T, v(q) = T, v(r) = F$. We constructed the tree for this formula on page 24 and reproduce it below for convenience.

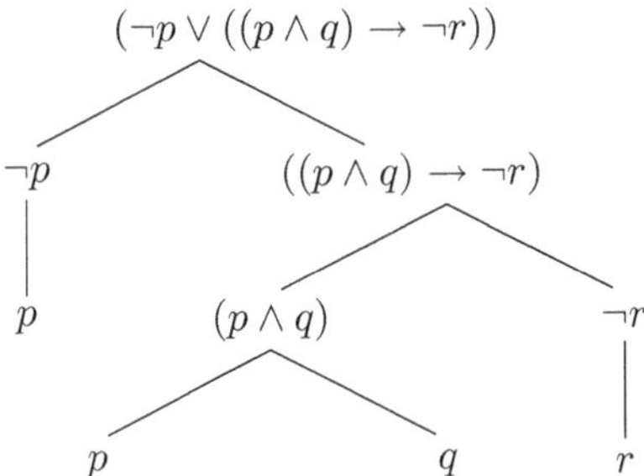

For the given v, we work out $v(\psi)$ for all subformulas ψ of ϕ, starting from the simplest subformulas, namely the propositional variables appearing in ϕ, and work our way up the tree till we get up to $v(\phi)$. Here this gives

$$v(p) = T$$
$$v(q) = T$$
$$v(r) = F$$
$$v(\neg p) = F$$
$$v((p \wedge q)) = T$$
$$v(\neg r) = T$$
$$v(((p \wedge q) \to \neg r)) = T$$
$$v((\neg p \vee ((p \wedge q) \to \neg r))) = T.$$

In practice, when working out $v(\phi)$ for a given v and ϕ, all one uses are the subformulas ψ of ϕ, rather than the whole tree, which saves some paper! But the principle is the same: starting with the values of $v(p)$ for the propositional variables that appear in ϕ, work out the values of $v(\psi)$ for progressively more complicated subformulas of ϕ until one obtains the value of $v(\phi)$ itself.

Of course, once one is familiar with the game, one can often take some shortcuts. In this example, once one has worked out that $v(\neg r) = T$, the truth table for $\to$ gives that $v(((p \wedge q) \to \neg r))$ has to equal T, irrespective of the value of $v((p \wedge q))$. Then, thanks to the truth table for $\vee$, $v((\neg p \vee ((p \wedge q) \to \neg r)))$ has to equal T, irrespective of the value of $v(\neg p)$. But be warned that such shortcuts cannot always be taken.

Exercise 2.17 ————————————————————————————————————

Let v be the truth assignment defined on the set of propositional variables $\{p, q, r\}$ by $v(p) = T$, $v(q) = F$, $v(r) = F$. Find the truth value under v of each of the following formulas.

(a) $\neg q$

(b) $(\neg p \vee r)$

(c) $(p \leftrightarrow (\neg r \to s))$

(d) $((q \wedge (r \to \neg r)) \vee ((p \vee r) \leftrightarrow \neg q))$

(e) $(p \to ((\neg r \to p) \to (\neg q \vee r)))$

Solution

(a) As $v(q) = F$, we have $v(\neg q) = T$.

(b) The subformulas here are p, r, $\neg p$, and finally the formula $(\neg p \vee r)$ itself. The values of these subformulas under v are

$$v(p) = T$$
$$v(r) = F$$
$$v(\neg p) = F$$
$$v((\neg p \vee r)) = F.$$

(c) Not a misprint for a change, but a trick question! As we have not specified the value of $v(s)$, we cannot work out the value of $v((p \leftrightarrow (\neg r \rightarrow s)))$.

(d) Working out the values of subformulas of

$$((q \wedge (r \rightarrow \neg r)) \vee ((p \vee r) \leftrightarrow \neg q)),$$

we have

$$v(p) = T$$
$$v(q) = F$$
$$v(r) = F$$
$$v(\neg r) = T$$
$$v((r \rightarrow \neg r)) = T$$
$$v((q \wedge (r \rightarrow \neg r))) = F$$
$$v((p \vee r)) = T$$
$$v(\neg q) = T$$
$$v(((p \vee r) \leftrightarrow \neg q)) = T$$
$$v(((q \wedge (r \rightarrow \neg r)) \vee ((p \vee r) \leftrightarrow \neg q))) = T.$$

Perhaps easier to say after the event than to spot beforehand, once one has spotted that $v(((p \vee r) \leftrightarrow \neg q)) = T$, then thanks to the truth table for $\vee$, we must have

$$v(((q \wedge (r \rightarrow \neg r)) \vee ((p \vee r) \leftrightarrow \neg q))) = T,$$

irrespective of the value of $v((q \wedge (r \rightarrow \neg r)))$. This would have been a bit of a shortcut, but we suspect that sometimes looking for a shortcut might take up time that could have been used working out the truth value the slow way!

(e) Not given.

From now on, we will often not show the intermediate steps in working out $v(\phi)$ and as your confidence increases you might find yourself doing the same.

One fact which we hope seems obvious is that the truth value of a formula ϕ under a truth assignment v does not depend on the values of $v(p)$ for propositional variables p which do not appear in ϕ. Despite being obvious, it's worth seeing how to prove the result, which can be regarded as a consequence of the following exercise. This exercise provides useful practice in proving a result for all formulas of a certain sort by mathematical induction on the length of a formula.

Exercise 2.18

Let v and v' be truth assignments which take the same values for all propositional variables except p, i.e. $v(p) \neq v'(p)$ and $v(q) = v'(q)$ for all other propositional variables q. Show that $v(\phi) = v'(\phi)$ for all formulas ϕ built up using the connectives $\neg, \wedge, \vee$ in which the propositional variable p does not appear.

Solution

We shall prove this by mathematical induction on the length of such formulas ϕ. The induction hypothesis is that $v(\phi) = v'(\phi)$ for all formulas ϕ in which the propositional variable p does not appear, where ϕ has length $\leq n$.

Recall that our preferred measure of the length of a formula is the number of occurrences of connectives in it, here $\neg, \wedge, \vee$.

If ϕ is a formula of the given form, namely one in which p does not appear, with length 0, then it has to be of the form q, where q is a propositional variable other than p. Then we have $v(q) = v'(q)$, that is, $v(\phi) = v'(\phi)$ for this ϕ.

Now suppose that the result holds for all ϕ of the given form with length $\leq n$. To prove from this the induction hypothesis for $n + 1$, it is enough to show that the result holds for any formula of the given form with length $n + 1$. Let ϕ be such a formula. As ϕ has length $n + 1 \geq 1$, ϕ contains at least one connective, so is of one of the forms $\neg\theta$, $(\theta \wedge \psi)$, $(\theta \vee \psi)$, for subformulas θ, ψ. The sum of the lengths of θ and ψ is n, so that both θ and ψ have length $\leq n$. Also as p does not appear in ϕ, it cannot appear in θ or ψ, so that the hypothesis can be used for both these subformulas. We must deal with each of the possible forms.

Case: ϕ is of the form $\neg\theta$

By the induction hypothesis $v(\theta) = v'(\theta)$, so that as v, v' are truth assignments

$$v(\neg\theta) = v'(\neg\theta),$$

that is, $v(\phi) = v'(\phi)$ for this form of ϕ.

Case: ϕ is of the form $(\theta \wedge \psi)$

By the induction hypothesis $v(\theta) = v'(\theta)$ and $v(\psi) = v'(\psi)$, so that as v, v' are truth assignments

$$v((\theta \wedge \psi)) = v'((\theta \wedge \psi)),$$

that is, $v(\phi) = v'(\phi)$ for this form of ϕ.

The case when ϕ is of the form $(\theta \vee \psi)$ is of course similar, completing the inductive step. The result for all $n \geq 0$, that is, for all formulas ϕ in which p does not appear, follows by mathematical induction.

The result of this last exercise confirms our intuition that the truth value of a formula ϕ under a truth assignment v depends only on the values v takes for the propositional variables in ϕ. This means that we can summarize the values ϕ can take under all possible truth assignments by looking only at the different truth assignments on the finitely many variables in it. How many of these latter assignments are there? This is the subject of the next exercise.

Exercise 2.19

(a) How many different truth assignments are there on the set of propositional variables $\{p, q, r\}$?

(b) How many different truth assignments are there on the set of propositional variables $\{p_1, p_2, \ldots, p_n\}$, where n is a positive integer?

Solution

(a) For a truth assignment v, there are two choices for the value of $v(p)$. For each such choice there are then two choices for $v(q)$. Also for each choice of $v(p)$ and $v(q)$, there are also two choices for $v(r)$, giving a total of $2 \times 2 \times 2 = 2^3 = 8$ different truth assignments.

(b) Extending the reasoning above, by an easy use of mathematical induction, we can show that there are 2^n different truth assignments.

We can now summarize the truth value of a formula ϕ under each of the different truth assignments on the variables in it by what is called the *truth table* of ϕ. This extends the terminology we used for the truth tables describing how to compute truth values for each of the connectives. If the propositional variables in ϕ are amongst $p_1, p_2, \ldots, p_n$, the table would have the form

p_1	p_2	$\cdots$	p_n	ϕ
T	T	$\ldots$	T	?
T	T	$\ldots$	F	?
$\vdots$	$\vdots$		$\vdots$	$\vdots$
F	F	$\ldots$	F	?

Normally we would only list variables that are used in ϕ. But there are occasions when it is useful to give the table using more variables than actually appear in ϕ, hence our use of the word 'amongst'.

where each row represents a truth assignment v giving particular truth values to each of $p_1, p_2, \ldots, p_n$ and then gives the corresponding value of $v(\phi)$. As there are 2^n different truth assignments on the n propositional variables, the table would have 2^n rows.

Our first example is the formula $(\neg(p \vee q) \to (p \wedge q))$ using the propositional variables p, q, rather than p_1, p_2.

p	q	$(\neg(p \vee q) \to (p \wedge q))$
T	T	T
T	F	T
F	T	T
F	F	F

For instance, the second row of the truth table says that when p is given the value T and q the value F, the formula $(\neg(p \vee q) \to (p \wedge q))$ has the value T.

It often helps one to record the truth values of the subformulas of ϕ in the table to enable one to compute the final values of ϕ itself. In the example

Equivalently, the second row says that if v is the truth assignment such that $v(p) = T$ and $v(q) = F$, then $v((\neg(p \vee q) \to (p \wedge q))) = T$.

above, this could have been recorded as follows.

p	q	$(p \lor q)$	$\neg(p \lor q)$	$(p \land q)$	$(\neg(p \lor q) \to (p \land q))$
T	T	T	F	T	T
T	F	T	F	F	T
F	T	T	F	F	T
F	F	F	T	F	F

This analysis of the subformulas could also be expressed in a more succinct form as follows.

$(\neg$	$(p$	$\lor$	$q)$	$\to$	$(p$	$\land$	$q))$
F	T	T	T	T	T	T	T
F	T	T	F	T	T	F	F
F	F	T	T	T	F	F	T
T	F	F	F	F	F	F	F
③	①	②	①	④	①	②	①

The circled numbers indicate the order in which the columns were filled in. They are not part of the truth table and they could be left out. The values in column ④ tell us the truth value of the entire formula corresponding to the truth values of the propositional variables p, q in the columns labelled ①. Note that column ④ is that in which the principal connective, here $\to$, of the entire formula occurs.

Strictly speaking, the truth table is the simpler one just giving the final value of ϕ for each truth assignment, although this conceals many complicated and tedious computations.

Exercise 2.20 ____

Give the truth table of the formula $((p \to (q \land r)) \leftrightarrow \neg(p \lor r))$ using the propositional variables p, q, r.

Solution

One way of presenting the table is as follows.

p	q	r	$((p \to (q \land r)) \leftrightarrow \neg(p \lor r))$
T	T	T	F
T	T	F	T
T	F	T	T
T	F	F	T
F	T	T	F
F	T	F	T
F	F	T	F
F	F	F	T

Our rough work involved producing the following table giving truth values of subformulas.

$((p$	$\rightarrow$	$(q$	$\wedge$	$r))$	$\leftrightarrow$	$\neg$	$(p$	$\vee$	$r))$
T	T	T	T	T	F	F	T	T	T
T	F	T	F	F	T	F	T	T	F
T	F	F	F	T	T	F	T	T	T
T	F	F	F	F	T	F	T	T	F
F	T	T	T	T	F	F	F	T	T
F	T	T	F	F	T	T	F	F	F
F	T	F	F	T	F	F	F	T	T
F	T	F	F	F	T	T	F	F	F
①	③	①	②	①	④	③	①	②	①

There is no firm rule about the order in which one arranges the different truth assignments into rows, but it pays to be systematic – imagine having to write down the truth table of a formula using 10 variables and wanting to be sure that one has correctly listed all 2^{10} truth assignments somewhere in it! Our preferred method is to organize the rows so that all those assignments under which p_1 takes the value T appear in the top half of the table and all those in which it takes the value F appear in the bottom half. Then, having settled on a particular value of p_1, list all assignments in which p_2 takes the value T above those in which it takes the value F. Having settled on particular values of p_1 and p_2, repeat the process for p_3, and so on for further variables.

Exercise 2.21

Use the outline procedure above to list the rows of a table for a formula ϕ involving the variables p_1, p_2, p_3, p_4.

Solution

p_1	p_2	p_3	p_4	ϕ
T	T	T	T	?
T	T	T	F	?
T	T	F	T	?
T	T	F	F	?
T	F	T	T	?
T	F	T	F	?
T	F	F	T	?
T	F	F	F	?
F	T	T	T	?
F	T	T	F	?
F	T	F	T	?
F	T	F	F	?
F	F	T	T	?
F	F	T	F	?
F	F	F	T	?
F	F	F	F	?

Exercise 2.22

For each of the following formulas, give its truth table (where $p, q, r, p_1, p_2, p_3, p_4$ are propositional variables).

(a) $(p \wedge \neg p)$

(b) $(p \rightarrow \neg(q \leftrightarrow \neg p))$

(c) $((r \vee (q \wedge p)) \vee (\neg(q \leftrightarrow \neg r) \rightarrow p))$

(d) $(p_1 \rightarrow (p_3 \rightarrow (\neg p_4 \rightarrow p_2)))$

Solution

No solutions are given, but we hope that you found that there was only one row in the table for (d) in which the truth value of the formula was F!

We often describe a formula in terms of other formulas, not all of which are propositional variables, for instance $(\theta \rightarrow (\psi \vee \neg\theta))$. Thinking of the construction tree for this formula, we can regard its basic building blocks as being the subformulas θ and ψ. In such a case, we extend the idea of a truth table by giving the value of the whole formula for all possible combinations of truth values of these building blocks, in this case as follows:

θ	ψ	$(\theta \rightarrow (\psi \vee \neg\theta))$
T	T	T
T	F	F
F	T	T
F	F	T

Of course, if we knew more about the formulas θ and ψ, it could be that some of the rows above could never arise. For instance, if θ was the formula $(p \wedge \neg p)$, which always takes the value F, the top two rows of the table would be irrelevant. However, potentially θ and ψ could also be distinct propositional variables, so that the whole table is potentially relevant.

Exercise 2.23

For each of the following formulas, give its truth table.

(a) $(\phi \rightarrow (\psi \rightarrow \phi))$

(b) $\neg(\neg\phi \vee \phi)$

(c) $((\theta \vee (\phi \leftrightarrow \theta)) \rightarrow \neg(\psi \wedge \neg\phi))$

We trust that your solution to Exercise 2.23(a) showed that the formula $(\phi \rightarrow (\psi \rightarrow \phi))$ is true for all possible combinations of truth values of the subformulas ϕ and ψ, so that it is true under all truth assignments. Such a formula is called a *tautology*. Likewise your solution to Exercise 2.23(b) should have shown that the formula $\neg(\neg\phi \vee \phi)$ is false whatever the truth value of the subformula ϕ, so that it is false under all truth assignments. Such a formula is called a *contradiction*. Tautologies and contradictions will prove to be of special importance in much of the rest of the course.

Simple examples of tautologies are

$$(\phi \vee \neg\phi), \quad (\phi \rightarrow \phi) \quad \text{and} \quad (\neg\neg\phi \leftrightarrow \phi),$$

Strictly speaking, we should say that a tautology is true under all truth assignments which are defined on a set of propositional variables including those appearing in the formula. However, here and elsewhere we shall simply talk about 'all truth assignments' as a shorthand for this fuller description.

where ϕ is any formula. It is clear that they are tautologies by phrasing them using the intended interpretations of the connectives, 'ϕ or not ϕ' and so on, and verifying that they are tautologies by constructing their truth tables is very straightforward. Likewise $(\phi \wedge \neg\phi)$ is a pretty memorable contradiction – it corresponds well to the way we use the word 'contradiction' in everyday language.

It's well worth actively remembering these particular tautologies and contradictions.

Exercise 2.24

Show that each of the formulas $(\phi \vee \neg\phi)$, $(\phi \to \phi)$ and $(\neg\neg\phi \leftrightarrow \phi)$ is a tautology and that $(\phi \wedge \neg\phi)$ is a contradiction.

Exercise 2.25

Which, if any, of the following formulas is a tautology or a contradiction?

(a) $(p \to (p \to p))$

(b) $((p \to p) \to p)$

(c) $((p \to \neg p) \leftrightarrow (\neg p \to p))$

As an aside, the philosopher Ludwig Wittgenstein (1889–1951) described a tautology as a statement which conveys no information. Indeed, taking ϕ to be the statement 'it will rain', asserting ϕ gives useful information, for instance influencing one to take an umbrella when going outdoors. But asserting the tautology $(\phi \vee \neg\phi)$ is totally unhelpful in this regard!

The last exercise provides a reminder that there are formulas which are neither a tautology nor a contradiction. Tautologies and contradictions are of particular interest in the rest of the book, but don't forget that in general formulas don't have to fall into one of these categories.

Exercise 2.26

Let ϕ be a formula. Show that ϕ is a tautology if and only if $\neg\phi$ is a contradiction.

Solution

In conversation in a class, we would probably accept an informal (but convincing!) argument based on the observation that if the value of ϕ on each row of its truth table is T, then the value of $\neg\phi$ on the corresponding row of its truth table must be F, and vice versa. However, as other problems of this sort might not yield as easily to this sort of analysis, we shall record a more formal way of presenting this solution in case such an approach is needed elsewhere.

We shall show that if ϕ is a tautology then $\neg\phi$ is a contradiction and that if $\neg\phi$ is a contradiction then ϕ is a tautology.

First let us suppose that ϕ is a tautology. We need to show that $\neg\phi$ is false under all truth assignments. So let v be any truth assignment. As ϕ is a tautology we have $v(\phi) = T$. Thus $v(\neg\phi) = F$ (as a truth assignment respects the truth table of $\neg$, but by this stage of your study, you no longer need to say this). Hence $v(\neg\phi) = F$ for all truth assignments v, so that $\neg\phi$ is a contradiction, as required.

Conversely, suppose that $\neg\phi$ is a contradiction. Then for any truth assignment v we have $v(\neg\phi) = F$, so that, as v is a truth assignment, we can only have $v(\phi) = T$. Thus $v(\phi) = T$ for all truth assignments v, so that ϕ is a tautology.

Exercise 2.27

Let ϕ, ψ be formulas.

(a) Show that if ϕ and $(\phi \to \psi)$ are tautologies, then ψ is a tautology.

(b) Is it the case that if ϕ and ψ are tautologies, then $(\phi \to \psi)$ is a tautology?

(c) Is it the case that if $(\phi \to \psi)$ and ψ are tautologies, then ϕ is a tautology?

Now that we have the basic ideas of how to interpret the formal language, we can start to investigate relationships between one formula and others, to work towards our goal of representing mathematical arguments in a formal way. We shall begin with the idea of *logically equivalent* formulas in the next section.

Further exercises

Exercise 2.28

For a formula ϕ built up using the connectives $\neg, \wedge, \vee$, let ϕ^* be constructed by replacing each propositional variable in ϕ by its negation.

So if ϕ is the formula
$$((q \vee p) \wedge \neg p),$$
ϕ^* is
$$((\neg q \vee \neg p) \wedge \neg\neg p).$$

(a) For any truth assignment v, let v^* be the truth assignment which gives each propositional variable the opposite value to that given by v, i.e.

$$v^*(p) = \begin{cases} T, & \text{if } v(p) = F, \\ F, & \text{if } v(p) = T, \end{cases}$$

for all propositional variables p. Show that $v(\phi) = v^*(\phi^*)$. [*Hint*: This is really a statement about all formulas ϕ of a certain sort, so what is the likely method of proof?]

(b) (i) Use the result of part (a) to show that ϕ is a tautology if and only if ϕ^* is a tautology.

(ii) Is it true that ϕ is a contradiction if and only if ϕ^* is a contradiction? Explain your answer.

Exercise 2.29

Suppose that we are given a set S of truth assignments with an odd number of elements. Let $\mathbf{D}$ be the set of formulas of the language which a majority of the truth assignments in S makes true. Which of the following statements is always true? Give reasons in each case.

(a) For any well-formed formula ϕ, either ϕ or $\neg\phi$ belongs to $\mathbf{D}$.

(b) If ϕ belongs to $\mathbf{D}$ and $(\phi \to \theta)$ is a tautology, then θ belongs to $\mathbf{D}$.

(c) If ϕ and $(\phi \to \theta)$ belong to $\mathbf{D}$, then θ belongs to $\mathbf{D}$.

Exercise 2.30

Prove that any formula built up from $\neg$ and $\to$ in which no propositional variable occurs more than once cannot be a tautology.

Exercise 2.31

Let ϕ, ψ be formulas.

(a) If ψ is a contradiction, under what circumstances, if any, is $(\phi \rightarrow \psi)$ a contradiction?

(b) If ϕ is a contradiction, under what circumstances, if any, is $(\phi \rightarrow \psi)$ a contradiction?

(c) If $(\phi \rightarrow \psi)$ is a contradiction, must either of ϕ, ψ be contradictions?

2.4 Logical equivalence

Formulas can be of great complexity and it is often very valuable to see whether the statements they represent could have be rephrased in a simpler but equivalent way. For instance, in normal language one would normally simplify the statement 'it isn't the case that it's not raining' into 'it is raining', which conveys the same information. The corresponding formal concept is that of logical equivalence, which we define below, and one of the recurring themes in the book is whether, given a formula, one can find a 'simpler' formula logically equivalent to it.

Definition *Logically equivalent*

Formulas ϕ and ψ are said to be *logically equivalent*, which we write as $\phi \equiv \psi$, if for all truth assignments v, $v(\phi) = v(\psi)$.

As ever, we should strictly speaking talk about all truth assignments v on a set of propositional variables including all of those appearing in ϕ or ψ.

For example, we have $(\theta \leftrightarrow \chi)$ is logically equivalent to $(\neg\theta \leftrightarrow \neg\chi)$ for any formulas θ, χ, as can be seen by comparing their truth tables:

θ	χ	$(\theta \leftrightarrow \chi)$
T	T	T
T	F	F
F	T	F
F	F	T

θ	χ	$(\neg\theta \leftrightarrow \neg\chi)$
T	T	T
T	F	F
F	T	F
F	F	T

By writing the tables so that the rows giving the different combinations of truth values for θ and χ are in the same order, it is easy to see that the tables match, so that for all truth assignments v, $v((\theta \leftrightarrow \chi)) = v((\neg\theta \leftrightarrow \neg\chi))$, showing that

$$(\theta \leftrightarrow \chi) \equiv (\neg\theta \leftrightarrow \neg\chi).$$

The definition of logical equivalence could be expressed in these terms, saying that the truth tables of ϕ and ψ match. However, when not all the basic building blocks of one of the formulas appears in the other, as is the case with the logically equivalent formulas θ and $((\theta \wedge \chi) \vee (\theta \wedge \neg\chi))$, the truth tables would have to be constructed in terms of truth values of the same set of subformulas, in this case θ and χ. The definition in terms of truth assignments is more elegant!

Exercise 2.32

Show that $\theta \equiv ((\theta \wedge \chi) \vee (\theta \wedge \neg\chi))$, for any formulas θ, χ.

Solution

Let v be any truth assignment.

If $v(\theta) = F$, then

$$v((\theta \wedge \chi)) = v((\theta \wedge \neg\chi)) = F,$$

regardless of whether $v(\chi)$ is true or false, so that

$$v(((\theta \wedge \chi) \vee (\theta \wedge \neg\chi))) = F = v(\theta).$$

If $v(\theta) = T$, then regardless of whether $v(\chi) = T$ or $v(\chi) = F$ (in which case $v(\neg\chi) = T$), exactly one of $v((\theta \wedge \chi))$ and $v((\theta \wedge \neg\chi))$ equals T, so that

$$v(((\theta \wedge \chi) \vee (\theta \wedge \neg\chi))) = T = v(\theta).$$

Thus for all truth assignments v, $v(\theta) = v(((\theta \wedge \chi) \vee (\theta \wedge \neg\chi)))$, so that

$$\theta \equiv ((\theta \wedge \chi) \vee (\theta \wedge \neg\chi)).$$

Alternatively, we could show that the truth tables match. The truth table of $((\theta \wedge \chi) \vee (\theta \wedge \neg\chi))$ is

θ	χ	$((\theta \wedge \chi) \vee (\theta \wedge \neg\chi))$
T	T	T
T	F	T
F	T	F
F	F	F

To help compare the truth tables, we regard θ as constructed from the subformulas θ, χ, giving the table

θ	χ	θ
T	T	T
T	F	T
F	T	F
F	F	F

which matches the first.

We list a number of very useful simple logical equivalences involving $\neg, \wedge, \vee$ in the following theorem. Some will seem very obvious. Some describe ways in which the connectives interact with each other. All are worth remembering and you will often find it helpful to exploit them.

Theorem 2.3

The following are all logical equivalences.

(a) $(\phi \wedge \psi) \equiv (\psi \wedge \phi)$ (commutativity of $\wedge$)

(b) $(\phi \vee \psi) \equiv (\psi \vee \phi)$ (commutativity of $\vee$)

(c) $(\phi \wedge \phi) \equiv \phi$ (idempotence of $\wedge$)

(d) $(\phi \vee \phi) \equiv \phi$ (idempotence of $\vee$)

(e) $(\phi \wedge (\psi \wedge \theta)) \equiv ((\phi \wedge \psi) \wedge \theta)$ (associativity of $\wedge$)

(f) $(\phi \vee (\psi \vee \theta)) \equiv ((\phi \vee \psi) \vee \theta)$ (associativity of $\vee$)

(g) $\neg\neg\phi \equiv \phi$ (law of double negation)

(h) $\neg(\phi \wedge \psi) \equiv (\neg\phi \vee \neg\psi)$ (De Morgan's Law)

(i) $\neg(\phi \vee \psi) \equiv (\neg\phi \wedge \neg\psi)$ (De Morgan's Law)

(j) $(\phi \wedge (\psi \vee \theta)) \equiv ((\phi \wedge \psi) \vee (\phi \wedge \theta))$ (distributivity of $\wedge$ over $\vee$)

(k) $(\phi \vee (\psi \wedge \theta)) \equiv ((\phi \vee \psi) \wedge (\phi \vee \theta))$ (distributivity of $\vee$ over $\wedge$))

(l) $(\phi \wedge (\psi \vee \phi)) \equiv \phi$ (absorption law for $\wedge$)

(m) $(\phi \vee (\psi \wedge \phi)) \equiv \phi$ (absorption law for $\vee$)

A more accurate description of part (a) would be that it demonstrates the commutativity of $\wedge$ *under logical equivalence*. The normal use of commutativity is with a binary operation $*$ on a set S which has the property that $a * b$ equals, rather than is *equivalent* to, $b * a$ for all $a, b \in S$. Hence the phrase 'under logical equivalence' is, strictly speaking, needed for this and other parts of this theorem.

Proof

We shall give an argument for part (f) and leave the rest to you as a straightforward exercise.

There are several acceptable ways to show that $(\phi \vee (\psi \vee \theta)) \equiv ((\phi \vee \psi) \vee \theta)$. One easy way is to write down the truth tables and show that these match. A perhaps more elegant method is to argue using truth assignments as follows.

Let v be any truth assignment.

If $v((\phi \vee (\psi \vee \theta))) = F$, then $v(\phi) = v((\psi \vee \theta)) = v(\psi) = v(\theta) = F$, so that $v((\phi \vee \psi)) = F$, giving

$$v(((\phi \vee \psi) \vee \theta)) = F.$$

Similarly, if $v(((\phi \vee \psi) \vee \theta)) = F$, then $v(\phi) = v(\psi) = v(\theta) = F$, so that

$$v((\phi \vee (\psi \vee \theta))) = F.$$

Thus for all truth assignments v,

$$v((\phi \vee (\psi \vee \theta))) = F \text{ if and only if } v(((\phi \vee \psi) \vee \theta)) = F,$$

so that for all truth assignments v,

$$v((\phi \vee (\psi \vee \theta))) = v(((\phi \vee \psi) \vee \theta)),$$

giving that $(\phi \vee (\psi \vee \theta)) \equiv ((\phi \vee \psi) \vee \theta)$. ∎

It would also have been acceptable to show that $v((\phi \vee (\psi \vee \theta))) = T$ if and only if $v(((\phi \vee \psi) \vee \theta)) = T$, for all truth assignments v, but for these formulas involving $\vee$, this would have involved more work.

You may well have noticed similarities between many of the equivalences in this theorem involving $\neg, \wedge, \vee$ and set identities involving set complement (written as $\backslash$), intersection ($\cup$) and union ($\cap$). For instance, the De Morgan Law

$$\neg(\phi \vee \psi) \equiv (\neg\phi \wedge \neg\psi)$$

is very similar to the set identity, for sets A, B regarded as subsets of a set X,

$$X \setminus (A \cup B) = (X \setminus A) \cap (X \setminus B).$$

This is one of the laws introduced by the English mathematician Augustus De Morgan (1806–1871) who made many important contributions to the growth of modern logic.

This is no coincidence, and a moment's thought about how to express what it means to be a member of the sets on each side of the set identity will convince you of the strong connection between the set operators and the corresponding logical connectives.

The sort of connection to which we refer is that $x \in C \cap D$ if and only if $x \in C$ *and* $x \in D$.

Exercise 2.33

Prove the remaining parts of Theorem 2.3.

There are very tempting connections between some pairs of the logical equivalences in Theorem 2.3. For instance, the logical equivalence

$$(\phi \vee (\psi \wedge \theta)) \equiv ((\phi \vee \psi) \wedge (\phi \vee \theta))$$

of part (k) corresponds to interchanging the occurrences of $\wedge$ and $\vee$ in the equivalence

$$(\phi \wedge (\psi \vee \theta)) \equiv ((\phi \wedge \psi) \vee (\phi \wedge \theta))$$

of part (j). These connections are made precise in what is called the Principle of Duality, which can be found in Exercise 2.44 at the end of this section.

We hope that it is pretty obvious that $\phi \equiv \psi$ if and only if $(\phi \leftrightarrow \psi)$ is a tautology. This means that each of the logical equivalences in the theorem above corresponds to a tautology involving $\leftrightarrow$. We think that the logical equivalences are more memorable than the corresponding tautologies!

Exercise 2.34

Show that $\phi \equiv \psi$ if and only if $(\phi \leftrightarrow \psi)$ is a tautology, for all formulas ϕ, ψ.

Much of what we expect the word 'equivalent' to convey about formulas is given by the results in the following exercise.

Exercise 2.35

Show each of the following, for all formulas ϕ, ψ, θ.

(a) $\phi \equiv \phi$

(b) If $\phi \equiv \psi$ then $\psi \equiv \phi$.

(c) If $\phi \equiv \psi$ and $\psi \equiv \theta$, then $\phi \equiv \theta$.

Exercise 2.35 shows that logical equivalence is what is called an *equivalence relation* on the set of all formulas of the underlying language. Logically equivalent formulas, while usually looking very different from each other, are the 'same' in terms of their truth under different interpretations. The set of all formulas logically equivalent to a given formula ϕ is called the *equivalence class* of ϕ under this relation and a natural question is whether each such class contains a formula which is in some way nice. We shall look at an example of one way of answering such a question later in the section.

We shall look at the theory of equivalence relations in Chapter 4.

In Theorem 2.5.

Also of great use are the following logical equivalences involving $\rightarrow$ and $\leftrightarrow$, especially those connecting $\rightarrow$ with $\neg, \wedge, \vee$.

Theorem 2.4

The following are logical equivalences.

(a) $(\phi \rightarrow \psi) \equiv (\neg\phi \vee \psi) \equiv \neg(\phi \wedge \neg\psi) \equiv (\neg\psi \rightarrow \neg\phi)$

(b) $(\phi \leftrightarrow \psi) \equiv ((\phi \rightarrow \psi) \wedge (\psi \rightarrow \phi))$

The formula $(\neg\psi \rightarrow \neg\phi)$ is called the *contrapositive* of $(\phi \rightarrow \psi)$.

Note that we have exploited the results of Exercise 2.35 to write the result of Theorem 2.4(a) on one line saying that four formulas are logically equivalent, rather than writing down several equivalences showing that each pair of formulas is logically equivalent. These logical equivalences give an indication of how some connectives can be expressed in terms of others – an idea which we shall take further in the next section when we look at the idea of an *adequate* set of connectives, that is, a set of connectives from which we can generate all conceivable connectives, not just $\rightarrow$ and $\leftrightarrow$. However, $\rightarrow$ and $\leftrightarrow$ are such important connectives in terms of expressing normal mathematical statements that you should not get the impression that their use is somehow to be avoided by the use of Theorem 2.4!

Exercise 2.36 ————————————————————————————

Prove Theorem 2.4.

——

Further nice and very useful results about logical equivalence are given in the following exercise.

Exercise 2.37 ————————————————————————————

Suppose that $\phi \equiv \phi'$ and $\psi \equiv \psi'$. Show each of the following.

(a) $\neg\phi \equiv \neg\phi'$

(b) $(\phi \wedge \psi) \equiv (\phi' \wedge \psi')$

(c) $(\phi \vee \psi) \equiv (\phi' \vee \psi')$

(d) $(\phi \to \psi) \equiv (\phi' \to \psi')$

(e) $(\phi \leftrightarrow \psi) \equiv (\phi' \leftrightarrow \psi')$

Solution

We give the solution to part (b) and leave the rest to you.

Let v be any truth assignment.

Suppose that $v((\phi \wedge \psi)) = T$, so that $v(\phi) = v(\psi) = T$. We then have

$$v(\phi') = v(\phi) \quad (\text{as } \phi \equiv \phi')$$
$$= T$$

and

$$v(\psi') = v(\psi) \quad (\text{as } \psi \equiv \psi')$$
$$= T,$$

so that

$$v((\phi' \wedge \psi')) = T.$$

Similarly we can show that if $v((\phi' \wedge \psi')) = T$ then

$$v((\phi \wedge \psi)) = T.$$

Thus for all truth assignments v,

$$v((\phi \wedge \psi)) = T \text{ if and only if } v((\phi' \wedge \psi')) = T,$$

so that $(\phi \wedge \psi) \equiv (\phi' \wedge \psi')$.

——

The results of this exercise can be generalised to show that if θ is a formula containing occurrences of ϕ as a subformula and all these occurrences are replaced by a formula ϕ' where $\phi \equiv \phi'$ to turn θ into the formula θ', then $\theta \equiv \theta'$.

The set of all propositions in a language using the set of connectives $\{\neg, \wedge, \vee\}$ with logical equivalence taking the place of $=$ is an example of what is called a *Boolean algebra.*

The proof of such a result would involve induction on the length of the formula θ and we would have to be more specific about the connectives being used. We leave an example of such a proof for you as Exercise 2.46.

We shall look at Boolean algebras in Section 4.4 of Chapter 4.

The essentially algebraic results of Theorems 2.3 and 2.4 and Exercises 2.35 and 2.37 provide an alternative way of showing formulas are logically equivalent to that of working directly from the definition of logical equivalence. For instance, take the logical equivalence

$$(\phi \wedge (\phi \vee \neg\phi)) \equiv \phi.$$

Using first principles, we could argue as follows. Let v be any truth assignment.

If $v(\phi) = T$, then $v((\phi \vee \neg\phi)) = T$, so that $v((\phi \wedge (\phi \vee \neg\phi))) = T = v(\phi)$.

If $v(\phi) = F$, then $v((\phi \wedge \psi)) = F$ for any formula ψ, so that in particular $v((\phi \wedge (\phi \vee \neg\phi))) = F = v(\phi)$.

Thus $(\phi \wedge (\phi \vee \neg\phi)) \equiv \phi$.

Alternatively, using the algebraic results, as $(\phi \vee \neg\phi) \equiv (\neg\phi \vee \phi)$ (by Theorem 2.3(b)), we have

$$(\phi \wedge (\phi \vee \neg\phi)) \equiv (\phi \wedge (\neg\phi \vee \phi)) \quad \text{(by Exercise 2.37(c))}$$

and by Theorem 2.3(l) we have

$$(\phi \wedge (\neg\phi \vee \phi)) \equiv \phi,$$

so that by Exercise 2.35(c) we have

$$(\phi \wedge (\phi \vee \neg\phi)) \equiv \phi.$$

Exercise 2.38 ____

Establish each of the following equivalences, where ϕ, ψ, θ and all the θ_i are formulas. You are welcome to do this from first principles or by exploiting the results of Theorems 2.3 and 2.4 and Exercises 2.35 and 2.37.

(a) $((\phi \wedge \psi) \vee \neg\theta) \equiv ((\phi \vee \neg\theta) \wedge (\psi \vee \neg\theta))$

(b) $(\phi \to \neg\psi) \equiv (\psi \to \neg\phi)$

(c) $(\theta \to (\phi \vee \psi)) \equiv (\phi \vee (\psi \vee \neg\theta))$

(d) $((\theta_1 \wedge \theta_2) \wedge (\theta_3 \wedge \theta_4)) \equiv (\theta_1 \wedge ((\theta_2 \wedge \theta_3) \wedge \theta_4))$

Solution

We shall give a solution to (a) and leave the rest to you.

$$((\phi \wedge \psi) \vee \neg\theta) \equiv (\neg\theta \vee (\phi \wedge \psi)) \quad \text{(by Theorem 2.3(b))}$$
$$\equiv ((\neg\theta \vee \phi) \wedge (\neg\theta \vee \psi)) \quad \text{(by Theorem 2.3(k)}$$
$$\text{and Exercise 2.35(c)).}$$

But

$$(\neg\theta \vee \phi) \equiv (\phi \vee \neg\theta) \text{ and } (\neg\theta \vee \psi) \equiv (\psi \vee \neg\theta)$$

by Theorem 2.3(b), so by Exercise 2.37(b)

$$((\neg\theta \vee \phi) \wedge (\neg\theta \vee \psi)) \equiv ((\phi \vee \neg\theta) \wedge (\psi \vee \neg\theta)).$$

Then by Exercise 2.35(c),

$$((\phi \wedge \psi) \vee \neg\theta) \equiv ((\phi \vee \neg\theta) \wedge (\psi \vee \neg\theta)).$$

We have shown in Theorem 2.3(e) that $(\phi \wedge (\psi \wedge \theta)) \equiv ((\phi \wedge \psi) \wedge \theta)$ and in Exercise 2.38(d) you were asked to show that

$$((\theta_1 \wedge \theta_2) \wedge (\theta_3 \wedge \theta_4)) \equiv (\theta_1 \wedge ((\theta_2 \wedge \theta_3) \wedge \theta_4)).$$

These equivalences are most easily established by noticing that each of the relevant formulas are true precisely for those truth assignments which make each of the subformulas θ_i true. A generalization of these logical equivalences is given in the following exercise.

Exercise 2.39 ____

Suppose that the formula ϕ is constructed by taking subformulas $\theta_1, \theta_2, \ldots, \theta_n$ in that order and joining them together only using the connective $\wedge$, with brackets inserted in such a way as to make ϕ a formula. Show that ϕ is true precisely for those truth assignments which make each of the subformulas $\theta_1, \theta_2, \ldots, \theta_n$ true. Deduce that any two such formulas (using the same $\theta_1, \theta_2, \ldots, \theta_n$ in that order) are logically equivalent. [*Hint*: Use the version of mathematical induction with the hypothesis that the result holds for all $k \leq n$ where $n \geq 1$.]

Examples for $n = 4$ include $((\theta_1 \wedge \theta_2) \wedge (\theta_3 \wedge \theta_4))$ and $(\theta_1 \wedge ((\theta_2 \wedge \theta_3) \wedge \theta_4))$. There is a general result for associative binary operations similar to the ultimate conclusion of this exercise which we invite you to look at in Exercise 2.45.

Exercise 2.40 ____

State and prove (by any preferred method) a result similar to that in Exercise 2.39 for formulas built up from n subformulas using $\vee$ rather than $\wedge$.

____

Given the results of these exercises, it will be convenient for us to introduce shorthand notations for a formula which is a successive conjunction of more than two subformulas and for a formula which is a successive disjunction of more than two subformulas.

Notation

If a formula is constructed by conjunction of $\theta_1, \theta_2, \ldots, \theta_n$ so that they appear in that order joined by $\wedge$s and suitably placed brackets, we shall write it as

$$(\theta_1 \wedge \theta_2 \wedge \ldots \wedge \theta_n),$$

that is, we ignore all the brackets around the θ_is except the outermost pair. A further shorthand is to write this as

$$\bigwedge_{i=1}^{n} \theta_i,$$

further ignoring these outermost brackets and using $\bigwedge$ to represent lots of $\wedge$s. Similarly if the formula is constructed by disjunction, i.e. using $\vee$ rather than $\wedge$, we shall write it as

$$(\theta_1 \vee \theta_2 \vee \ldots \vee \theta_n)$$

and use the shorthand

$$\bigvee_{i=1}^{n} \theta_i.$$

Note that several different formulas get represented by the same shorthand. For instance, $((p \wedge q) \wedge (p \wedge r))$ and $(p \wedge ((q \wedge p) \wedge r))$ both get represented as $(p \wedge q \wedge p \wedge r)$. In the contexts where we shall use these shorthands, this won't matter – all that will matter is that the formulas so represented are all logically equivalent.

Many of the useful basic logical equivalences can be extended to cover conjunctions and disjunctions of more than two subformulas, as you will see in the following exercise.

You might like to think about how many different formulas $(p \wedge q \wedge p \wedge r)$ represents in this way and, more generally, how many are represented by $\bigwedge_{i=1}^{n} \theta_i$. The answer will be one of what are called *Catalan* numbers.

Exercise 2.41 ____

Establish the following logical equivalences.

(a) $\displaystyle \neg \bigwedge_{i=1}^{n} \theta_i \equiv \bigvee_{i=1}^{n} \neg \theta_i$

(b) $\displaystyle \neg \bigvee_{i=1}^{n} \theta_i \equiv \bigwedge_{i=1}^{n} \neg \theta_i$

(c) $\displaystyle \left(\bigwedge_{i=1}^{n} \theta_i \right) \vee \left(\bigwedge_{j=1}^{m} \psi_j \right) \equiv \bigwedge_{i=1}^{n} \bigwedge_{j=1}^{m} (\theta_i \vee \psi_j)$

(d) $\displaystyle \left(\bigvee_{i=1}^{n} \theta_i \right) \wedge \left(\bigvee_{j=1}^{m} \psi_j \right) \equiv \bigvee_{i=1}^{n} \bigvee_{j=1}^{m} (\theta_i \wedge \psi_j)$

We are being somewhat casual about brackets in these formulas, in the cause of comprehensibility we hope! The same philosophy will pervade our solutions.

Solution

(a) We shall use mathematical induction on $n \geq 1$. For $n = 1$ the result is trivially true. For the inductive step, we suppose that the result holds for $n \geq 1$ and must show that it holds for $n + 1$. We have

$$\neg \bigwedge_{i=1}^{n+1} \theta_i \equiv \neg \left(\left(\bigwedge_{i=1}^{n} \theta_i \right) \wedge \theta_{n+1} \right)$$

$$\equiv \left(\neg \left(\bigwedge_{i=1}^{n} \theta_i \right) \vee \neg \theta_{n+1} \right) \qquad (\text{as } \neg(\phi \wedge \psi) \equiv (\neg\phi \vee \neg\psi)).$$

By the induction hypothesis, $\neg \bigwedge_{i=1}^{n} \theta_i \equiv \bigvee_{i=1}^{n} \neg \theta_i$, so by Exercise 2.37(c) we have

$$\left(\neg \left(\bigwedge_{i=1}^{n} \theta_i \right) \vee \neg \theta_{n+1} \right) \equiv \left(\left(\bigvee_{i=1}^{n} \neg \theta_i \right) \vee \neg \theta_{n+1} \right)$$

$$\equiv \bigvee_{i=1}^{n+1} \neg \theta_i,$$

This is a typical use of Exercise 2.37 to replace one subformula by an equivalent subformula.

so that

$$\neg \bigwedge_{i=1}^{n+1} \theta_i \equiv \bigvee_{i=1}^{n+1} \neg \theta_i,$$

as required. The result follows by mathematical induction.

(b) Not given.

(c) This one could be done by mathematical induction on both $n \geq 1$ and $m \geq 1$: first show that the result holds for $n = 1$ and all $m \geq 1$ using induction on m, and then assume that the result holds for some $n \geq 1$ and all m and show that it holds for $n + 1$ and all m. However, there is a shortcut which we will take.

First we fix $m = 1$ and show the result then holds for all $n \geq 1$. We shall write ϕ rather than ψ_1 for a reason which will be revealed later! The result holds trivially for $n = 1$. If the result holds for some $n \geq 1$, we then have

As forecast, we shall be a bit casual about brackets!

$$\left(\bigwedge_{i=1}^{n+1} \theta_i\right) \vee \phi \equiv \left(\left(\bigwedge_{i=1}^{n} \theta_i\right) \wedge \theta_{n+1}\right) \vee \phi$$

$$\equiv \left(\left(\bigwedge_{i=1}^{n} \theta_i\right) \vee \phi\right) \wedge (\theta_{n+1} \vee \phi)$$

$$(\text{as } ((\psi \wedge \theta) \vee \phi) \equiv ((\psi \vee \phi) \wedge (\theta \vee \phi)))$$

$$\equiv \left(\bigwedge_{i=1}^{n}(\theta_i \vee \phi)\right) \wedge (\theta_{n+1} \vee \phi)$$

(using the induction hypothesis and

Exercise 2.37(b))

$$\equiv \bigwedge_{i=1}^{n+1}(\theta_i \vee \phi),$$

as required. So by mathematical induction, we have

$$\left(\bigwedge_{i=1}^{n} \theta_i\right) \vee \phi \equiv \bigwedge_{i=1}^{n}(\theta_i \vee \phi),$$

for all $n \geq 1$.

Now to prove the required result, we replace ϕ in the result above by $\bigwedge_{j=1}^{m} \psi_j$ to obtain

$$\left(\bigwedge_{i=1}^{n} \theta_i\right) \vee \left(\bigwedge_{j=1}^{m} \psi_j\right) \equiv \bigwedge_{i=1}^{n}\left(\theta_i \vee \bigwedge_{j=1}^{m} \psi_j\right). \tag{$*$}$$

As $(\theta \vee \phi) \equiv (\phi \vee \theta)$, the subsidiary result for $m = 1$ gives

$$\phi \vee \bigwedge_{i=1}^{n} \theta_i \equiv \bigwedge_{i=1}^{n}(\phi \vee \theta_i),$$

which, by replacing n by m, the θ_is for $i = 1, \ldots, n$ by ψ_j for $j = 1, \ldots, m$, and ϕ by θ_i gives

$$\theta_i \vee \bigwedge_{j=1}^{m} \psi_j \equiv \bigwedge_{j=1}^{m} (\theta_i \vee \psi_j).$$

Substituting this in $(*)$ gives the required result, namely

$$\left(\bigwedge_{i=1}^{n} \theta_i \right) \vee \left(\bigwedge_{j=1}^{m} \psi_j \right) \equiv \bigwedge_{i=1}^{n} \bigwedge_{j=1}^{m} (\theta_i \vee \psi_j),$$

for all $n, m \geq 1$.

(d) Not given.

We are about to state a result involving a quite complicated description of a particular sort of formula, as follows:

> ψ is a disjunction of formulas which are conjunctions of propositional variables and/or negated propositional variables.

What does this mean? An example of what we mean is

$$((p \wedge \neg q \wedge p) \vee q \vee (\neg r \wedge s) \vee \neg s).$$

Recall our shorthand for ignoring brackets in long conjunctions and long disjunctions.

Here each of $(p \wedge \neg q \wedge p)$, q, $(\neg r \wedge s)$ and $\neg s$ are conjunctions of propositional variables and/or negated propositional variables – OK, you might not like it, but each of the q and $\neg s$ is a conjunction of just one thing! Also these conjunctions are joined together by $\vee$s to form the disjunction. Such a formula is said to be in *disjunctive form*. A more general version of such a form is

$$\bigvee_{i=1}^{n} \left(\bigwedge_{j=1}^{k_i} q_{i,j} \right),$$

When $n = 1$, no $\vee$ actually appears, and an example of the sort of formula you get is $(p \wedge q \wedge \neg r)$.

where each $q_{i,j}$ is a propositional variable or its negation.

The result will also involve a corresponding form where the roles of $\wedge$ and $\vee$ are interchanged. This is called a *conjunctive form*, which is a conjunction of formulas which are disjunctions of propositional variables and/or negated propositional variables, i.e. of the form

$$\bigwedge_{i=1}^{n} \left(\bigvee_{j=1}^{k_i} q_{i,j} \right),$$

where each $q_{i,j}$ is a propositional variable or its negation. An example of this is

$$(p \wedge (\neg q \vee r) \wedge (r \vee \neg p \vee q \vee p)).$$

As with disjunctive form, there are some fairly trivial formulas which are in conjunctive form, like each of q, $\neg p$, $(p \wedge q)$ (for which the k_is in the general form above all equal 1) and $(q \vee r \vee \neg p)$ (for which the n in the general form equals 1). Actually all these trivial formulas are simultaneously in both conjunctive and disjunctive form. The result we are leading towards says that for any given formula ϕ using connectives in the set $\{\wedge, \vee, \neg, \rightarrow, \leftrightarrow\}$, there are logically equivalent formulas, one in disjunctive form and one in conjunctive form, but usually these latter formulas are not the same. For instance, $((p \to q) \to r)$ is logically equivalent to

$$((p \wedge \neg q) \vee r)$$

which is in disjunctive form and

$$((p \vee r) \wedge (\neg q \vee r))$$

which is in conjunctive form. Now for the theorem! This result tells us that any formula, however complicated a jumble of variables and connectives it appears to be, is logically equivalent to a formula with a nice, orderly shape.

Theorem 2.5

Let ϕ be a formula using connectives in the set $\{\wedge, \vee, \neg, \to, \leftrightarrow\}$. Then ϕ is logically equivalent to a formula $\phi^{\vee}$ in disjunctive form and a formula $\phi^{\wedge}$ in conjunctive form.

Proof

First we remove all occurrences of $\leftrightarrow$ in ϕ by replacing all subformulas of the form $(\theta \leftrightarrow \psi)$ by $((\theta \to \psi) \wedge (\psi \to \theta))$. In the resulting formula, we then remove all occurrences of $\to$ by replacing all subformulas of the form $(\theta \to \psi)$ by $(\neg\theta \vee \psi)$. In this way we have produced a formula logically equivalent to ϕ which only uses the connectives $\wedge, \vee, \neg$. If we can prove the result for all formulas of this type, then the result holds for the original ϕ.

These are logical equivalences in Theorem 2.4.

So let's now suppose that the connectives in ϕ are in the set $\{\wedge, \vee, \neg\}$. We have to prove the result for all formulas ϕ of this type and to cope with formulas of arbitrary complexity we shall use induction on the length of ϕ, where, as before, our preferred measure of length is the number of connectives in ϕ.

For a formula ϕ of this type of length 0, ϕ can only be a propositional variable p, which is already in both disjunctive and conjunctive form.

So $p^{\vee} = p^{\wedge} = p$.

Now suppose that the result holds for all formulas using connectives in the set $\{\wedge, \vee, \neg\}$ of length $\leq n$, where $n \geq 0$, and that ϕ is a formula of this type with $n + 1$ connectives. As ϕ has at least one connective, it must have one of the following three forms:

$$(\theta \wedge \psi), \quad (\theta \vee \psi), \quad \neg\theta,$$

where crucially the subformulas θ and ψ have length $\leq n$, so that the inductive hypothesis applies to them. We must consider each of these three forms separately and will leave some of the details to you. We'll make heavy use of some of the results of Exercise 2.41 and Exercise 2.37 to replace subformulas by logically equivalent formulas.

Case: ϕ is of the form $(\theta \wedge \psi)$

By the inductive hypothesis θ and ψ are logically equivalent to $\theta^\wedge$ and $\psi^\wedge$ respectively, both in conjunctive form. Then

$$\phi \equiv (\theta \wedge \psi) \equiv (\theta^\wedge \wedge \psi^\wedge).$$

But $(\theta^\wedge \wedge \psi^\wedge)$ is in conjunctive form as both $\theta^\wedge$ and $\psi^\wedge$ are in this form. So we can take $\phi^\wedge$ to be $(\theta^\wedge \wedge \psi^\wedge)$.

That was rather easy, but what about $\phi^\vee$? For this, first note that by the inductive hypothesis θ and ψ are logically equivalent to $\theta^\vee$ and $\psi^\vee$ respectively, both in disjunctive form. So

$$\phi \equiv (\theta \wedge \psi) \equiv (\theta^\vee \wedge \psi^\vee).$$

The $\wedge$ as principal connective of $(\theta^\vee \wedge \psi^\vee)$ means that it is not usually in disjunctive form, so some more work is needed. The formulas $\theta^\vee$ and $\psi^\vee$ are of the forms $\bigvee_{i=1}^{n} \theta_i$ and $\bigvee_{j=1}^{m} \psi_j$ respectively where each θ_i and ψ_j is a conjunction of propositional variables and/or negated propositional variables. So

$$\phi \equiv \left(\bigvee_{i=1}^{n} \theta_i \right) \wedge \left(\bigvee_{j=1}^{m} \psi_j \right)$$

and by Exercise 2.41 part (d), the formula on the right is logically equivalent to

$$\bigvee_{i=1}^{n} \bigvee_{j=1}^{m} (\theta_i \wedge \psi_j).$$

As each θ_i and ψ_j is a conjunction of propositional variables and/or negated propositional variables, so is each $(\theta_i \wedge \psi_j)$. That means that $\bigvee_{i=1}^{n} \bigvee_{j=1}^{m} (\theta_i \wedge \psi_j)$ is in disjunctive form, so that we can take this formula as $\phi^\vee$.

Case: ϕ is of the form $(\theta \vee \psi)$

This is left as an exercise for you.

Case: ϕ is of the form $\neg\theta$

By the inductive hypothesis, θ is logically equivalent to $\theta^\wedge$ in conjunctive form, which we can write as $\bigwedge_{i=1}^{n} \theta_i$, where each θ_i is a disjunction of propositional variables and/or their negations. Using Exercise 2.41 part (a), we have

$$\phi \equiv \neg\theta \equiv \neg \bigwedge_{i=1}^{n} \theta_i$$
$$\equiv \bigvee_{i=1}^{n} \neg\theta_i.$$

Each θ_i is a disjunction of propositional variables and/or their negations, so of the form $\bigvee_{j=1}^{n_i} q_{i,j}$, where each $q_{i,j}$ is a propositional variable or its negation.

By Exercise 2.41 part (b), $\neg\theta_i$ is logically equivalent to $\bigwedge_{i=1}^{n} \neg q_{i,j}$. Each $q_{i,j}$ is of the form p or $\neg p$, where p is a propositional variable. So $\neg q_{i,j}$ is of the form $\neg p$ or $\neg\neg p$. In the latter case $\neg q_{i,j}$ is logically equivalent to p. For each i, j put

$$r_{i,j} = \begin{cases} \neg p, & \text{if } q_{i,j} \text{ is a propositional variable } p, \\ p, & \text{if } q_{i,j} \text{ is } \neg p \text{ where } p \text{ is a propositional variable,} \end{cases}$$

so that $r_{i,j}$ is always a propositional variable or its negation and is logically equivalent to $\neg q_{i,j}$. At last we have

$$\phi \equiv \bigvee_{i=1}^{n} \neg\theta_i$$
$$\equiv \bigvee_{i=1}^{n} \neg \bigvee_{j=1}^{n_i} q_{i,j}$$
$$\equiv \bigvee_{i=1}^{n} \bigwedge_{j=1}^{n_i} \neg q_{i,j}$$
$$\equiv \bigvee_{i=1}^{n} \bigwedge_{j=1}^{n_i} r_{i,j},$$

which is in disjunctive form and can thus be taken as $\phi^{\vee}$.

We shall leave it as an exercise for you to find a suitable $\phi^{\wedge}$ logically equivalent to ϕ in this case. $\blacksquare$

Exercise 2.42

(a) Explain how to construct a suitable $\phi^{\wedge}$ and $\phi^{\vee}$ in the case when ϕ is of the form $(\theta \vee \psi)$ in the proof of Theorem 2.5.

(b) Explain how to construct a suitable $\phi^{\wedge}$ in the case when ϕ is of the form $\neg\theta$ in the proof of Theorem 2.5.

Exercise 2.43

For each of the following formulas, follow the method used in the proof of Theorem 2.5 to find a disjunctive form and a conjunctive form equivalent to it.

(a) $((p_1 \vee (p_2 \leftrightarrow \neg p_1)) \rightarrow (\neg p_1 \wedge p_3))$

(b) $(p \rightarrow (q \rightarrow (r \vee \neg p)))$

(c) $(p \leftrightarrow \neg p)$

A given formula will in general be logically equivalent to several different formulas in disjunctive form. For instance,

$$(p \vee (\neg p \wedge q)) \equiv (p \vee q),$$

where both are in disjunctive form. We shall see in the next section that amongst these different disjunctive forms there are some that fit a standard format, and so can be said to be in a *normal* form. The same thing goes for conjunctive forms.

In this section we have looked at propositional formulas built up from the connectives $\neg$, $\wedge$, $\vee$, $\rightarrow$ and $\leftrightarrow$. But these are not the only connectives we might have introduced based on normal language. For instance, we might have introduced connectives for the likes of 'unless' and 'neither ... nor'. In the next section we shall look at further possible connectives in a very general way. Then we shall discover a remarkable fact about how even very complicated connectives can always be built up from some of the very simple ones we have met in this section.

Further exercises

Exercise 2.44

Let ϕ be a formula built up using the connectives $\neg, \wedge, \vee$. The *dual* ϕ' of ϕ is the formula obtained from ϕ by replacing all occurrences of $\wedge$ by $\vee$, of $\vee$ by $\wedge$, and all propositional variables by their negations.

For instance, if ϕ is the formula
$$(\neg p \wedge ((p \wedge q) \vee r)),$$
then ϕ' is
$$(\neg\neg p \vee ((\neg p \vee \neg q) \wedge \neg r)).$$

(a) Show that ϕ' is logically equivalent to $\neg\phi$. (This is called the *Principle of Duality*.) [*Hint*: Use induction on the length of ϕ.]

(b) Hence, using Theorem 2.4, show that if ϕ, ψ are formulas built up using the connectives $\neg, \wedge, \vee$, then

$$(\phi \rightarrow \psi) \equiv (\psi' \rightarrow \phi')$$

and

$$(\phi \leftrightarrow \psi) \equiv (\phi' \leftrightarrow \psi').$$

(c) Use the following method to show that Theorem 2.3(k) follows from Theorem 2.3(j). Theorem 2.3(j) states that

$$(\phi \wedge (\psi \vee \theta)) \equiv ((\phi \wedge \psi) \vee (\phi \wedge \theta)),$$

so that by the result of Exercise 2.34,

$$((\phi \wedge (\psi \vee \theta)) \leftrightarrow ((\phi \wedge \psi) \vee (\phi \wedge \theta)))$$

is a tautology. Use the result of part (b) above and the result of Exercise 2.28(b)(i) in Section 2.3 to show that

$$((\phi \vee (\psi \wedge \theta)) \leftrightarrow ((\phi \vee \psi) \wedge (\phi \vee \theta)))$$

is a tautology. Hence, by Exercise 2.34,

$$(\phi \vee (\psi \wedge \theta)) \equiv ((\phi \vee \psi) \wedge (\phi \vee \theta)),$$

which is Theorem 2.3(k).

(d) Identify other logical equivalences in Theorem 2.3 which are related in the same way using the method of part (c).

Exercise 2.45

Let X be a non-empty set and suppose that $*\colon X^2 \longrightarrow X$ is a function with the associative property, that is,

$$(x * (y * z)) = ((x * y) * z), \quad \text{for all } x, y, z \in X,$$

where we write the image under the function of the pair (a, b) in X^2 as $(a * b)$. Let $x_1, x_2, \ldots, x_n$ be elements of X and suppose that brackets and $*$s are inserted into the string of symbols $x_1 x_2 \ldots x_n$ to give an expression which can be computed using the function $*$ to give an element of X. Show that the computations of all such expressions (for the same $x_1 x_2 \ldots x_n$ in that order) will result in the same element of X.

For instance, for $n = 4$, each of the expressions

$$(x_1 * (x_2 * (x_3 * x_4)))$$
$$(x_1 * ((x_2 * x_3) * x_4))$$
$$((x_1 * x_2) * (x_3 * x_4))$$
$$(((x_1 * x_2) * x_3) * x_4)$$
$$((x_1 * (x_2 * x_3)) * x_4)$$

give the same element of X.

[*Hints*: Use induction on n with hypothesis that each such expression equals both of

$$(x_1 * (x_2 * (\ldots * (x_{n-1} * x_n) \ldots)))$$

and

$$(((\ldots (x_1 * x_2) * \ldots) * x_{n-1}) * x_n).]$$

Exercise 2.46

Let θ be a formula built up using the connectives $\neg, \wedge$ and let ϕ be one of its subformulas. We shall write $\theta[\phi'/\phi]$ for the formula obtained by replacing all occurrences of the subformula ϕ in θ by the formula ϕ'. Show that if $\phi \equiv \phi'$ then $\theta \equiv \theta[\phi'/\phi]$. [*Hints*: Fix the formulas ϕ and ϕ' and do an induction on the length of θ. But first be much more specific about the meaning of $\theta[\phi'/\phi]$ by defining it as follows:

For instance if θ is the formula

$$((p \wedge p) \wedge (q \wedge \neg(p \wedge p))),$$

ϕ is the formula $(p \wedge p)$ and ϕ' is the formula $\neg\neg p$, then $\theta[\phi'/\phi]$ is the formula

$$(\neg\neg p \wedge (q \wedge \neg\neg\neg p)).$$

$$\theta[\phi'/\phi] = \begin{cases} \theta, & \text{if } \phi \text{ does not occur as a subformula of } \theta, \\ \phi', & \text{if } \phi \text{ is the subformula } \theta \text{ of } \theta, \\ \neg\psi[\phi'/\phi], & \text{if } \phi \text{ occurs as a subformula of } \theta \text{ (with } \phi \neq \theta) \text{ and } \theta \text{ is of the form } \neg\psi, \\ (\psi[\phi'/\phi] \wedge \chi[\phi'/\phi]), & \text{if } \phi \text{ occurs as a subformula of } \theta \text{ (with } \phi \neq \theta) \text{ and } \theta \text{ is of the form } (\psi \wedge \chi). \end{cases}$$

Making the description of $\theta[\phi'/\phi]$ makes the problem much easier to solve!]

2.5   The expressive power of connectives

So far we have looked at formulas built up using the connectives $\wedge$, $\vee$, $\neg$, $\rightarrow$ and $\leftrightarrow$, which have intended interpretations corresponding to uses in everyday language, conveyed by their standard truth tables. Surely there are other connectives which might arise from everyday language, in which case we might ask the following questions.

(i) How many different connectives are there?

(ii) Can some connectives be expressed in terms of others?

(iii) Is there any 'best' set of connectives?

It's not too hard to think of some further everyday connectives, although it might be harder to settle on reasonable truth tables for them. For instance, there are the exclusive 'or' (meaning '...or ...but not both'), 'is implied by' and 'unless'.

The correspondence between the symbols and everyday language has already required some firm, and even tough, decisions, like 'or' being taken as inclusive rather than exclusive and 'implies' carrying with it the convention that $\phi \rightarrow \psi$ is true when ϕ is false.

Exercise 2.47

Suggest truth tables for each of 'ϕ or ψ' with the exclusive 'or', 'ϕ is implied by ψ' and 'ϕ unless ψ'.

These extra connectives are all binary – they connect two propositions. How about connectives requiring more than two propositions? They are perhaps less everyday than the likes of 'and' and 'implies', but they do exist. Take, for instance, 'at least two of the following statements are true: ...'. This doesn't fit in well with the sort of construction rules we've had for well-formed formulas if we leave open how many statements do follow. But we can nail things down by specifying a natural number n (with $n > 2$ to make things interesting!) and modifying this as 'at least two of the following n statements are true: ...'.

Exercise 2.48

(a) (i) How many rows would you need for the truth table of the proposition 'at least two of the following 3 statements are true: ϕ, ψ, θ'?

(ii) Write down the truth table for this proposition.

(b) (i) How many rows would you need for the truth table of the proposition 'at least two of the following n statements are true: $\phi_1, \phi_2, \ldots, \phi_n$', where $n \geq 2$?

(ii) On how many of these rows would you expect the proposition to be true?

Solution

(a) (i) Each of the three propositions ϕ, ψ and θ could take the values T or F independent of the values taken by the other two. Thus there are $2 \times 2 \times 2 = 2^3 = 8$ different combinations of truth values to be taken into account in the table, so that the latter needs 8 rows.

(ii)

We've set out the rows in what we hope is a transparently systematic basis!

ϕ	ψ	θ	at least two of ϕ, ψ, θ are true
T	T	T	T
T	T	F	T
T	F	T	T
T	F	F	F
F	T	T	T
F	T	F	F
F	F	T	F
F	F	F	F

(b) (i) As each of the propositions $\phi_1, \phi_2, \ldots, \phi_n$ could potentially be true or false independent of the values taken by the others, the table would normally require

$$\underbrace{2 \times 2 \times \ldots \times 2}_{n} = 2^n$$

rows.

(ii) Not given.

Rather than producing even more baroque examples of vaguely natural connectives, we'll simply state that we've opened the floodgates of connectives, laying open the possibility that there are lots of connectives of n arguments, for arbitrarily large natural numbers n. We shall leave behind the everyday language descriptions of connectives and concentrate on what characterizes them within 2-valued logic, namely their truth tables. Each truth table in essence describes the values of a function, as in the following definition.

A given truth table might have several descriptions: e.g. 'if … then' and 'implies' have the same truth table.

Definition   Truth function

A *truth function of n arguments* is any function $f \colon \{T, F\}^n \longrightarrow \{T, F\}$.

$\{T, F\}^n$ means
$$\underbrace{\{T, F\} \times \{T, F\} \times \ldots \times \{T, F\}}_{n}.$$

A truth function can often be described nicely by its rule, for instance the function of 3 arguments we saw in Exercise 2.48(a) which takes the value T when at least two of its arguments take the value T. But as often as not, the only way of describing a truth function is by giving its full table of values, in the form:

x_1	x_2	$\ldots$	x_n	$f(x_1, x_2, \ldots, x_n)$
T	T	$\ldots$	T	?
T	T	$\ldots$	F	?
$\vdots$	$\vdots$		$\vdots$	$\vdots$
F	F	$\ldots$	F	?

Exercise 2.49

(a) If f is a truth function of n arguments, how many rows are there in its table of values?

(b) Explain why the number of truth functions of n arguments is 2^{2^n}.

Clearly there are infinitely many truth functions, allowing for all possible values of n. If we want to represent all of these within our formal language, do we need a special connective symbol for each one? Or can we represent some of them in terms of others? We have already seen how some of the more basic connectives are interrelated, in terms of logical equivalence. For instance, we have $(\phi \to \psi) \equiv (\neg\phi \vee \psi)$, so that any use of the connective $\to$ in a formula could be replaced by a construction involving $\neg$ and $\vee$. Likewise we can talk about a formula in the formal language representing a particular truth function, as in the following definition.

More precisely there are countably infinitely many truth functions.

Definition   Representing a formula

Let f be a truth function of n arguments $x_1, x_2, \ldots, x_n$ and let ϕ be a formula of the formal language involving propositional variables out of the set $\{p_1, p_2, \ldots, p_n\}$. We shall say that ϕ *represents* f if the table of values of f

x_1 x_2 $\ldots$ x_n	$f(x_1, x_2, \ldots, x_n)$
T $\ T$ $\ldots\ T$	?
T $\ T$ $\ldots\ F$	?
$\vdots$ $\ \vdots$ $\quad\vdots$	$\vdots$
F $\ F$ $\ldots\ F$	?

matches the truth table of ϕ in an obvious way:

p_1 p_2 $\ldots$ p_n	ϕ
T $\ T$ $\ldots\ T$	$f(T, T, \ldots, T)$
T $\ T$ $\ldots\ F$	$f(T, T, \ldots, F)$
$\vdots$ $\ \vdots$ $\quad\vdots$	$\vdots$
F $\ F$ $\ldots\ F$	$f(F, F, \ldots, F)$

or, more formally, for any truth assignment v, if $v(p_i) = x_i$ for $i = 1, 2, \ldots, n$, then $v(\phi) = f(x_1, x_2, \ldots, x_n)$.

For instance, the truth function $f_\to$ of 2 arguments given by $f_\to(x_1, x_2) = F$ if and only if $x_1 = T$ and $x_2 = F$ is represented by the formula $(p_1 \to p_2)$. Also the truth function of 3 arguments corresponding to the connective introduced in Exercise 2.48(a) is represented by the formula

$f_\to$ is just the truth function corresponding to 'implies'.

$$((p_1 \wedge p_2) \vee (p_2 \wedge p_3) \vee (p_3 \wedge p_1)).$$

If we were interested in whether both these truth functions could be represented in a language using a limited set of connectives, say consisting of just $\neg$ and $\wedge$, then it so happens it can be done, by exploiting logical equivalences, first $(\phi \to \psi) \equiv (\neg\phi \lor \psi)$ (for appropriate ϕ, ψ) to eliminate the uses of $\to$ and then $(\phi \lor \psi) \equiv \neg(\neg\phi \land \neg\psi)$ to eliminate the uses of $\lor$. But perhaps these are just nicely behaved truth functions. With how uncomplicated a set of connectives can we represent all truth functions?

Pretty remarkably, there are very uncomplicated and small sets of connectives with which one can represent all truth functions. This property merits a definition.

> ### Definition *Adequate set of connectives*
>
> A set S of connectives is *adequate* if all truth functions can be represented by formulas using connectives from this set.

In many ways the nicest (to the author!) such adequate set is $\{\neg, \land, \lor\}$, as is shown in the following theorem.

> ### Theorem 2.6
>
> The set of connectives $\{\neg, \land, \lor\}$ is adequate.

Proof

Let $f: \{T, F\}^n \longrightarrow \{T, F\}$ be a truth function of n arguments. We shall use the table of values of f to construct a formula ϕ using $\neg, \land, \lor$ representing f.

$x_1\ x_2\ \ldots\ x_n$	$f(x_1, x_2, \ldots, x_n)$
$T\ \ T\ \ldots\ T$	?
$T\ \ T\ \ldots\ F$	?
$\vdots\ \ \vdots\ \quad\ \vdots$	$\vdots$
$F\ \ F\ \ldots\ F$	?

First we deal with the case when $f(x_1, x_2, \ldots, x_n) = F$ on all 2^n rows of the table. Simply take ϕ to be the formula $(p_1 \land \neg p_1)$ – the truth table of this formula, regarded as involving variables out of the set $\{p_1, p_2, \ldots, p_n\}$, will give the value F on all lines.

Much more interesting, and needing some real effort, is the case when the table of values of f has some rows for which $f(x_1, x_2, \ldots, x_n) = T$. For each such row, coded by a particular n-tuple $\langle x_1, x_2, \ldots, x_n \rangle$ in $\{T, F\}^n$, construct the formula $\theta_{\langle x_1, x_2, \ldots, x_n \rangle}$ as follows:

$$(q_1 \land q_2 \land \ldots \land q_n) \quad \text{where } q_i = \begin{cases} p_i, & \text{if } x_i = T, \\ \neg p_i & \text{if } x_i = F, \end{cases} \text{ for } i = 1, 2, \ldots, n.$$

The key property of this formula is that the only truth assignment v that makes it true is the one corresponding to the row coded by $\langle x_1, x_2, \ldots, x_n \rangle$ in $\{T, F\}^n$, i.e. defined by $v(p_i) = x_i$, for each $i = 1, 2, \ldots, n$.

Now let ϕ be the disjunction of all the $\theta_{\langle x_1, x_2, \ldots, x_n \rangle}$ which arise for the truth function f. We claim that ϕ represents f.

Of course we could have used the logical equivalence $(\phi \to \psi) \equiv \neg(\phi \land \neg\psi)$ to eliminate the $\to$ directly, without going via the use of $\lor$.

This proof has the bonus that it gives an explicit construction of a formula ϕ using $\neg, \land, \lor$ representing a given truth function f and that this formula has a helpful standard shape.

If you really want a formula which involves all of the variables $p_1, p_2, \ldots, p_n$, then take the formula $(p_1 \land \neg p_1) \land \ldots \land (p_n \land \neg p_n)$.

Check this!

Take any truth assignment v, where $v(p_i) = x_i$ for each $i = 1, 2, \ldots, n$. We must show that $v(\phi) = f(x_1, x_2, \ldots, x_n)$ and to do this it's enough to show that $v(\phi) = T$ if and only if $f(x_1, x_2, \ldots, x_n) = T$. If $v(\phi) = T$, then as ϕ is a disjunction of formulas $\theta_{\langle y_1, y_2, \ldots, y_n \rangle}$, one of the latter is true under v. But then one of these θs must be $\theta_{\langle x_1, x_2, \ldots, x_n \rangle}$. By the construction of ϕ, this can only be included when $f(x_1, x_2, \ldots, x_n) = T$, which is what we require. Conversely, if $f(x_1, x_2, \ldots, x_n) = T$, then $\theta_{\langle x_1, x_2, \ldots, x_n \rangle}$ is one of the disjuncts of ϕ; and as the truth assignment v makes $\theta_{\langle x_1, x_2, \ldots, x_n \rangle}$ true, it must then make ϕ true. ∎

Let's look at the ϕ which the proof constructs for the truth function f of 3 variables given by the following table:

x_1	x_2	x_3	$f(x_1, x_2, x_3)$
T	T	T	T
T	T	F	T
T	F	T	F
T	F	F	F
F	T	T	F
F	T	F	T
F	F	T	F
F	F	F	F

There are three rows on which $f(x_1, x_2, x_3) = T$, corresponding to the triples $\langle T, T, T \rangle$, $\langle T, T, F \rangle$ and $\langle F, T, F \rangle$. The corresponding θs are

$$\theta_{\langle T,T,T \rangle} : \quad (p_1 \wedge p_2 \wedge p_3),$$
$$\theta_{\langle T,T,F \rangle} : \quad (p_1 \wedge p_2 \wedge \neg p_3),$$
$$\theta_{\langle F,T,F \rangle} : \quad (\neg p_1 \wedge p_2 \wedge \neg p_3).$$

Thus f is represented by the formula ϕ obtained by taking the disjunction of these three formulas:

$$((p_1 \wedge p_2 \wedge p_3) \vee (p_1 \wedge p_2 \wedge \neg p_3) \vee (\neg p_1 \wedge p_2 \wedge \neg p_3)).$$

Note that this formula is in what we previously called a disjunctive form, namely a disjunction of formulas each of which is a conjunction of variables, possibly negated. Furthermore these latter conjunctions and the similar conjunctions obtained for almost all other truth functions f all involve the same variables, $p_1, p_2, \ldots, p_n$, which merits saying that the formula is in a 'normal form', to which we give a special name, as in the following definition.

We met this in Theorem 2.5 in Section 2.4.

The exception being the case when $f(x_1, x_2, \ldots, x_n) = F$ for all $x_1, x_2, \ldots, x_n$; but even then we could represent f by the formula $(p_1 \wedge \neg p_1 \wedge \ldots \wedge p_n \wedge \neg p_n)$, which uses all of $p_1, p_2, \ldots, p_n$.

Definition   *Disjunctive normal form*

A formula is said to be in *disjunctive normal form*, often abbreviated as *dnf*, if for some $n \geq 1$ it is a disjunction of formulas of the form

$$(q_1 \wedge q_2 \wedge \ldots \wedge q_n),$$

where for each $i = 1, 2, \ldots, n$, q_i is one of p_i and $\neg p_i$. As an exceptional case, we shall also say that the formula $(p_1 \wedge \neg p_1 \wedge \ldots \wedge p_n \wedge \neg p_n)$ is in disjunctive normal form.

To avoid redundancy in a formula in dnf, we shall also ban any of these conjunctions from appearing more than once in the formula.

Thus a benefit of this proof of Theorem 2.6 is that we know not only that each truth function can be represented by a formula using $\wedge$, $\vee$ and $\neg$, but how to construct such a formula with a nice shape, namely one in disjunctive normal form.

A further benefit of Theorem 2.6 is that it gives a more general result than Theorem 2.5 in Section 2.4. The latter theorem told us that any formula built up using connectives in the set $\{\wedge, \vee, \neg, \rightarrow, \leftrightarrow\}$ was logically equivalent to a formula in disjunctive form. But now we can say that any formula ψ built up using *any* connectives, not just these familiar ones, is logically equivalent to a formula in dnf. All we do is take the truth function f represented by ψ and then apply Theorem 2.6 to get a formula ϕ which represents f, and hence has to be logically equivalent to ψ. We hope that after having seen the proof of Theorem 2.6, the construction of the formulas in the dnf for ψ now seems very natural – we just represent each truth assignment making ψ true by a corresponding conjunction of propositional variables and/or their negations, and then join together the relevant conjunctions by $\vee$s.

Exercise 2.50 ____

Write down formulas in dnf representing each of the following:

(a) the truth function of 2 arguments represented by the formula $(p_1 \leftrightarrow p_2)$;

(b) the truth function of 3 arguments represented by the formula $\neg(p_1 \vee p_3)$, regarded as built up from variables out of the set $\{p_1, p_2, p_3\}$;

(c) the truth function f of 3 arguments where $f(x_1, x_2, x_3)$ is true if at least two of x_1, x_2, x_3 is false.

Solution

We shall give a solution to (b) and leave the others to you.

The formula $\neg(p_1 \vee p_3)$, regarded as built up from variables out of the set $\{p_1, p_2, p_3\}$, represents a truth function f whose table of values matches the truth table of $\neg(p_1 \vee p_3)$ relative to these 3 variables, thus with 8 rows. The rows on which $f(x_1, x_2, x_3) = T$ correspond to the truth assignments making both p_1 and p_3 false, with p_2 given any truth value. So the relevant triples for the dnf are $\langle F, T, F \rangle$ and $\langle F, F, F \rangle$, giving the dnf

$$((\neg p_1 \wedge p_2 \wedge \neg p_3) \vee (\neg p_1 \wedge \neg p_2 \wedge \neg p_3)).$$

____

There is a similar result to Theorem 2.6 in terms of conjunctive forms, which we also met in Theorem 2.5 of Section 2.4. A formula θ is said to be in *conjunctive normal form*, abbreviated as *cnf*, if, for some n, it is a conjunction of disjunctions of propositional variables and their negations, where each variable in the set $\{p_1, p_2, \ldots, p_n\}$ appears exactly once in each disjunction, for example

$$((p_1 \vee \neg p_2 \vee p_3) \wedge (p_1 \vee \neg p_2 \vee \neg p_3) \wedge (\neg p_1 \vee p_2 \vee \neg p_3)).$$

In the next exercise, we invite you to prove in two different ways that every truth function f can be represented by a formula in cnf.

Exercise 2.51

(a) Prove that any truth function f can be represented by a formula ϕ in cnf by negating a formula in dnf representing a suitably chosen truth function related to f and manipulating the resulting formula.

(b) Prove that any truth function f can be represented by a formula ϕ in cnf by an adaptation of the proof of Theorem 2.6, using the table of values of the function f. [*Hints*: Which rows should you look at? Also for each row you look at, try to construct in a systematic way a formula of the form $(q_1 \vee q_2 \vee \ldots \vee q_n)$, where each q_i is one of p_i and $\neg p_i$.]

(c) Write down formulas in cnf representing each of the truth functions in Exercise 2.50.

As a check, our cnf for $(p_1 \leftrightarrow p_2)$ is
$$((\neg p_1 \vee p_2) \wedge (p_1 \vee \neg p_2)).$$

Once we have one adequate set of connectives, it's easy to show that some other set S is adequate – all one has to do is show that each of $\neg, \wedge, \vee$ can be represented by a formula using connectives in S. For instance, as we have $(\phi \vee \psi) \equiv \neg(\neg\phi \wedge \neg\psi)$, so that $\vee$ can be represented by a formula using $\neg$ and $\wedge$, the set $\{\neg, \wedge\}$ is adequate.

Exercise 2.52

Show that each of the following sets of connectives is adequate.

(a) $\{\neg, \vee\}$

(b) $\{\neg, \rightarrow\}$

(c) $\{|\}$ where $|$ is a two-place connective with truth table:

p	q	$(p\|q)$
T	T	F
T	F	T
F	T	T
F	F	T

The symbol $|$ is often called 'nand' or 'not ($\ldots$ and $\ldots$)' – you have to say it so that the bracketing is unambiguous! – or the *Sheffer stroke*, and $(\phi|\psi)$ is logically equivalent to $\neg(\phi \wedge \psi)$.

The following definition introduces a couple of new symbols which we shall use occasionally later in the book.

Definitions Propositional constants

The symbol $\bot$ is called the *propositional constant for falsity*. The symbol $\top$ is called the *propositional constant for truth*. These propositional constants are used in the construction of formulas as though they are propositional variables, but for every truth assignment v,

$$v(\bot) = F \text{ and } v(\top) = T.$$

What distinguishes a propositional constant from a variable is that its truth value is the same for every truth assignment v.

Thus

$$(\bot \rightarrow (p \rightarrow \bot)), \ \bot, \ \neg\top, \ (q \rightarrow \top)$$

are formulas, while $\bot p$ and $\top\bot$ are not. Also for all truth assignments v,

$$v((\bot \wedge p)) = F, \quad v((\top \wedge p)) = v(p), \quad v((\top \vee \bot)) = T.$$

***Exercise* 2.53**

Show that the set $\{\rightarrow, \bot\}$ is adequate.

One of the benefits of knowing that a set of connectives is adequate comes when devising a formal proof system for propositional calculus. A reasonable aim for such a system is that one should be able to prove within it all tautologies, while at the same time not involving too many connectives for which the formal system would have to give rules or axiom schemes. An adequate set of connectives will do nicely!

How might one show that a set S of connectives is *not* adequate? This turns out to be less straightforward and requires a variety of tricks, depending on the connectives involved. For instance, the set $\{\wedge, \vee\}$ is not adequate. Obviously, if we could find a formula involving these connectives that represented $\neg$, the set would be adequate. A spot of experimenting with shortish formulas using $\wedge$ and $\vee$ will not come up with such a formula, but how do we know that there isn't some very long formula that does the job? One general principle to answer this is to look for some special property possessed by *all* formulas of a particular sort built up using connectives in S which thereby rules out that all truth functions can be represented. In the case of $S = \{\wedge, \vee\}$, one such property is the following.

> Any formula built up using a single variable p (as many times as you like!) with $\wedge$ and $\vee$ always takes the value T under the truth assignment v such that $v(p) = T$.

This means that such a formula could not represent one of the (two) truth functions $f \colon \{T, F\} \longrightarrow \{T, F\}$ such that $f(T) = F$, so that the set S is not adequate.

It is pretty easy to see that this property holds, but it is a useful discipline to prove it properly. A key feature of the proof is that it has to encompass all formulas ϕ of the sort described, and one of the standard ways of doing this is by induction on some measure of the length of ϕ. Have a go at the proof in the next exercise. In our solution we shall as ever use the number of connectives in ϕ as the measure of its length.

There are other such properties that would do, for instance that for any ϕ built up from $p_1, p_2, \ldots, p_n$ with $\wedge$ and $\vee$, if v is the truth assignment such that $v(p_i) = T$ for all $i = 1, 2, \ldots, n$, then $v(\phi) = T$. That would mean that no truth function f of n arguments such that $f(T, T, \ldots, T) = F$ can be represented.

The essence of why the property holds is that the truth tables of both $(\theta \wedge \psi)$ and $(\theta \vee \psi)$ give the value T when ϕ and ψ take the value T.

***Exercise* 2.54**

Prove that if ϕ is built up using the variable p with $\wedge$ and $\vee$, and v is the truth assignment such that $v(p) = T$, then $v(\phi) = T$.

Solution

We use induction on the number n of connectives in ϕ.

The base case is $n = 0$. The only formula of the given type with no connectives is p and the required property, namely that if $v(p) = T$ then $v(\phi) = T$, holds trivially.

For the inductive step assume that the property holds for all the formulas of this special form with $\leq n$ connectives, and that ϕ is a formula of this type with $n + 1$ connectives. Then ϕ has to be one of the forms $(\theta \wedge \psi)$ and $(\theta \vee \psi)$, where, as θ and ψ have the same form and at most n connectives, both have the required property, i.e. if $v(p) = T$ then $v(\theta) = v(\psi) = T$. Then whichever form ϕ has, the truth tables of $\wedge$ and $\vee$ ensure that $v(\phi) = T$, as required.

The result follows by mathematical induction.

As usual with such arguments, it is almost always essential to frame the property of formulas ϕ of the special sort so that it includes the case when ϕ is a propositional variable, which contains no connectives – hence $n = 0$. For the inductive step, the $(n + 1)$th connective, in this case, joins two formulas whose combined connective length is n – we don't know how many connectives are in each, but we do know that in each case it's n or less.

Use your ingenuity to resolve the following problems and provide suitably convincing arguments!

Exercise 2.55

Show that none of the following sets of connectives is adequate.

(a) $\{\neg\}$

(b) $\{\rightarrow\}$

(c) $\{\vee, \bot\}$, where $\bot$ is the propositional constant for falsity.

Exercise 2.56

Let $S = \{\neg, \leftrightarrow\}$.

(a) Show that every truth function of one argument can be represented by a formula using connectives in $\{\neg, \leftrightarrow\}$.

(b) By finding a property possessed by all formulas built up from two propositional variables p and q using $\neg$ and $\leftrightarrow$ (and verifying that this property does indeed hold), show that the set $\{\neg, \leftrightarrow\}$ is not adequate.

Exercise 2.57

There are 16 different possible truth tables for a two-place connective $*$. For which of these is $\{*\}$ an adequate set of connectives? In each case explain why it gives or does not give (as appropriate) an adequate set.

An alternative approach to showing that a set S of connectives is not adequate is to investigate, for each non-negative integer n, how many truth functions of n arguments can be represented using S. If for some n this number is less than 2^{2^n}, then S is not adequate. Meanwhile, knowing how many, and which, truth functions can be represented by S is of interest in its own right.

Exercise 2.58

How many truth functions of n arguments can be represented using the set $\{\wedge\}$?

Solution

Our method is to investigate whether there is some sort of 'normal' form for formulas ϕ built up using variables in the set $\{p_1, p_2, \ldots, p_n\}$ and connectives out of the set S, in this case just the connective $\wedge$. It would be nice if each ϕ was logically equivalent to just one formula in this normal form, so that counting the number of formulas in normal form gives the number of different truth functions which can be represented. In this case, there are some very useful logical equivalences involving $\wedge$ which lead to such a normal form. They are as follows:

$$(\phi \wedge (\psi \wedge \theta)) \equiv ((\phi \wedge \psi) \wedge \theta) \quad \text{(associativity)},$$
$$(\phi \wedge \psi) \equiv (\psi \wedge \phi) \quad \text{(commutativity)},$$
$$(\phi \wedge \phi) \equiv \phi \quad \text{(idempotency)}.$$

Recall that the correct terminology here is that $\wedge$ is associative, commutative and idempotent under logical equivalence. The associativity and commutativity are particularly helpful, because using these we can show that given any formula ϕ built up just using $\wedge$, any rearrangement of the variables in ϕ gives a formula logically equivalent to ϕ. In particular we can rearrange the variables in ϕ so that all the p_i, for a given i, are in the same subformula θ_i, and obtain a formula logically equivalent to ϕ of the form

$$(\theta_{i_1} \wedge (\theta_{i_2} \wedge (\ldots \wedge \theta_{i_k}) \ldots))),$$

where the propositional variables appearing in ϕ are $p_{i_1}, p_{i_2}, \ldots, p_{i_k}$, with $i_1 < i_2 < \ldots < i_k$. For example, the formula

$$((p_3 \wedge p_1) \wedge (p_3 \wedge ((p_4 \wedge p_1) \wedge p_3)))$$

is logically equivalent to the formula

$$((p_1 \wedge p_1) \wedge ((p_3 \wedge (p_3 \wedge p_3)) \wedge p_4)).$$

The idempotency of $\wedge$ under logical equivalence then gives us that each component θ_i is logically equivalent to just a single p_i. Thus a given ϕ is logically equivalent to a normal form which is a simple conjunction of just the p_is appearing in ϕ, that is,

$$(p_{i_1} \wedge p_{i_2} \wedge \ldots \wedge p_{i_k}),$$

where the propositional variables appearing in ϕ are $p_{i_1}, p_{i_2}, \ldots, p_{i_k}$ with $i_1 < i_2 < \ldots < i_k$. In the example above, this would be the formula

$$(p_1 \wedge p_3 \wedge p_4).$$

The number of distinct formulas using variables in the set $\{p_1, p_2, \ldots, p_n\}$ which are in this normal form equals the number of non-empty subsets of this set, namely $2^n - 1$.

The same sort of method pays dividends in many parts of the following exercises.

Logically equivalent formulas, regarded as using variables out of the set $\{p_1, p_2, \ldots, p_n\}$, represent the same truth function. So to count the different truth functions representable, we just want one formula out of each class of logically equivalent formulas.

Not all of the n propositional variables in $\{p_1, p_2, \ldots, p_n\}$ might appear in a given ϕ.

Back to being casual about brackets, thanks to associativity!

The set $\{p_1, p_2, \ldots, p_n\}$ has n elements, so has 2^n subsets, including the empty set.

Exercise 2.59

How many truth functions of n variables can be represented by using each of the following sets of connectives?

(a) $\{\vee\}$

(b) $\{\neg\}$

(c) $\{\neg, \wedge, \vee\}$

(d) $\{\leftrightarrow\}$

(e) $\{\neg, \leftrightarrow\}$

> The result of this gives a nice alternative way of showing that the set $\{\neg, \leftrightarrow\}$ is not adequate.

Exercise 2.60

(a) Let f be a truth function of n arguments such that $f(T, T, \ldots, T) = T$. Show that f can be represented by a formula using connectives in the set $\{\wedge, \vee, \rightarrow\}$. [*Hints:* f can be represented by a formula ϕ using $\{\neg, \wedge, \vee\}$ in dnf (or cnf, whichever you prefer). The fact that $f(T, T, \ldots, T) = T$ gives just enough information about ϕ to enable all the occurrences of $\neg$ to be eliminated using $\wedge$, $\vee$ and $\rightarrow$, with the aid of logical equivalences such as $(\neg\theta \vee \psi) \equiv (\theta \rightarrow \psi)$.]

(b) Hence show that the number of truth functions of n arguments representable using $\{\wedge, \vee, \rightarrow\}$ is $2^{2^n - 1}$ and also find the number when using $\{\wedge, \rightarrow\}$.

You may be surprised to know that there is no known nice formula for the number of truth functions of n arguments representable using such a simple set as $\{\wedge, \vee\}$, which has been sought for a variety of applications, but for which only not very good upper and lower bounds have been found.

> The problem for $\{\wedge, \vee\}$ is equivalent to ones in the contexts of Boolean algebra and sets, e.g. how many different sets can be created from up to n sets by taking unions and intersections.

Further exercises

Exercise 2.61

Let $*$ be a ternary connective with the following truth table.

p	q	r	$*(p, q, r)$
T	T	T	F
T	T	F	T
T	F	T	T
T	F	F	T
F	T	T	T
F	T	F	T
F	F	T	T
F	F	F	F

(a) Prove that $\{*\}$ is not adequate.

(b) Is $\{*, \rightarrow\}$ adequate? Prove your answer.

Exercise 2.62 ——————————————————————————————

Suppose that ϕ is a formula in disjunctive normal form (but not necessarily involving all the propositional variables in each disjunct). Prove that ϕ is a contradiction if and only if for each disjunct ψ of ϕ there is some propositional variable p such that both p and $\neg p$ appear in ψ.

Exercise 2.63 ——————————————————————————————

The result that $\{\wedge, \vee, \neg\}$ is an adequate set of connectives can be restated by saying that all truth functions $f \colon \{T, F\}^n \longrightarrow \{T, F\}$, for all $n \geq 0$, can be obtained by (repeated) composition of the three truth functions $f_\wedge$, $f_\vee$ and $f_\neg$ (of respectively 2, 2 and 1 arguments) corresponding to the connectives $\wedge$, $\vee$ and $\neg$. Ignoring the particular interpretation of T and F as truth values, we can conclude from this that the set of all functions from X^n to X, where X is a two element set and $n \geq 0$, can be obtained by composition of a finite number (three!) of relatively simple functions. Show that the same applies for any finite set X, that is, that there is a finite subset of the set of all functions from X^n to X for all $n \geq 0$ from which any function in the set can be obtained by composition.

2.6 Logical consequence

In this section we shall start looking at arguments involving propositional formulas, which will be our first step towards modelling mathematical proof. One's expectation of a (correct!) mathematical argument is that it should involve statements which follow from previous statements in some sort of convincing way, right up to the desired concluding result. It is a tall order trying to nail down all the ways in which mathematicians are convinced by an argument, so in this section we shall concentrate on just one requirement of a convincing argument, as follows. One measure of a statement ϕ following from statements in a set Γ is that whenever all the statements in Γ are true then ϕ must also be true. So, for instance, in everyday maths it follows from the statement that the function f from $\mathbb{R}$ to $\mathbb{R}$ is differentiable, along with all sorts of other tacit assumptions about the arithmetic of $\mathbb{R}$, that f is continuous. It is not the case that every function f is differentiable, but whenever one does have an f that is differentiable, then it must also be continuous. We shall capture this general idea by the following definition.

> Think of the statements in Γ as the assumptions underlying the argument.

Definitions   Logical consequence

Let Γ be a set of formulas and ϕ a formula involving propositional variables in a set P. Then ϕ is a *logical consequence* of Γ, or equivalently Γ *logically implies* ϕ, when for all truth assignments v on P, if $v(\gamma) = T$ for all $\gamma \in \Gamma$, then $v(\phi) = T$. We write this as $\Gamma \vDash \phi$.

In the case where Γ is the empty set, we write $\vDash \phi$ to say that for all truth assignments v, $v(\phi) = T$, i.e. ϕ is a tautology.

When ϕ is not a logical consequence of Γ, we write $\Gamma \nvDash \phi$. Similarly when ϕ is not a tautology, we write $\nvDash \phi$.

> Another description of $\Gamma \vDash \phi$ is that every truth assignment satisfying Γ also satisfies ϕ. Informally, ϕ is true whenever Γ is true.

So we have

$$\models ((p \wedge q) \rightarrow p),$$

as the formula is a tautology. We have

$$\{q, (r \rightarrow \neg p)\} \models (q \vee r)$$

as for each of the truth assignments v satisfying both q and $(r \rightarrow \neg p)$, v also satisfies $(q \vee r)$. Thus we have

$$\{q, (r \rightarrow \neg p)\} \not\models (q \rightarrow r)$$

as there is a truth assignment v satisfying both q and $(r \rightarrow \neg p)$ which does not satisfy $(q \rightarrow r)$, for instance v defined by

$$v(p) = v(q) = T, \ v(r) = F.$$

Three of the eight different truth assignments on $\{p, q, r\}$ satisfy both q and $(r \rightarrow \neg p)$.

Exercise 2.64

Decide which of the following logical implications hold.

(a) $\{p, \neg r\} \models (q \rightarrow (r \rightarrow \neg p))$

(b) $\{p, (q \leftrightarrow r)\} \models (q \rightarrow (r \rightarrow \neg p))$

(c) $\models ((p \rightarrow q) \rightarrow p)$

(d) $\{(p \vee q)\} \models ((p \rightarrow q) \rightarrow q)$

(e) $\{p_{2i} : i \in \mathbb{N}\} \models ((p_{17} \rightarrow p_{14}) \rightarrow p_{87})$

(f) $\{(p_{2i} \rightarrow p_i) : i \in \mathbb{N}\} \models ((p_{34} \vee p_{17}) \rightarrow p_{17})$

Solution

(a) In any truth assignment v satisfying all the formulas in the set $\{p, \neg r\}$, we must have $v(p) = T$ and $v(r) = F$. Then $v((r \rightarrow \neg p)) = T$, so regardless of the value of $v(q)$, we have $v((q \rightarrow (r \rightarrow \neg p))) = T$. Thus it is the case that $\{p, \neg r\} \models (q \rightarrow (r \rightarrow \neg p))$.

(b) The truth assignment v defined by $v(p) = v(q) = v(r) = T$ satisfies all the formulas in $\{p, (q \leftrightarrow r)\}$, but $v((q \rightarrow (r \rightarrow \neg p))) = F$. So it is not the case that $\{p, (q \leftrightarrow r)\} \models (q \rightarrow (r \rightarrow \neg p))$.

Equivalently, we have shown that $\{p, (q \leftrightarrow r)\} \not\models (q \rightarrow (r \rightarrow \neg p))$.

(c) Not given.

(d) Suppose that the truth assignment v satisfies $(p \vee q)$. If $v(q) = T$, then $v(((p \rightarrow q) \rightarrow q)) = T$. If $v(q) = F$, so that $v(p) = T$, we have $v((p \rightarrow q)) = F$, so that $v(((p \rightarrow q) \rightarrow q)) = T$.

Thus in all cases where $v((p \vee q)) = T$, we have $v(((p \rightarrow q) \rightarrow q)) = T$, so that $\{(p \vee q)\} \models ((p \rightarrow q) \rightarrow q)$.

(e) Not given.

(f) Not given.

Exercise 2.65

Is $\Gamma \not\models \phi$ equivalent to saying $\Gamma \models \neg\phi$?

When we extend the definition of logical consequence to the more complicated, and mathematically more useful, predicate languages in Chapter 4, you will see that the idea does capture something of great importance. To give you a foretaste, the set Γ might give axioms for an interesting theory, for instance group theory, and $\Gamma \vDash \phi$ will then mean that in every structure which makes all of Γ true, i.e. in every group, the formula ϕ is also true. Then the property of groups that statement ϕ represents holds for *all* groups.

In, for instance, our solution to Exercise 2.64(d), we establish a logical consequence $\Gamma \vDash \phi$ by direct appeal to the definition, by looking at all truth assignments which satisfy Γ. But this is quite far from how we usually infer statements from others within a mathematical proof. For instance, while it is the case that a function f being continuous is a logical consequence of f being differentiable, this is normally established by a sequence of several non-trivial steps. In general, we tend to use quite small steps in proofs and in the exercise below we give logical consequences corresponding to some very small such steps. From now on in this section we shall concentrate on inferences involving propositional formulas. Later in the book we shall look at inferences involving a richer language, closer to one usable for everyday mathematics.

Notation

We shall sometimes cheat on set notation for the Γ in $\Gamma \vDash \phi$ by dropping some of the set brackets { }, writing e.g.

$$\theta, \psi \vDash \phi \text{ instead of } \{\theta, \psi\} \vDash \phi,$$
$$\Gamma, \theta \vDash \phi \text{ instead of } \Gamma \cup \{\theta\} \vDash \phi$$
$$\text{and } \Gamma, \Delta \vDash \phi \text{ instead of } \Gamma \cup \Delta \vDash \phi.$$

We hope that the context will make it clear what is meant.

None of the parts of the next exercise should be very challenging, but even if you don't attempt them all, do read all the parts of the exercise as they are potentially more important than their simplicity suggests.

Exercise 2.66

Let ϕ, ψ, θ be formulas. Show each of the following.

(a) $(\phi \wedge \psi) \vDash \phi$

(b) $(\phi \wedge \psi) \vDash \psi$

(c) $\phi, \psi \vDash (\phi \wedge \psi)$

(d) $\phi \vDash (\phi \vee \psi)$

(e) $\phi \vDash (\psi \vee \phi)$

(f) If $\phi \vDash \theta$ and $\psi \vDash \theta$, then $(\phi \vee \psi) \vDash \theta$.

With this and later exercises, turn the symbols into natural language when you think about the problem. For example, you might think of part (c) as saying that any assignment making both the formulas ϕ and ψ true must make the formula $(\phi \wedge \psi)$ true.

Solution

We shall give a solution only to part (a), to give you an idea of how simple a convincing explanation can be!

If a truth assignment v satisfies $(\phi \wedge \psi)$, then from the truth table of $\wedge$ we must have $v(\phi) = T \ (= v(\psi))$.

These simple logical consequences are important because they illustrate how we infer using the connectives 'and' and 'or'. Here are some more simple logical consequences of similar importance in representing small steps of inference.

Exercise 2.67

Let ϕ, ψ, θ, χ be formulas. Show each of the following.

(a) $\phi, (\phi \rightarrow \psi) \vDash \psi$

(b) If $\phi \vDash \psi$ then $\neg\psi \vDash \neg\phi$.

(c) If $\phi \vDash \psi$ and $\theta \vDash \chi$, then $(\phi \wedge \theta) \vDash (\psi \wedge \chi)$.

(d) If $\phi \vDash \psi$ and $\theta \vDash \chi$, then $(\phi \vee \theta) \vDash (\psi \vee \chi)$.

The result of the first part of this exercise, inferring ψ from ϕ and $(\phi \rightarrow \psi)$, has been regarded as so important that it has been given a special name, *Modus Ponens*. It is plainly a crucial feature of how to infer with $\rightarrow$. When we come to our formal proof system in the next chapter, we shall adopt a formal rule of inference corresponding to Modus Ponens. One yardstick of a formal system will be whether it can mirror other simple logical consequences – if it cannot, what hope for deriving those that are more complicated!

We hope that it comes as no surprise that there are strong connections between logical consequence and the connective $\rightarrow$ which represents implication. One such connection is given by the following theorem.

Theorem 2.7

Let Γ be a set of formulas and ϕ, ψ be formulas. Show that

$$\Gamma, \phi \vDash \psi \quad \text{if and only if} \quad \Gamma \vDash (\phi \rightarrow \psi).$$

We shall be very keen to match this result within our formal system.

Exercise 2.68

Prove Theorem 2.7.

Exercise 2.69

Let $\gamma_1, \gamma_2, \ldots, \gamma_n$ be finitely many formulas and ϕ a formula. Show that

$$\gamma_1, \gamma_2, \ldots, \gamma_n \vDash \phi \quad \text{if and only if} \quad \vDash ((\gamma_1 \wedge \gamma_2 \wedge \ldots \wedge \gamma_n) \rightarrow \phi).$$

There are useful connections between logical consequence and logical equivalence, as you can show, we hope straightforwardly, in the next exercise.

Exercise 2.70

Let ϕ, ψ, θ be formulas.

(a) Show that $\phi \equiv \psi$ if and only if $\phi \vDash \psi$ and $\psi \vDash \phi$.

(b) Suppose that $\phi \equiv \psi$. Show that

    (i) if $\phi \vDash \theta$, then $\psi \vDash \theta$;

    (ii) if $\theta \vDash \phi$, then $\theta \vDash \psi$.

The next exercise involves a very straightforward and useful result about cascading logical consequences. Again, this illustrates further connections between $\vDash$ and $\rightarrow$.

Exercise 2.71

(a) Let ϕ, ψ, θ be formulas. Show that if $\phi \vDash \psi$ and $\psi \vDash \theta$, then $\phi \vDash \theta$.

(b) Let Γ be a set of formulas and $\phi_1, \phi_2, \ldots, \phi_n$ finitely many formulas such that

$$\Gamma \vDash \phi_1, \quad \phi_1 \vDash \phi_2, \quad \phi_2 \vDash \phi_3, \quad \ldots, \quad \phi_{n-1} \vDash \phi_n.$$

Show that $\Gamma \vDash \phi_n$.

One way of using the result of the last exercise is to combine simple logical consequences to establish a more complicated logical consequence. But not every complicated logical consequence would necessarily get broken down in such a linear fashion. There might be other routes, for instance as in the following exercise.

Exercise 2.72

Let Γ, Δ be sets of formulas and ϕ, ψ, θ formulas. Suppose that $\Gamma \vDash (\phi \vee \psi)$, $\Delta, \phi \vDash \theta$ and $\psi \vDash \theta$. Show that $\Gamma, \Delta \vDash \theta$.

We have previously introduced the symbols $\bot$ and $\top$ as the propositional constants for falsity and truth, respectively. Here's an easy exercise involving them.

Recall from Section 2.5 that these propositional constants are used in formulas just like propositional variables, but that for every truth assignment v, $v(\bot) = F$ and $v(\top) = T$.

Exercise 2.73

Let ϕ, ψ be any formulas. Show the following.

(a) $(\phi \vee \neg\phi) \equiv \top$

(b) $(\phi \wedge \neg\phi) \equiv \bot$

(c) $\psi \vDash \top$

(d) $\bot \vDash \psi$

(e) $\top \vDash \psi$ if and only if ψ is a tautology.

(f) $\psi \vDash \bot$ if and only if ψ is a contradiction.

So $\top$ is a tautology and $\bot$ is a contradiction.

Parts (b) and (d) are of particular interest. If we have a set of formulas Γ and a formula ϕ for which both ϕ and $\neg\phi$ are logical consequences of Γ, it is easy to show that Γ logically implies the contradiction $(\phi \wedge \neg\phi)$, so that using (b) and (d) (and essentially the result of Exercise 2.71(b)), we can show that all formulas ψ are logical consequences of Γ. Such sets Γ most certainly exist – for instance, just take Γ to be the set $\{p, \neg p\}$ for a propositional variable p – but they are somehow rather undiscriminating when it comes to investigating their logical consequences! In fact, for such a set Γ, there are no truth assignments making all of the formulas of Γ true. We shall ask you to show this, phrasing the result using the following new terminology.

Definition   Satisfiable

The set Γ of formulas is *satisfiable* if there is some truth assignment v which satisfies Γ, i.e. $v(\gamma) = T$ for all $\gamma \in \Gamma$.

Plainly the issue of whether there are circumstances in which all statements in a given set can be made simultaneously true, i.e. are satisfiable, will be of interest, for instance when the statements attempt to axiomatize a mathematical theory.

Exercise 2.74

Let Γ be a set of formulas. Show that $\Gamma \models \bot$ if and only if Γ is not satisfiable.

Exercise 2.75

Let Γ be a set of formulas and ϕ a formula such that both $\Gamma \models \phi$ and $\Gamma \models \neg\phi$.   The converse result holds trivially!
Show that $\Gamma \models \psi$, for all formulas ψ.

Exercise 2.76

Let Γ be a set of formulas and ϕ a formula. Show that $\Gamma \cup \{\neg\phi\}$ is satisfiable if and only if $\Gamma \not\models \phi$.

Solution

We shall show that if $\Gamma \not\models \phi$, then $\Gamma \cup \{\neg\phi\}$ is satisfiable, and we leave the rest of the proof to you!

Suppose that $\Gamma \not\models \phi$. So it is not the case that all truth assignments v which satisfy Γ also satisfy ϕ. So there is some truth assignment v which satisfies Γ and doesn't satisfy ϕ. As $v(\phi) \neq T$, this means $v(\phi) = F$, so that $v(\neg\phi) = T$. Therefore this v satisfies $\Gamma \cup \{\neg\phi\}$, which is thus satisfiable.

Exercise 2.77

Let Γ be a set of formulas. Show that Γ is satisfiable if and only $\Gamma \not\models \phi$ for some formula ϕ.

It might be tempting to think that a set of formulas Γ for which $\Gamma \vDash \perp$ is in some sense silly. But such a set is often of great value within a mathematical proof, in the context of what is called *proof by contradiction*. You should have met this before and, if you are anything like the author, have been so excited by this method of proof that you spent a long period trying to use it in every mathematical argument! A classical argument is that found in Euclid (at around 300 BC) to show that there are infinitely many primes. A modern version of this proof goes as follows.

> This has the grander Latin name of *reductio ad absurdum*.

Suppose that there are only finitely many primes, listed as $p_1, p_2, \ldots, p_n$. Consider the number $N = p_1 p_2 \ldots p_n + 1$. As division by any p_i leaves remainder 1, none of $p_1, p_2, \ldots, p_n$ divides N. But N can be factorized as a product of primes, so there is another prime dividing N not equal to one of $p_1, p_2, \ldots, p_n$. This contradicts that $p_1, p_2, \ldots, p_n$ lists all the primes. Thus there are infinitely many primes.

For the purposes of this section, the underlying structure of this proof by contradiction is as follows: to prove that ϕ follows from the set of formulas Δ, we assume the negation of ϕ, i.e. $\neg\phi$, and from this and Δ derive a contradiction. Hey presto! This means that from Δ we can infer the hoped for ϕ. Formally we have the following theorem.

> In the proof above, ϕ says that there are infinitely many primes while Δ in some way gives more fundamental properties of the integers.

Theorem 2.8   *Proof by contradiction*

Let Δ be a set of formulas and ϕ a formula. If $\Delta, \neg\phi \vDash \perp$ then $\Delta \vDash \phi$.

Proof

Suppose that $\Delta \cup \{\neg\phi\} \vDash \perp$. Then by the result of Exercise 2.74, there are no truth assignments making all of the formulas in $\Delta \cup \{\neg\phi\}$ true. This means that if the truth assignment v makes all the formulas of Δ true, $v(\neg\phi)$ must be false, so that $v(\phi)$ must be true; that is, every truth assignment satisfying Δ also satisfies ϕ. Thus $\Delta \vDash \phi$. ∎

Exercise 2.78 ——————————————————————————

Show that the converse of the theorem above holds, that is, if Δ is a set of formulas and ϕ a formula such that $\Delta \vDash \phi$, then $\Delta \cup \{\neg\phi\} \vDash \perp$.

Exercise 2.79 ——————————————————————————

Let Δ be a set of formulas and ϕ a formula. Show that if $\Delta, \phi \vDash \perp$ then $\Delta \vDash \neg\phi$.

Exercise 2.80 ——————————————————————————

Let ϕ, θ be formulas such that $\phi, \theta \vDash \perp$. Show that $\phi \vDash \neg\theta$ and $\theta \vDash \neg\phi$. Is it necessarily the case that $\vDash (\neg\phi \wedge \neg\theta)$?

When we extend our formalisation of language to predicate languages and write down some set of formulas Γ axiomatizing something of real mathematical significance, like the theory of groups or of linear orders, we shall not only be interested in whether a particular formula ϕ is a consequence of Γ but in finding all formulas which are consequences of Γ. One tempting possibility is that for a given a set of formulas Γ and any formula ϕ, we have that $\Gamma \vDash \phi$ or $\Gamma \vDash \neg\phi$.

Exercise 2.81 ————————————————————————

Let Γ be the set $\{(p \vee q)\}$, where p, q are propositional variables. Is it the case that for all formulas ϕ in the language using just these two propositional variables that $\Gamma \vDash \phi$ or $\Gamma \vDash \neg\phi$?

Solution

No. Consider the formula ϕ given by p. Then taking the truth assignment v defined by $v(p) = F$, $v(q) = T$, we have $v((p \vee q)) = T$ and $v(p) = F$, so that $\Gamma \nvDash \phi$. Taking the truth assignment u defined by $u(p) = T$, $u(q) = T$, we have $u((p \vee q)) = T$ and $u(\neg p) = F$ (here $\neg\phi$ is $\neg p$), so that $\Gamma \nvDash \neg\phi$.

____

So in general, given a set Γ, we should not expect that for all formulas ϕ, $\Gamma \vDash \phi$ or $\Gamma \vDash \neg\phi$. There are some sets Γ for which this holds, besides those for which there is no truth assignment making all formulas of Γ true, and they are described as *complete*. We shall discuss these some more in the next chapter.

Some comments on decidability

The question of *how* we can tell whether ϕ is a logical consequence of the set Γ is rather interesting. One of the hopes of those who developed the ideas of describing interesting parts of mathematics using axioms was that the logical consequences of these axioms would be *decidable*, meaning that there is an algorithmic procedure which would, after a finite number of steps, say whether or not a given formula ϕ is a logical consequence of a set of formulas Γ.

We shall have to wait till Chapter 5 for the predicate languages which we might use to axiomatize some interesting mathematics. For the moment we shall just discuss the decidability of $\Gamma \vDash \phi$ for propositional languages.

If Γ is the empty set, there is a very straightforward algorithmic procedure for deciding whether $\vDash \phi$, i.e. whether ϕ is a tautology – just construct its truth table and check whether it takes the value T on all the finitely many rows of this table.

What can be done when Γ is non-empty? If the set Γ is finite, we can create the formula $(\bigwedge_{\gamma \in \Gamma} \gamma \to \phi)$, look at its truth table and exploit the result of Exercise 2.69. The formula is a tautology exactly when $\Gamma \vDash \phi$ and checking its truth table involves just finitely many steps. For the mean-minded there is then the further question of whether the method involving the construction of a truth table is practicable. Checking whether a formula is a tautology involves a finitely long process which is good news for many purposes. But if there is a large number n of propositional variables involved, the number of

We might even have that both hold, so that by the result of Exercise 2.75, $\Gamma \vDash \psi$ for all formulas ψ. But then by Exercise 2.74, there are no truth assignments making Γ true.

For an example giving many such sets Γ, see Exercise 2.87 later.

See Davis [10] for some of the history, leading up to precise definitions of 'decidable' and, indeed, of 'algorithmic procedure'. For the exciting theory stemming from these ideas, see Cutland [9], Enderton [12], Epstein and Carnielli [14] or Kleene [22].

We hope that you are willing to agree that one can give a finite set of simple instructions for producing the truth table of a formula ϕ that can be undertaken in finitely many steps.

As for what is practicable, you may have observed the author's reluctance to produce truth tables of long formulas or ones involving more than 8 rows!

rows in the truth table, 2^n, will be so large as to make the process impracticable. So there is an interest in thinking about other ways of checking whether $\Gamma \vDash \phi$.

When Γ is an infinite set, it is highly unlikely that $\Gamma \vDash \phi$ will be decidable. For instance, suppose that the language has propositional variables in the set $\{p_i : i \in \mathbb{N}\}$. An attempt by brute force of checking all truth assignments v to see first whether v satisfies Γ and, if so, whether v also satisfies ϕ, is likely to be far from an algorithmic procedure, as there are uncountably infinitely many different truth assignments on the variables $\{p_i : i \in \mathbb{N}\}$, so the full check couldn't happen in finitely many steps. But sometimes an infinite set Γ has a simple enough structure to make checking whether $\Gamma \vDash \phi$ practicable. For instance, if Γ is the set

$$\{p_0\} \cup \{(p_i \rightarrow p_{i+1}) : i \in \mathbb{N}\},$$

there is actually only the one truth assignment making all the formulas of Γ true, namely v where $v(p_i) = T$ for all $i \in \mathbb{N}$. So checking whether $\Gamma \vDash \phi$ for a given formula ϕ is simply a matter of working out the truth value $v(\phi)$ for this v and seeing whether this is T! There are more challenging examples of this positive behaviour – see for example Exercise 2.86 below. But normally with an infinite set Γ, we are doomed. For instance, it can be shown that there are subsets I of the set $\mathbb{N}$ for which there is no algorithmic procedure for deciding whether or not a given natural number is in I. Take such a set I and let Γ be the set $\{p_i : i \in I\}$. Then we can't even decide whether $\Gamma \vDash p_n$ for each $n \in \mathbb{N}$ – this is equivalent to deciding whether $n \in I$ – let alone whether $\Gamma \vDash \phi$ for more complicated ϕ.

A rather clever question which one can ask when Γ is infinite and $\Gamma \vDash \phi$ is whether all of the infinite amount of information coded in Γ is needed to logically imply ϕ. Perhaps there is some finite subset Δ of Γ for which $\Delta \vDash \phi$.

There's a bit of a reminder about the set-theoretic background and the theory of infinite sets in Section 6.4 of Chapter 6. The ideas of checking whether a truth assignment satisfies infinitely many formulas in Γ and then whether the possibly infinitely many truth assignments satisfying Γ also satisfy ϕ are meaningful to most modern mathematicians. But the practicalities of doing this checking are another matter!

We shall address this question for infinite Γ using the soundness and completeness theorems in Section 3.3 of Chapter 3.

Exercise 2.82 ───────────────────────────────────

Let Γ and Δ be sets of formulas with $\Delta \subseteq \Gamma$, and let ϕ be a formula.

(a) Show that if $\Delta \vDash \phi$ then $\Gamma \vDash \phi$.

(b) Give a counterexample to show that the converse of (a) is false.

In general the given subset Δ won't be suitable. But this doesn't mean that if $\Gamma \vDash \phi$ there might not be *some* non-trivial subset Δ of Γ for which $\Delta \vDash \phi$.

Within the predicate calculus, the logical consequences ϕ of a set Γ will be of much greater mathematical interest than those we have been looking at in this chapter. For instance, if Γ axiomatizes group theory, its logical consequences will be all statements that must be true for all groups. We shall be able to axiomatize group theory with finitely many axioms Γ. But checking whether $\Gamma \vDash \phi$ for this finite set Γ cannot be done by the same finite process as we gave above for dealing with formulas. For a finite set of propositional formulas Γ and a given ϕ, there are essentially only finitely many different truth assignments needed to check to see whether $\Gamma \vDash \phi$. But for the axioms of group theory, there are infinitely many groups making these axioms true for which we would have to check whether ϕ is also true. This cannot give a finite process for checking whether $\Gamma \vDash \phi$. For predicate calculus we thus have to investigate an alternative way of establishing logical consequence. Our chosen route is to look more closely at how we actually prove results within mathematics and it is both wise and very revealing to look first at how we might formally handle proofs using the relatively simple propositional formulas we have to hand right now. This is what we shall do in the next chapter. Bear in mind that our main aim will be to produce a formal proof system within which derivations correspond to logical consequences. That such proof systems can be found represents a considerable achievement!

Further exercises

Exercise 2.83 ____________________________________

Which of the following sets of formulas are satisfiable? (p, q, r and the p_i, $i \in \mathbb{N}$, are propositional variables.)

(a) $\{(p \rightarrow q), (q \rightarrow r), (r \rightarrow p)\}$

(b) $\{(p \vee (q \leftrightarrow \neg p)), \neg(p \vee q)\}$

(c) $\{(p_i \leftrightarrow \neg p_j) : i < j,\ i, j \in \mathbb{N}\}$

Exercise 2.84 ____________________________________

Three individuals, Green, Rose and Scarlet, are suspected of a crime. They testify under oath as follows.

> GREEN: Rose is guilty and Scarlet is innocent.
> ROSE: If Green is guilty, then so is Scarlet.
> SCARLET: I am innocent, but at least one of the others is guilty.

This is based on an exercise in Kleene [22] attributed to H. Jerome Keisler.

(a) Could all the suspects be telling the truth?

(b) The testimony of one of the suspects follows from that of another. Identify which!

(c) Assuming that all three are innocent of the crime, who has committed perjury?

(d) Assuming that everyone's testimony is true, who is innocent and who is guilty?

(e) Assuming that those who are innocent told the truth and those who are guilty told lies, who is innocent and who is guilty?

Exercise 2.85 ____________________________________

Suppose that L is a propositional language which, besides the usual connectives, also includes constants $\top$ (for true) and $\bot$ (for false). For any formula ϕ of L and any propositional variable p, write for short:

$$(\phi/p) \text{ for } (\phi[\top/p] \vee \phi[\bot/p]),$$

where $\phi[\psi/p]$ is the result of substituting the formula ψ for the variable p throughout ψ. Prove the following.

(a) $\phi \vDash (\phi/p)$.

(b) If $\phi \vDash \theta$ and p does not occur in the formula θ, then $(\phi/p) \vDash \theta$.

(c) If $\phi \vDash \theta$, then there is a formula ψ involving at most the propositional variables in common to ϕ and θ such that $\phi \vDash \psi$ and $\psi \vDash \theta$.

(d) If $\phi \vDash \theta$ and ϕ and θ have no propositional variables in common, then either $\neg\phi$ or θ is a tautology.

The desirable result of part (c), that the logical implication of θ from ϕ should somehow depend only on propositional variables in common to the formulas, is called *Craig's interpolation lemma*, after the American logician Bill Craig. The formula ψ is called the *interpolant* between ϕ and θ.

Exercise 2.86

Consider a propositional language with variables p_i for each $i \in \mathbb{N}$. Let Γ be the set of formulas

$$\{(p_n \to p_m) : n, m \in \mathbb{N}, n < m\}.$$

(a) Which of the following sets of formulas is satisfiable? In each case, justify your answer.

   (i) $\Gamma \cup \{p_0\}$

   (ii) $\Gamma \cup \{\neg p_0, p_1\}$

   (iii) $\Gamma \cup \{\neg p_2, p_1\}$

   (iv) $\Gamma \cup \{\neg p_1, p_2\}$

(b) Describe all the truth assignments which satisfy Γ, explaining why they satisfy Γ and how you know that you have found all possible assignments.

(c) For which pairs (m, n) is the formula $(p_n \to p_m)$ a logical consequence of Γ and for which pairs is this not the case?

(d) Find and describe an algorithmic procedure which, for any formula ϕ, decides whether $\Gamma \vDash \phi$. (For a given ϕ, such a procedure would have to decide within a finite number of steps whether or not $\Gamma \vDash \phi$.)

Exercise 2.87

Suppose that the language has the set $\{p_i : i \in \mathbb{N}\}$ of propositional variables and let Γ be the set $\{q_i : i \in \mathbb{N}\}$, where q_i is either p_i or $\neg p_i$ for each $i \in \mathbb{N}$. Show that for any formula ϕ in this language, exactly one of ϕ and $\neg \phi$ is a logical consequence of Γ.

Exercise 2.88

Suppose that Γ is an infinite set of formulas in the language with the finite set of propositional variables $\{p_1, p_2, \ldots, p_n\}$. Is the following argument correct?

Every formula in Γ is logically equivalent to one of the 2^{2^n} formulas in dnf using these variables, so that there is a finite set Σ of these latter formulas such that, for all formulas ϕ in the language

$$\Gamma \vDash \phi \quad \text{if and only if} \quad \Sigma \vDash \phi.$$

As Σ is finite, there is an algorithmic procedure for deciding whether, for any formula ϕ, $\Sigma \vDash \phi$. Therefore there is an algorithmic procedure which, for any formula ϕ, decides whether $\Gamma \vDash \phi$.

3 FORMAL PROPOSITIONAL CALCULUS

3.1 Introduction

It is time to look at a formal description of at least part of what constitutes a mathematical argument. To some extent we have captured what we expect as the outcome of such an argument by the idea of *logical consequence*. If $\Gamma \vDash \phi$, then the truth of ϕ under a given set of circumstances stems from the truth of the formulas in Γ. But the process of checking whether all truth assignments satisfying Γ also satisfy ϕ does not correspond to the way we produce arguments in mathematics. Typically a mathematical argument, aiming to prove something, will consist of various statements, arranged in some sort of progression and usually with accompanying justification or explanation. The statements could have a variety of statuses. For instance, they might be assumptions (like 'suppose that f is differentiable', 'suppose that $n < 0$'): or they might be axioms, agreed principles, about the subject matter (like 'multiplication of real numbers is commutative'); or, normally the most interesting bits of an argument, they follow on from previous statements, via some sort of justification.

When we get to the more general set-up of the predicate calculus, we can write down, for instance, a set Γ of axioms for group theory. Then asking for this Γ whether $\Gamma \vDash \phi$ amounts to asking whether all groups satisfy ϕ. Plainly investigating the truth of ϕ in all groups, one by one, is a tall order. So we turn to something like a proof from axioms.

The range of acceptable mathematical arguments seems dauntingly large and to make any progress we must initially take a restricted view of which features of everyday argument to formalize. In this chapter we shall work within the very limited context of arguments involving formulas like those we have dealt with so far. Although such formulas are a relatively dull, though necessary, component of interesting statements within mathematics, any attempt to formalize arguments must incorporate them and indeed must be the bedrock for any formalization.

In this chapter, as already noted, we shall only work with statements that are well-formed formulas built up from propositional variables using various connectives. We will forgo the brevity of many elegant mathematical arguments by breaking proofs down into very small steps, with only a limited number of ways of justifying each. We are going to try to capture the idea of a formula ϕ being derived within a formal proof system S from assumptions out of a set Γ, where each step of the derivation, including ϕ itself, arises in one of the following ways:

(i) as one of the assumptions out of the set Γ;

(ii) as an axiom, that is, a formula previously agreed as allowable in *any* proof within the system S (unlike the particular assumptions of this derivation);

(iii) as the consequence of applying a rule of inference of the system S to formulas already derived.

Although formal systems are about to be specified in abstract terms of allowable strings of symbols and manipulations involving them, for the sake of our sanity we describe the system using terminology which suggests their intended meanings. Hence our use of 'assumption' instead of member of Γ, 'rule of inference' instead of construction rule and so on.

Into all this will be built all sorts of finiteness conditions, for instance that formulas are finitely long strings of symbols and that derivations consist of finitely many steps, so that derivations are finitely rather than infinitely long. This may strike you as entirely reasonable from your own experience! But sense can be made of dropping such finiteness conditions. More importantly, one of the main reasons for undertaking this analysis was to skirt round the problems caused by incautious use of infinity within mathematics, by building in finiteness where possible, though without completely ruling out the use of infinity. As a consequence of this, once the axioms and the rules of the system S have been specified, a derivation will turn out to be essentially a mechanical process involving strings of symbols of a given formal language. No regard will be given to any intended meaning of the symbols or components of the system. Only the shape and interrelationships of the symbols, their *syntax*, will matter. The correctness of a derivation would in principle be checkable by a machine armed with the construction rules of the system, and under suitable conditions on the 'size' of the system, a machine might be able to generate all derivations of the system. This mechanical aspect is one reason for describing this mathematical edifice of statements or propositions as a calculus.

What might we hope for from our formal system? It would be desirable that it should match logical consequence, so that whenever $\Gamma \vDash \phi$, there should be some sort of corresponding formal proof of ϕ from the assumptions in Γ. Likewise, when formulas are formally proved, they should in some sense or other be true. We shall make all these requirements more precise later in this chapter.

The shape of this chapter is as follows. In Section 3.2 we define and explore a formal proof system for handling the propositional statements of the previous chapter. One of the points of this formal system is that with it we should be able to derive all tautologies and be unable to derive statements which aren't tautologies; and this is dealt with in Section 3.3. The detailed work in these two sections will have a pay-off later in the book when we discuss the predicate calculus, which is much more important and complex than the propositional calculus, but builds on the work done in these sections. Section 3.4 could almost certainly be omitted if one was trying to get to the predicate calculus as quickly as possible. In this section we look at some interesting and rather tricky issues to do with formal proof systems in general, which are much easier to discuss in the context of propositional calculus than predicate calculus, precisely because the former is less complex.

3.2   A formal system for propositional calculus

We are about to describe a formal proof system and say what is meant by
a formal derivation of a formula. Our aim is that the formal system should
match logical consequence. For a set Γ of formulas and a formula ϕ, we write
$\Gamma \vDash \phi$ to express that ϕ is a logical consequence of Γ. We can regard Γ as a
set of assumptions from which ϕ follows. We shall use the similar notation
$\Gamma \vdash \phi$ to express that there is a formal derivation of ϕ from Γ. As you will
see from the following definition of our main formal proof system S used in
this chapter, a formal derivation exploits only the shape of formulas, not any
consideration of their truth or falsity.

So the symbol $\vdash$ refers to formal derivations of formulas and $\vDash$ refers to their interpretation using truth assignments.

From here on in this chapter, we'll use the letter S for this particular system.

Definitions   The formal system S

The system will manipulate (well-formed) formulas in the language L
consisting of countably many propositional variables $p_0, p_1, p_2, \ldots, p_n, \ldots$
and the connectives $\rightarrow, \neg$. For simplicity we shall use lower case letters
$p, q, r, \ldots$ to stand for these propositional variables.

When it's important to emphasize that the language has countably infinitely many variables, we'll use the p_is.

Let Γ be a set (possibly empty) of formulas and let ϕ be a formula. A
derivation of $\Gamma \vdash \phi$ within S is a finite sequence of formulas

$$\phi_1, \phi_2, \phi_3, \ldots, \phi_n,$$

where the final formula ϕ_n in the sequence is the formula ϕ and the
inclusion of each formula ϕ_i can be explained in one of the following
ways:

(i) $\phi_i \in \Gamma$;

(ii) ϕ_i is a formula of one of the following forms:

   (Ax 1) $(\phi \rightarrow (\psi \rightarrow \phi))$,

   (Ax 2) $((\phi \rightarrow (\psi \rightarrow \theta)) \rightarrow ((\phi \rightarrow \psi) \rightarrow (\phi \rightarrow \theta)))$,

   (Ax 3) $((\neg\phi \rightarrow \neg\psi) \rightarrow (\psi \rightarrow \phi))$,

   where ϕ, ψ, θ are any formulas of L;

(iii) there are two previous formulas in the sequence, ϕ_j and ϕ_k with
   $j, k < i$, where ϕ_k is the formula $(\phi_j \rightarrow \phi_i)$.

When there is such a derivation, we shall say that $\Gamma \vdash \phi$ is a *formal
theorem* of the system. We shall also say that $\Gamma \vdash \phi$ is *derivable* and
that ϕ is *derivable from* Γ. If Γ is empty, we shall also say that ϕ *is
derivable*.

We shall use

$$\Gamma \vdash \phi$$

as a shorthand for 'there is a derivation of $\Gamma \vdash \phi$'.

Although 'formal theorem' is a bit of a mouthful (compared to e.g. 'theorem'), it will occasionally help us distinguish between the results of derivations within the system and theorems *about* derivations, seen from outside the system.

Our first example of a derivation is one of

$$\{p, r, (s \rightarrow q)\} \vdash (q \rightarrow p).$$

To make this and all future derivations easier to follow, we shall write the
sequence of formulas constituting the derivation as a succession of lines down

the page, with the first formula in the sequence (ϕ_1) at the top. The lines are numbered to help keep track of how each line is explained. The explanations of each line use a shorthand we'll explain.

$$
\begin{array}{lll}
(1) & p & \text{Ass} \\
(2) & (p \to (q \to p)) & \text{Ax}\,1 \\
(3) & p & \text{Ass} \\
(4) & (q \to p) & \text{MP}, 1, 2
\end{array}
$$

The formula on both lines 1 and 3 (ϕ_1 and ϕ_3 in the sequence) is p. The explanation for both these lines is that the formula p is an element of the set $\{p, r, (s \to q)\}$ – recall that a derivation of $\Gamma \vdash \phi$ can include a formula which is an element of the set Γ, that's (i) in the definition above. We have used the abbreviation 'Ass' as a shorthand for the formula on that line being in the relevant set Γ, because the intended interpretation of the Γ in $\Gamma \vdash \phi$ is as a set of assumptions from which we can derive ϕ. Hence a use of (i) within a derivation is akin to saying that a formula is an assumption; and (i) is called the *Rule of Assumptions*.

The formula on line 2 (ϕ_2 in the sequence) is an instance of one of the forms Ax 1, Ax 2 and Ax 3 above. Our shorthand explanation of Ax 1 just says which of these has been used. Formulas of these forms which can be slapped down anywhere in a derivation, and so have a privileged status, are called *axioms* of the system.

The formula on line 4 (ϕ_4 in the sequence) is explained by (iii) in the definition above. The relevant previous lines are p (which is ϕ_1 in the sequence) and $(p \to (q \to p))$ (ϕ_2 in the sequence), the latter being of the form $(\phi_1 \to \phi_4)$. We have used the abbreviation 'MP, 1,2' – the 1 and 2 refer to the previous lines being exploited, while MP stands for *Modus Ponens*, the inference of ψ from ϕ and 'if ϕ then ψ', which we met in Exercise 2.67 in the last section. The formal version of Modus Ponens, as described in (iii), is called a *rule of inference* of the system.

We promised to comment on line 3 of the proof, which is just the same as line 1. It really is redundant! There's nothing we can derive from it that we couldn't have derived from line 1. But its inclusion doesn't break any of the rules for a derivation. (In real-life proofs one occasionally repeats lines superfluously too!) Dropping line 3 from this derivation gives an alternative derivation of $\{p, r, (s \to q)\} \vdash (q \to p)$ – a derivation with 3 lines rather than 4. In general a formal theorem has infinitely many derivations (as we ask you to show in Exercise 3.4 below) and in this book we shan't attempt to make judgments whether any one of them is in some sense better than the others!

Another feature of the derivation is that it makes use of the element p of the set $\{p, r, (s \to q)\}$ with the Rule of Assumptions, but makes no use of the elements r and $(s \to q)$. A derivation of $\Gamma \vdash \phi$ isn't obliged to make use of all the elements of Γ or indeed any of them. In this case, as the derivation only uses p, it is also a derivation of

$$\{p\} \vdash (q \to p),$$

and indeed of

$$\Gamma \vdash (q \to p),$$

From now on we shall refer to the *lines* or *steps* of a derivation. The ith line will correspond to the ith formula in the sequence.

If you think that line 3 looks unnecessary, don't worry! You are right and we'll comment on this later. But it's not incorrect.

It is the only rule of inference of this system. Later we shall look at alternative systems with more than one rule of inference.

for any set of formulas Γ with $p \in \Gamma$. This reasoning leads us to a straightforward pair of general results about this formal system, as follows.

Theorem 3.1

(i) Suppose that a derivation of $\Gamma \vdash \phi$ involves uses of the Rule of Assumptions only with the formulas $\theta_1, \theta_2, \ldots, \theta_k$ from Γ. Then there is also a derivation of

$$\{\theta_1, \theta_2, \ldots, \theta_k\} \vdash \phi.$$

(ii) Suppose that there is a derivation of $\Gamma \vdash \phi$ and that Δ is a set of formulas with $\Gamma \subseteq \Delta$. Then there is also a derivation of $\Delta \vdash \phi$.

This is called the *thinning* rule.

Proof

In both cases exactly the same derivation as the original one of $\Gamma \vdash \phi$ provides the required derivation, as any use of the Rule of Assumptions involves a formula out of the new set. ∎

Note that like many general results about derivations which follow, the proof of Theorem 3.1 is *constructive* in the sense that it gives a recipe for a new derivation in terms of given ones. This is a desirable feature which will not, however, be achieved in all such general results.

Theorem 3.1 is a theorem *about* a formal proof system, rather than a formal theorem *of* it. To help distinguish between these two uses of the word 'theorem', a result about a formal system is often called a *metatheorem*.

This is similar to our use of the word metalanguage for the language used to talk about a formal system to distinguish it from the formal language L used within the system.

Exercise 3.1 ─────────────────────────

Suppose that the sequence $\phi_1, \phi_2, \phi_3, \ldots, \phi_n$ is a derivation of $\Gamma \vdash \phi$ (so that ϕ_n is ϕ). Show that for each $i = 1, 2, \ldots, n-1$ there is also a derivation of $\Gamma \vdash \phi_i$.

Exercise 3.2 ─────────────────────────

The following sequences of formulas lack explanations for each line. For which could explanations be added to turn the sequence into a derivation of

$$(\neg\neg q \to \neg p) \vdash (p \to \neg q)?$$

(a) $\quad (\neg\neg q \to \neg p)$

$\quad\quad ((\neg\neg q \to \neg p) \to (p \to \neg q))$

$\quad\quad (p \to \neg q)$

(b) $\quad ((p \to \neg q) \to ((\neg\neg q \to \neg p) \to (p \to \neg q)))$

$\quad\quad ((\neg\neg q \to \neg p) \to (p \to \neg q))$

$\quad\quad (p \to \neg q)$

(c) $\quad ((\neg\neg q \to \neg p) \to (p \to \neg q))$

$\quad\quad (\neg p \to (\neg\neg\neg q \to \neg p))$

$\quad\quad (\neg\neg q \to \neg p)$

$\quad\quad (p \to \neg q)$

Exercise 3.3 ____

Say something interesting about derivations that make no use of the rule of Modus Ponens! Do likewise for derivations that make no use of either the Rule of Assumptions or instances of any of the axioms.

Exercise 3.4 ____

Suppose that $\Gamma \vdash \phi$ (i.e. that this is a formal theorem). Show that it has infinitely many derivations.

So far in this section, we have presented this formal system as being a particular way of shuffling symbols around, following various rules – this is an important part of the exercise. Derivations can be produced and checked in a mechanical way, without any regard to whether the symbols or process as a whole mean anything in the real world. But of course we did really want to formalize mathematical proof and we have dropped strong hints about the ways in which features of the system have intended meanings. We want this formal proof system to be capable of deriving $\Gamma \vdash \phi$ exactly when $\Gamma \vDash \phi$. Recall that $\Gamma \vDash \phi$ says that every truth assignment that makes all the formulas in Γ true also makes ϕ true. The formal language uses the symbols $\rightarrow, \neg$ and with their usual interpretations: these connectives form an adequate set, so that the expressive power of the language is as large as we need for this aim. It's thus no surprise that we have chosen the axioms Ax 1, Ax 2 and Ax 3 of the formal system to be tautologies, and that the rule of Modus Ponens matches a strong form of the logical consequence $\{\theta, (\theta \rightarrow \psi)\} \vDash \psi$ (as you can check in Exercise 3.5 below). We shall show later (very straightforwardly) that if $\Gamma \vdash \phi$, then $\Gamma \vDash \phi$. But the reverse also holds, namely that the formal system can derive any logical consequence $\Gamma \vDash \phi$ (expressed in terms of the connectives $\rightarrow$ and $\neg$). We think that this is pretty surprising – the system has so few axioms and rules! Likewise we shall prove this result later, though the proof is not at all straightforward!

In particular, taking Γ as the empty set, the system can derive $\vdash \phi$ exactly when ϕ is a tautology.

A psychological attraction of the adequate set $\{\rightarrow, \neg\}$, as opposed to say $\{\vee, \neg\}$ is that 'implication' seems central to drawing conclusions in an argument.

The first of these results for the proof system is called the *soundness* or *correctness* theorem and the second is called the *completeness* or *adequacy* theorem.

Exercise 3.5 ____

(a) Show that all instances of Ax 1, Ax 2 and Ax 3 are tautologies.

(b) Show that if $\Gamma \vDash \theta$ and $\Gamma \vDash (\theta \rightarrow \psi)$, then $\Gamma \vDash \psi$.

One way of describing the result of part (b) is by saying that Modus Ponens is a *valid rule of inference*, in some sense preserving truth when we look at the interpretation of formulas. When we look later at further rules of inference in other formal systems, we shall usually want them to be valid in the same sort of way as this.

Exercise 3.6

Show that the thinning rule (Theorem 3.1(ii)) is valid; that is, show that for sets of formulas Γ, Δ with $\Gamma \subseteq \Delta$, if $\Gamma \vDash \phi$ then $\Delta \vDash \phi$.

It's probably good for your soul and appreciation of later results about the system to have a go now at some derivations, using bare hands. Later results will provide some merciful shortcuts about derivations and you might even think that the completeness theorem spares you from ever doing any formal derivations, as you could instead check for logical consequences – however, this would be a misconception of the benefits and drawbacks of two very different concepts.

Notation

As with logical consequence, $\Gamma \vDash \phi$, we shall sometimes cheat on set notation for the Γ in $\Gamma \vdash \phi$ by dropping some of the set brackets { }, writing e.g.

$$\theta, q \vdash \phi \text{ instead of } \{\theta, q\} \vdash \phi,$$
$$\Gamma, \theta \vdash \phi \text{ instead of } \Gamma \cup \{\theta\} \vdash \phi$$
$$\text{and } \Gamma, \Delta \vdash \phi \text{ instead of } \Gamma \cup \Delta \vdash \phi.$$

Also recall that the set Γ in $\Gamma \vdash \phi$ could be empty, in which case we'd write $\vdash \phi$.

We hope that the context will make it clear what is meant.

Exercise 3.7

Give derivations of each of the following. [Some might be hard!]

(a) $\phi, (\phi \to \psi), (\psi \to \theta) \vdash \theta$

(b) $p, (p \to q), (p \to (q \to r)) \vdash r$

(c) $\vdash (p \to p)$

(d) $(\psi \to \theta), (\phi \to \psi) \vdash (\phi \to \theta)$

(e) $(\neg\neg\psi \to \neg\neg\phi) \vdash (\psi \to \phi)$

(f) $\neg\psi \vdash (\psi \to \phi)$

(g) $\vdash (\psi \to (\neg\psi \to \phi))$

Solution

(a) We have lots of assumptions to play with and it turns out that we can avoid using any instances of axioms.

$$
\begin{array}{lll}
(1) & \phi & \text{Ass} \\
(2) & (\phi \to \psi) & \text{Ass} \\
(3) & \psi & \text{MP}, 1, 2 \\
(4) & (\psi \to \theta) & \text{Ass} \\
(5) & \theta & \text{MP}, 3, 4
\end{array}
$$

If we seriously had to live and work with this proof system, then of course we would develop lots of proof strategies to cope with derivations such as these. For instance, we would systematically record all our derivations, perhaps giving some of them memorable names, so that we could exploit them in future. We don't think that it is worthwhile for the purposes of studying this book. There is, however, personal satisfaction from creating successful derivations and understanding better the formal system.

(b) Not given. Our private solution uses no axioms.

(c) We'll give this one as it's a key lemma in a later result. In as much as there are strategies for derivations, the chances are that one is looking for an instance of one of the axioms that ends in '$\dots \to (p \to p)$' for which the bits before the $\to$ also look like instances of axioms, which get eliminated by use of Modus Ponens – the allowed set of assumptions is empty, so there's not much to play with! As no $\neg$s are involved in what we are trying to derive, the chances are that we won't need to use an instance of Ax 3.

$$
\begin{array}{llll}
(1) & (p \to ((p \to p) \to p)) & & \text{Ax 1} \\
(2) & ((p \to ((p \to p) \to p)) \to ((p \to (p \to p)) \to (p \to p))) & & \text{Ax 2} \\
(3) & ((p \to (p \to p)) \to (p \to p)) & & \text{MP}, 1, 2 \\
(4) & (p \to (p \to p)) & & \text{Ax 1} \\
(5) & (p \to p) & & \text{MP}, 3, 4
\end{array}
$$

(d) Not given. Our private solution includes instances of Ax 1 and Ax 2.

(e) Not given. Our private solution includes instances of Ax 3.

(f) Not given. Our private solution includes instances of Ax 1 and Ax 3.

(g) Not given. This one might be pretty hard. We've not even tried ....! But we shall show it is derivable later.

So far we've either given you or, more commonly, asked you to give various derivations. Derivations can get very complicated and impenetrable, and could distract us from what is usually much more interesting, namely whether a derivation exists. In the next exercise we start our move towards a variety of results of the latter sort, not necessarily requiring us to give full derivations, but telling us that derivations exist.

Exercise 3.8 ──────────────

In this exercise we ask you to show how to exploit given derivations to create another derivation.

(a) Show that if $\Gamma \vdash (\phi \to \psi)$, then $\Gamma, \phi \vdash \psi$.

(b) Show that if $\Gamma \vdash \phi$ and $\Delta, \phi \vdash \psi$, then $\Gamma, \Delta \vdash \psi$.

(c) Show that if, for some ψ, $\Gamma \vdash \psi$ and $\Gamma \vdash \neg\psi$, then $\Gamma \vdash \phi$, for any formula ϕ. [*Hint*: Use a result from Exercise 3.7.]

Solution

We shall give a solution to part (a) and leave (b) and (c) to you.

Suppose that $\Gamma \vdash (\phi \to \psi)$ and that we have a derivation of it using k lines (so that the formula on the kth line is $(\phi \to \psi)$). This can then be turned into a derivation of $\Gamma, \phi \vdash \psi$ by adding lines as follows.

$$
\begin{array}{lll}
\vdots \quad \vdots & \quad \vdots & \\
(k) \quad (\phi \to \psi) & \cdots & \\
(k+1) \quad \phi & \text{Ass} & \\
(k+2) \quad \psi & \text{MP}, k, k+1 &
\end{array}
$$

Our experience is that students almost always produce cleverer and more elegant solutions than ours!

Somewhat curiously, there are formulas built up using only $\to$s and no $\neg$s for which Ax 3 is needed in any derivation. An example is the derivation of $\vdash (((p \to q) \to p) \to p)$ (Peirce's law). We shall investigate this subtlety in Section 3.4.

Recall that we are using the notation $\Gamma \vdash \phi$ as a shorthand for saying that a derivation of $\Gamma \vdash \phi$ exists.

If we can derive both $\Gamma \vdash \psi$ and $\Gamma \vdash \neg\psi$ for some ψ, then the set Γ is said to be *inconsistent*. This exercise shows that if Γ is inconsistent, then any formula can be derived from Γ.

Note that the solution we gave for part (a) above is constructive: it actually gives a recipe for creating an actual derivation from earlier ones, rather than merely showing that a derivation exists. We hope that your solutions to parts (b) and (c) have the same constructive character, which is highly desirable from all sorts of perspectives which we won't elaborate on here. Later in the book we shall encounter at least one important metatheorem about derivations for which there is no constructive proof.

The relevant result is the completeness theorem for predicate calculus.

How might we make it easier to find derivations? One way is as follows. In real maths, if you wish to prove a formula of the form 'if ϕ then ψ' (which is the intended meaning of $(\phi \to \psi)$), one assumes ϕ and attempts to prove ψ. If this is successful, we conclude that 'if ϕ then ψ'. The analogue of this for the formal system is that

For instance, to prove 'If the function f is differentiable, then f is continuous', one would normally start the proof with 'Suppose that f is differentiable' and then show that f is continuous.

$$\text{if } \Gamma, \phi \vdash \psi, \text{ then } \Gamma \vdash (\phi \to \psi).$$

This result does hold for the system and is usually called the *deduction theorem* for the system. (It is the converse of the result of Exercise 3.8(a) above.) To be honest, the statement of the deduction theorem doesn't give a derivation of $\Gamma \vdash (\phi \to \psi)$. It just says that one exists. Nevertheless it's no bad thing to be able to show that something is derivable. Take for instance the task of deriving $\vdash (\psi \to (\neg\psi \to \phi))$ (which is Exercise 3.7(g)). Observe that the principal connective of the formula $(\psi \to (\neg\psi \to \phi))$ is an $\to$, so that the deduction theorem could be of help: if one could derive

However, the proof of the deduction theorem, Theorem 3.3 below, will show how to turn a derivation of $\Gamma, \phi \vdash \psi$ into a derivation of $\Gamma \vdash (\phi \to \psi)$.

$$\psi \vdash (\neg\psi \to \phi),$$

then the deduction theorem says that one can derive

Bear in mind that a derivation with some assumptions available for use might be easier to spot than a derivation with none!

$$\vdash (\psi \to (\neg\psi \to \phi)).$$

But why stop there? The principal connective of $(\neg\psi \to \phi)$ is also a $\to$, so it is tempting to try to derive

$$\psi, \neg\psi \vdash \phi,$$

and, if successful, use the deduction theorem to conclude that one can derive

$$\psi \vdash (\neg\psi \to \phi),$$

leading to the desired result. Here is a derivation of $\psi, \neg\psi \vdash \phi$.

(1)	$\neg\psi$	Ass
(2)	$(\neg\psi \to (\neg\phi \to \neg\psi))$	Ax 1
(3)	$(\neg\phi \to \neg\psi)$	MP, 1, 2
(4)	$((\neg\phi \to \neg\psi) \to (\psi \to \phi))$	Ax 3
(5)	$(\psi \to \phi)$	MP, 3, 4
(6)	ψ	Ass
(7)	ϕ	MP, 5, 6

This derives $\psi, \neg\psi \vdash \phi$ so that two applications of the deduction theorem gives that $\vdash (\psi \to (\neg\psi \to \phi))$ is derivable.

We shall make use later of the result that $\psi, \neg\psi \vdash \phi$ is derivable. Essentially it says that every formula can be derived from a contradiction.

Before we prove the deduction theorem, have a go at using it for yourself.

Exercise 3.9

Using the deduction theorem if you like, show that the following are derivable.

(a) $\vdash ((\phi \to (\psi \to \theta)) \to (\psi \to (\phi \to \theta)))$

(b) $\vdash (\neg\neg p \to p)$

(c) $\vdash (p \to \neg\neg p)$

Solution

(a) Not given. Our private solution makes three uses of the deduction theorem.

(b) Not given. Our private solution includes use of Ax 1, Ax 3 and the deduction theorem.

(c) As we can derive $\vdash (\neg\neg p \to p)$, replacing both the occurrences of p by $\neg p$ means we can derive $\vdash (\neg\neg\neg p \to \neg p)$. All we now need do is add a couple of lines to this derivation, so assuming that we've obtained this last result after k lines, the rest of the derivation looks like:

$$
\begin{array}{lll}
\vdots \ \ \vdots & & \vdots \\
(k) & (\neg\neg\neg p \to \neg p) & \cdots \\
(k+1) & ((\neg\neg\neg p \to \neg p) \to (p \to \neg\neg p)) & \text{Ax 3} \\
(k+2) & (p \to \neg\neg p) & \text{MP}, k, k+1
\end{array}
$$

> We hope that this is an obvious result about the system: if we can derive $\Gamma \vdash \phi$ and we replace all occurrences of the variable p in ϕ and the formulas in the set Γ by a formula θ, turning ϕ into the formula ϕ' and the set Γ into the set Γ', then we can derive $\Gamma' \vdash \phi'$. We'll come back to this in Exercise 3.15.

We shall use the deduction theorem to obtain a formal analogue of proof by contradiction, mentioned in Section 2.6 of Chapter 2. To give you a break from the unremitting horrors of providing your own derivations in this marvellously compact, but inherently obscure, system, we shall actually give a proof!

Theorem 3.2 Proof by contradiction

If $\Gamma, \neg\phi \vdash \psi$ and $\Gamma, \neg\phi \vdash \neg\psi$, then $\Gamma \vdash \phi$.

Proof

We need the following lemma.

> *Lemma:* $(\neg\phi \to \phi) \vdash \phi$.

> *Proof of lemma:* Earlier we derived $\psi, \neg\psi \vdash \phi$, for any formulas ϕ, ψ. Taking the ϕ to be $\neg(\neg\phi \to \phi)$ and the ψ to be the formula ϕ, this means that we can derive

$$\phi, \neg\phi \vdash \neg(\neg\phi \to \phi).$$

One use of the deduction theorem gives

$$\neg\phi \vdash (\phi \to \neg(\neg\phi \to \phi)),$$

and a further use of the deduction theorem gives

$$\vdash (\neg\phi \to (\phi \to \neg(\neg\phi \to \phi))).$$

We shall turn this derivation into one of $(\neg\phi \to \phi) \vdash \phi$ by adding the following lines. (We'll simplify our line numbering by describing the line on which $\vdash (\neg\phi \to (\phi \to \neg(\neg\phi \to \phi)))$ is derived as line 0.)

$$
\begin{array}{lll}
(0) & (\neg\phi \to (\phi \to \neg(\neg\phi \to \phi))) & \cdots \\
(1) & ((\neg\phi \to (\phi \to \neg(\neg\phi \to \phi))) & \\
    & \quad \to ((\neg\phi \to \phi) \to (\neg\phi \to \neg(\neg\phi \to \phi)))) & \mathrm{Ax\,2} \\
(2) & ((\neg\phi \to \phi) \to (\neg\phi \to \neg(\neg\phi \to \phi))) & \mathrm{MP}, 0, 1 \\
(3) & (\neg\phi \to \phi) & \mathrm{Ass} \\
(4) & (\neg\phi \to \neg(\neg\phi \to \phi)) & \mathrm{MP}, 2, 3 \\
(5) & ((\neg\phi \to \neg(\neg\phi \to \phi)) \to ((\neg\phi \to \phi) \to \phi)) & \mathrm{Ax\,3} \\
(6) & ((\neg\phi \to \phi) \to \phi) & \mathrm{MP}, 4, 5 \\
(7) & \phi & \mathrm{MP}, 3, 6
\end{array}
$$

$\square$

Let us now prove the required theorem. We suppose that we have derivations of both $\Gamma, \neg\phi \vdash \psi$ and $\Gamma, \neg\phi \vdash \neg\psi$. Applying the deduction theorem to the second of these, we can derive

$$\Gamma \vdash (\neg\phi \to \neg\psi).$$

By adding the following lines to this derivation (as before labelling the final line of this derivation as line 0 for readability here),

$$
\begin{array}{lll}
(0) & (\neg\phi \to \neg\psi) & \cdots \\
(1) & ((\neg\phi \to \neg\psi) \to (\psi \to \phi)) & \mathrm{Ax\,3} \\
(2) & (\psi \to \phi) & \mathrm{MP}, 0, 1
\end{array}
$$

we turn the derivation into one of

$$\Gamma \vdash (\psi \to \phi).$$

By the result of Exercise 3.8(a), we can derive

$$\Gamma, \psi \vdash \phi.$$

As (by our supposition) we can also derive $\Gamma, \neg\phi \vdash \psi$ the result of Exercise 3.8(b) (with a suitable reinterpretation of the symbols used) tells us that we can derive

$$\Gamma, \neg\phi \vdash \phi,$$

so that by the deduction theorem we can derive

$$\Gamma \vdash (\neg\phi \to \phi).$$

Combining the derivation of this with that of $\vdash ((\neg\phi \to \phi) \to \phi)$ in our lemma above, one use of Modus Ponens gives us a derivation of

$$\Gamma \vdash \phi,$$

as required. $\blacksquare$

We can use this and other earlier results to obtain another metatheorem with claims to the title of 'proof by contradiction':

If $\Gamma, \phi \vdash \psi$ and $\Gamma, \phi \vdash \neg\psi$, then $\Gamma \vdash \neg\phi$.

We can justify this as follows. Suppose that we can derive both $\Gamma, \phi \vdash \psi$ and $\Gamma, \phi \vdash \neg\psi$. As we can derive $\vdash (\neg\neg\phi \to \phi)$ (using the result of Exercise 3.9(b), replacing p by ϕ), we can also derive $\neg\neg\phi \vdash \phi$ (using the result of Exercise 3.8(a)). Using the result of Exercise 3.8(b), we can thus derive both

$$\Gamma, \neg\neg\phi \vdash \psi$$

and

$$\Gamma, \neg\neg\phi \vdash \neg\psi.$$

By Theorem 3.2 we can then derive

$$\Gamma \vdash \neg\phi.$$

The use of proof by contradiction can be so subtle that it is worth showing you one worked example before letting you loose on your own.

Example 3.1

We shall show that $\neg(\theta \to \psi) \vdash \neg\psi$. Our strategy will be to take ψ as an extra assumption alongside $\neg(\theta \to \psi)$, derive both χ and $\neg\chi$ for some formula χ, and use the second version of proof by contradiction above to conclude that the negation of ψ is derivable from the single assumption $\neg(\theta \to \psi)$. Consider the following derivation.

$$
\begin{array}{lll}
(1) & \neg(\theta \to \psi) & \text{Ass} \\
(2) & \psi & \text{Ass} \\
(3) & (\psi \to (\theta \to \psi)) & \text{Ax}\,1 \\
(4) & (\theta \to \psi) & \text{MP}, 2, 3
\end{array}
$$

This derivation shows both that $\neg(\theta \to \psi), \psi \vdash \neg(\theta \to \psi)$ (using just the first line of the derivation) and that $\neg(\theta \to \psi), \psi \vdash (\theta \to \psi)$. So by the second form of proof by contradiction above, $\neg(\theta \to \psi) \vdash \neg\psi$.

Notice the cunning way we have exploited the assumption $\neg(\theta \to \psi)$, which begins with a negation symbol. The axioms and tules of the formal system don't immediately give us a way of exploiting this formula. However, by adding an extra assumption and using it to derive $(\theta \to \psi)$, the unnegated form of $\neg(\theta \to \psi)$, we obtain a contradiction, from which we can conclude something useful! $\qquad\qquad\blacklozenge$

Now it's your turn to prove some further theorems and metatheorems!

Exercise 3.10 ______________________________

Show each of the following, using the deduction theorem if you like, along with any other previous results in exercises and theorems.

(a) $\neg(\theta \to \psi) \vdash \theta$ [*Hints*: By Exercise 3.7(f), $\neg\theta \vdash (\theta \to \psi)$; and use the first form of proof by contradiction, in Theorem 3.2.]

(b) $\theta, \neg\psi \vdash \neg(\theta \to \psi)$

(c) If $\Gamma \vdash (\phi \to \psi)$ and $\Gamma \vdash (\psi \to \theta)$, then $\Gamma \vdash (\phi \to \theta)$.

(d) If $\Gamma, \neg\phi \vdash \neg\psi$, then $\Gamma, \psi \vdash \phi$.

(e) If $\Gamma, \phi \vdash \neg\psi$, then $\Gamma, \psi \vdash \neg\phi$.

(f) If $\Gamma, \phi \vdash \theta$ and $\Gamma, \neg\phi \vdash \theta$, then $\Gamma \vdash \theta$.

Our private solutions to most parts of this exercise exploit proof by contradiction. But not all do so – a common mistake of most mathematicians on meeting proof by contradiction for the first time is to get over-excited by the method and try to use it where it doesn't help!

Let us now prove the deduction theorem.

Theorem 3.3 Deduction theorem for this system

If $\Gamma, \phi \vdash \psi$, then $\Gamma \vdash (\phi \rightarrow \psi)$.

Proof

The method of proof is in many ways more significant than the detailed steps. It won't be sufficient simply to add steps to a derivation of $\Gamma, \phi \vdash \psi$ to turn it into one of $\Gamma \vdash (\phi \rightarrow \psi)$. Instead we look line by line at a derivation of $\Gamma, \phi \vdash \psi$ and show how to generate corresponding fragments of a derivation of $\Gamma \vdash (\phi \rightarrow \psi)$. A derivation of $\Gamma, \phi \vdash \psi$ is a sequence

$$\psi_1, \psi_2, \ldots, \psi_n = \psi,$$

where any use of the Rule of Assumptions involves a formula in the set $\Gamma \cup \{\phi\}$. We shall show by mathematical induction on $i = 1, 2, \ldots, n$ (where it helps to see i as coding the ith line of this derivation) that there is a derivation of

$$\Gamma \vdash (\phi \rightarrow \psi_i),$$

and will indeed give a recipe for constructing this derivation. We could describe the method of proof as *mathematical induction on the length of the derivation* of $\Gamma, \phi \vdash \psi$.

   basis of induction: show how to derive $\Gamma \vdash (\phi \rightarrow \psi_1)$;

      inductive step: assuming that, for $k > 1$, we have derived $\Gamma \vdash (\phi \rightarrow \psi_i)$

         for all $i < k$, show how to derive $\Gamma \vdash (\phi \rightarrow \psi_k)$.

Each step of this process usually replaces a single step in the derivation of $\Gamma, \phi \vdash \psi$ by several steps in the derivation, as you'll see.

Basis of induction

The first line (ψ_1) of the derivation of $\Gamma, \phi \vdash \psi$ could arise in two different ways: by a use of the Rule of Assumptions; and as an axiom, i.e. an instance of one of Ax 1, Ax 2 and Ax 3. We need to deal with each of these possibilities.

Suppose that ψ_1 is an assumption, so that it's in the set $\Gamma \cup \{\phi\}$. If it is in Γ then we can derive $\Gamma \vdash (\phi \rightarrow \psi_1)$ by

    (1) ψ_1 Ass
    (2) $(\psi_1 \rightarrow (\phi \rightarrow \psi_1))$ Ax 1
    (3) $(\phi \rightarrow \psi_1)$ MP, 1, 2.

Unfortunately this derivation won't work if ψ_1 isn't in Γ but is the assumption ϕ in $\Gamma \cup \{\phi\}$ – well, it will be a correct derivation, but not one of $\Gamma \vdash (\phi \rightarrow \psi_1)$ as it uses an assumption not in Γ. In the case that ψ_1 is ϕ itself, we can exploit the result of Exercise 3.7(c) using ϕ replacing p to get a derivation of $\Gamma \vdash (\phi \rightarrow \phi)$.

The other possibility is that ψ_1 is an axiom. We leave it for you to show that $\Gamma \vdash (\phi \rightarrow \psi_1)$ as an exercise.

We've written this derivation as a sequence rather than in lines just to save space!

Inductive step

We suppose that, for $k > 1$, we have derived $\Gamma \vdash (\phi \rightarrow \psi_i)$ for each i with $1 \leq i < k$. (You might like to imagine this as a derivation of $\Gamma \vdash (\phi \rightarrow \psi_1)$ followed by extra steps to turn it into one of $\Gamma \vdash (\phi \rightarrow \psi_2)$, and so on.) We now need to explain what to do with the kth step of the original derivation (deriving ψ_k), which of course will depend on how this arose. We can deal with use of the Rule of Assumptions or an instance of an axiom in the same way as we did for their use on the first line of a derivation. There is one extra possibility with which we now have to deal, namely that Modus Ponens was used to get this kth line. This means that the kth line is ψ_k where there are earlier lines ψ_i, ψ_j in the original derivation of $\Gamma, \phi \vdash \psi$ where ψ_j is the formula $\psi_i \rightarrow \psi_k$. By the induction hypothesis there are derivations of

$$\Gamma \vdash (\phi \rightarrow \psi_i)$$

and of

$$\Gamma \vdash (\phi \rightarrow (\psi_i \rightarrow \psi_k)).$$

All we need do to derive $\Gamma \vdash (\phi \rightarrow \psi_k)$ is to add extra lines to the new derivation so far as follows.

$$
\begin{array}{ll}
(n_i) & (\phi \rightarrow \psi_i) \\
\\
(n_j) & (\phi \rightarrow (\psi_i \rightarrow \psi_k)) \\
\\
(n_{k-1}) & (\phi \rightarrow \psi_{k-1}) \\
(n_{k-1} + 1) & ((\phi \rightarrow (\psi_i \rightarrow \psi_k)) \rightarrow ((\phi \rightarrow \psi_i) \rightarrow (\phi \rightarrow \psi_k))) \quad \text{Ax 2} \\
(n_{k-1} + 2) & ((\phi \rightarrow \psi_i) \rightarrow (\phi \rightarrow \psi_k)) \quad \text{MP}, n_j, n_{k-1} + 1 \\
(n_{k-1} + 3) & (\phi \rightarrow \psi_k) \quad \text{MP}, n_i, n_{k-1} + 2
\end{array}
$$

We've used n_i to denote the line of the new derivation on which the method has guaranteed to derive $(\phi \rightarrow \psi_i)$ and similarly with n_j and n_{k-1} – of course, one of i, j might actually be $k - 1$, in which case the layout of the derivation is a bit simpler.

This deals with the inductive step, so that the result follows by mathematical induction on $i = 1, 2, \ldots, n$. ∎

Exercise 3.11

Complete the proof of the deduction theorem by deriving $\Gamma \vdash (\phi \rightarrow \psi_k)$ in the case when ψ_k is an axiom.

Perhaps it's interesting that we needn't really worry here about which of Ax 1, Ax 2 and Ax 3 was used.

Solution

One straightforward derivation is as follows.

$$
\begin{array}{lll}
(1) & \psi_k & \text{Ax} \ldots \\
(2) & (\psi_k \rightarrow (\phi \rightarrow \psi_k)) & \text{Ax 1} \\
(3) & (\phi \rightarrow \psi_k) & \text{MP}, 1, 2
\end{array}
$$

Exercise 3.12

(a) Suppose that the recipe of our proof of the deduction theorem is slavishly
followed to create a derivation of $\Gamma \vdash (\phi \to \psi)$ from a given derivation of
$\Gamma, \phi \vdash \psi$. (This recipe includes our solutions to Exercises 3.7(c) and 3.11.)
Suppose that the latter derivation includes n_1 uses of the Rule of Assumptions involving ϕ, n_2 uses of the Rule of Assumptions involving formulas
in Γ (where we'll suppose that $\phi \notin \Gamma$), n_3 uses of axioms and n_4 uses of
Modus Ponens. Give a formula for the length of the resulting derivation
of $\Gamma \vdash (\phi \to \psi)$ in terms of n_1, n_2, n_3, n_4.

In any elegant derivation of
$\Gamma, \phi \vdash \psi$ you might expect n_1 to
equal 0 or 1. But don't suppose
that you are given such an elegant
derivation here!

(b) See if you can find an (easy!) example of a formal theorem for which you
can find a shorter derivation of $\Gamma \vdash (\phi \to \psi)$ than the one generated by
the recipe in our proof of the deduction theorem.

In this section we have looked at some of what can be derived within our particular formal proof system. The rules and axioms seem very restrictive and
we have begun to compensate for these restrictions by establishing metatheorems, like proof by contradiction and the deduction theorem, which make it
easier for us to establish what is derivable. A key feature of these metatheorems is that, while they give us a shortcut to showing something is derivable,
they all tell us how we could construct a full derivation within the original
restricted theorem, if we really wanted one. One point of Exercise 3.12 was
to emphasize this feature of the deduction theorem. Thus the metatheorems
about the formal system so far have the character of extending our knowledge
of when we can derive $\Gamma \vdash \phi$ by increasing our catalogue of actual derivations. But there is a completely different way of establishing whether $\Gamma \vdash \phi$,
involving the intended interpretation of the symbols using truth assignments
and logical consequence $\Gamma \vDash \phi$. This is what we shall investigate in the next
section.

Another point of Exercise 3.12 was
to try to improve your
understanding of the detail of the
proof of the deduction theorem!

At a more technical level, we shall use the method, seen in the proof of the
deduction theorem, of mathematical induction on the length of a derivation
again in later sections. For instance, we shall use it to prove metatheorems
about alternative formal systems to the one we have been looking at here.
We shall use this in the next section for one of the metatheorems about our
current system, relating the syntactic notion of derivability to the intended
interpretation in terms of validity.

Further exercises

Exercise 3.13

Suppose that Γ is a set of instances of axioms Ax 1, Ax 2 and Ax 3. Show that
for any formula ϕ, if $\Gamma \vdash \phi$ then $\vdash \phi$.

Exercise 3.14

Show each of the following. You may use any of the results in this section.

(a) $(\phi \to \psi), (\phi \to \neg\psi) \vdash \neg\phi$

(b) $\vdash (\phi \to (\psi \to (\theta \to \phi)))$

(c) $((p \to \neg p) \to p)$

(d) $\{(p_i \to p_{i+1}) : i \in \mathbb{N}\} \cup \{\neg p_5\} \vdash \neg p_2$

Exercise 3.15 ____

We mentioned earlier that we hope it is obvious that if we can derive $\Gamma \vdash \phi$ and we replace all occurrences of the variable p in ϕ and the formulas in the set Γ by a formula θ, turning ϕ into the formula ϕ' and the set Γ into the set Γ', then we can derive $\Gamma' \vdash \phi'$. Use the method of mathematical induction on the length of a derivation to prove this result.

3.3   Soundness and completeness

In this section we shall derive two very important metatheorems about the formal proof system S in the last section. These results show that the formal theorems of the system exactly match their intended interpretation as logical consequences, in the world of truth assignments. They thus connect two very different ideas of when one statement follows from others. They foreshadow similar results for predicate calculus later in the book, which are of much weightier significance than within the context of propositional calculus. The methods we shall use in this section to prove the results for propositional calculus will be very valuable to us when we get to the predicate calculus.

The first of these metatheorems essentially says that any formal theorem of the system S represents a logical consequence.

Theorem 3.4 Soundness theorem for S

Let ϕ be any formula and Γ a set of formulas. If $\Gamma \vdash \phi$, then $\Gamma \vDash \phi$. (In the case that the set Γ is empty, the result is to be read as saying that if $\vdash \phi$, then $\vDash \phi$, i.e. ϕ is a tautology.)

This is sometimes called the *correctness theorem.*

Proof

We shall suppose that $\Gamma \vdash \phi$ and show that for any truth assignment v, if v satisfies Γ (i.e. $v(\gamma) = T$ for all $\gamma \in \Gamma$), then $v(\phi) = T$.

Let

$$\phi_1, \phi_2, \ldots, \phi_n = \phi$$

be a derivation of $\Gamma \vdash \phi$ and v be a truth assignment satisfying Γ. We shall use induction on the length of the derivation to show that $v(\phi_i) = T$.

If Γ is empty, then our argument will show that $v(\phi) = T$ for all v. (We can regard any v as satisfying the empty set of formulas.)

Basis of induction

The first line (ϕ_1) of the derivation of $\Gamma \vdash \phi$ could arise in two different ways: by a use of the Rule of Assumptions; and as an axiom, i.e. an instance of one of Ax 1, Ax 2 and Ax 3. We need to deal with each of these possibilities.

Suppose that ϕ_1 is an assumption, so that it's in the set Γ. As v satisfies Γ, then in particular $v(\phi_1) = T$.

The other possibility is that ϕ_1 is an axiom. All instances of Ax 1, Ax 2 and Ax 3 are tautologies, as we asked you to show in Exercise 3.5(a) in Section 3.2. Thus $v(\phi_1) = T$.

Inductive step

We suppose that, for $k > 1$, we have shown $v(\phi_i) = T$ for each i with $1 \leq i < k$. We shall show that $v(\phi_k) = T$. The kth line of the derivation could have arisen from a use of the Rule of Assumptions or as an instance of an axiom. In these cases the same argument as used for the first line shows that $v(\phi_k) = T$. The one further possibility is that Modus Ponens was used to get the kth line. That means that the kth line is ϕ_k where there are earlier lines ϕ_i, ϕ_j in the original derivation of $\Gamma \vdash \phi$ and ϕ_j is the formula $(\phi_i \to \phi_k)$. By the induction hypothesis

$$v(\phi_i) = T \quad \text{and} \quad v((\phi_i \to \phi_k)) = T,$$

so that from the truth table of $\to$ we have $v(\phi_k) = T$.

This is essentially what you were asked to show in Exercise 3.5(b) in Section 3.2.

This deals with the inductive step, so that the result follows by induction. ∎

In theory one might derive $\Gamma \vdash \phi$ and apply the soundness theorem as a means of showing that $\Gamma \vDash \phi$. But for propositional formulas it's usually somewhat easier to test whether $\Gamma \vDash \phi$ directly using truth assignments than to find derivations in S. A more useful application of the soundness theorem is to show when something is not derivable. For instance, let's show that

By way of contrast, it might well *not* be easier to test whether $\Gamma \vDash \phi$ for formulas involving predicates, which are what we are aiming at, using our study of propositional formulas to help develop our ideas.

$$(p \to q) \vdash ((q \to p) \to q)$$

is not derivable. Take the truth assignment v such that $v(p) = v(q) = F$. We have

$$v((p \to q)) = T \quad \text{and} \quad v\big(((q \to p) \to q)\big) = F,$$

so that

$$(p \to q) \nvDash ((q \to p) \to q).$$

By the soundness theorem, $(p \to q) \vdash ((q \to p) \to q)$ cannot be derivable. It's handy to introduce the notation $\Gamma \nvdash \phi$ for '$\Gamma \vdash \phi$ is not derivable', so that a way of rephrasing the soundness theorem for this sort of application is as

This is similar to writing $\Gamma \nvDash \phi$ for ϕ not being a logical consequence of Γ.

$$\text{if } \Gamma \nvDash \phi \text{ then } \Gamma \nvdash \phi.$$

Exercise 3.16

Decide which of the following are derivable in S. Explain each of your answers in a way permitted by results established in the book so far. In particular, any explanation why one of these is derivable shouldn't exploit the completeness theorem, which hasn't yet been established – you should instead give a derivation.

(a) $\neg(q \to \neg p) \vdash ((\neg q \to p) \to \neg(q \to \neg p))$

(b) $\vdash ((p \to (p \to \neg r)) \to (q \to r))$

(c) $\{(p_i \to p_{i+1}) : i \in \mathbb{N}\} \vdash p_3$

(d) $\{(p_i \to p_{i+1}) : i \in \mathbb{N}\} \vdash (p_2 \to p_4)$

(e) $\{(p_i \to p_{i+1}) : i \in \mathbb{N}\} \vdash (p_4 \to p_2)$

There is another way of phrasing the soundness theorem in terms of *consistency*, a property of formal systems which is of critical importance.

Definitions   Consistency

Let Γ be a set of sentences. We say that Γ *is inconsistent* if there is a formula θ for which both

$$\Gamma \vdash \theta \quad \text{and} \quad \Gamma \vdash \neg\theta.$$

In the case that Γ is the empty set, we say that *the system S is inconsistent*.

We say that the set Γ *is consistent* if it is not inconsistent. In the case that Γ is the empty set, we say that *the system S is consistent*.

Of course, in the intended interpretation θ and $\neg\theta$ are contradictory statements.

For instance, the set $\{p, \neg p\}$ is inconsistent as there are very simple derivations of $\{p, \neg p\} \vdash \theta$ and $\{p, \neg p\} \vdash \neg\theta$, taking θ to be the formula p. Plainly, in general if the set Γ contains both a formula and its negation, it is inconsistent. Somewhat more effort is need to show that the set $\{\neg(p \to q), \neg(q \to r)\}$ is inconsistent.

Exercise 3.17 ________________________________

Show that the set $\{\neg(p \to q), \neg(q \to r)\}$ is inconsistent. [*Hint*: Show that $\neg(q \to r) \vdash q$ and $\neg(p \to q) \vdash \neg q$.]

An example of a consistent set is $\{\neg(p \to q), \neg(r \to q)\}$ as we shall show soon.

Before we rephrase the soundness theorem in terms of consistency, we shall explore some of the basic properties of the notion. Being told that a set Γ is inconsistent gives us some concrete information that there are derivations of a certain sort involving Γ, whereas being told that Γ is consistent gives the less tangible information denying that certain derivations exist. So it might not be surprising that many results about consistency are proved by first rephrasing them in terms of inconsistency, so that there are some concrete derivations to play with!

You may have encountered a similar situation if you have studied connectedness in topology. It is common to define 'disconnected' in terms of a concrete piece of information and then to define 'connected' as 'not disconnected'.

We have already derived some useful results about inconsistency in Section 3.2. One of these is as follows.

If, for some ψ, $\Gamma \vdash \psi$ and $\Gamma \vdash \neg\psi$, then $\Gamma \vdash \phi$, for any formula ϕ,

Exercise 3.8(c)

so that all formulas are derivable from an inconsistent set of assumptions. Clearly the converse of this result also holds: if $\Gamma \vdash \phi$ for any formula ϕ, then in particular both $\Gamma \vdash \psi$ and $\Gamma \vdash \neg\psi$. In terms of inconsistency, this result becomes

Γ is inconsistent if and only if $\Gamma \vdash \phi$ for any formula ϕ,

and by negating this appropriately we have proved the following theorem about consistency.

Theorem 3.5

Let Γ be a set of formulas (possibly empty). Then Γ is consistent if and only if $\Gamma \nvdash \phi$ for some ϕ.

One of the aims of formalizing mathematical argument is to provide a rigorous framework for deriving results of mathematical substance, e.g. about the integers or real numbers or geometry. We hope that it seems to you totally undesirable that within such a framework one could prove two contradictory statements. As most such formal frameworks in wider use incorporate a fragment equivalent to the formal system S, Theorem 3.5 gives a further reason why such a state of affairs would be undesirable: if the formal framework is inconsistent, then every statement can be derived from it, meaning that the mathematics has only trivial content. Being able to establish that sets of sentences describing a mathematical theory are consistent is thus of major importance.

The result of Exercise 3.8(c) is really the meaty bit here.

Later in this section we shall require some further technical results about consistency which we shall state as the following theorem.

Theorem 3.6

Let Γ be a set of formulas and ϕ a formula.
(a) $\Gamma \cup \{\neg\phi\}$ is consistent if and only if $\Gamma \nvdash \phi$.
(b) Suppose that Γ is consistent and $\Gamma \vdash \phi$. Then $\Gamma \cup \{\phi\}$ is consistent.

Proof
(a) To get a handle on this we need to turn the required result into a statement about inconsistency:

$\Gamma \cup \{\neg\phi\}$ is inconsistent if and only if $\Gamma \vdash \phi$.

We shall prove one half of this and leave the other for you as an exercise. We shall show that if $\Gamma \cup \{\neg\phi\}$ is inconsistent, then $\Gamma \vdash \phi$.

Suppose that $\Gamma \cup \{\neg\phi\}$ is inconsistent. Then for some ψ we can derive both of

$\Gamma, \neg\phi \vdash \psi$ and $\Gamma, \neg\phi \vdash \neg\psi$.

By Theorem 3.2 of Section 3.2, we can then derive $\Gamma \vdash \phi$.

(b) This is left as an exercise for you. ∎

Exercise 3.18 ____

(a) Do the other half of the proof of Theorem 3.6(a), namely that if $\Gamma \vdash \phi$, then $\Gamma \cup \{\neg\phi\}$ is inconsistent.

(b) Prove Theorem 3.6(b), namely that if Γ is consistent and $\Gamma \vdash \phi$, then $\Gamma \cup \{\phi\}$ is consistent. [We reckon that there is suitable machinery available in Section 3.2 to do this quite smoothly. What will you actually attempt to prove?]

As we are going to show that formal proof matches logical consequence, we should expect similarities between results about formal proofs and results about logical consequence. Results like Theorem 3.6 above which are expressed in terms of consistency can be seen to match results in Section 2.6 of Chapter 2 once we have established a link between consistency and satisfiability of a set of formulas. The following rephrasing of the soundness theorem gives part of this link.

Theorem 3.6(a) turns out to match the result of Exercise 2.76, which is that $\Gamma \cup \{\neg\phi\}$ is satisfiable if and only if $\Gamma \nvdash \phi$.

Theorem 3.7

The following general statements about the system S are equivalent.

(A) For all formulas ϕ and all sets of formulas Γ, if $\Gamma \vdash \phi$ then $\Gamma \vDash \phi$.

(B) For all sets of formulas Δ, if Δ is satisfiable then Δ is consistent.

As the soundness theorem holds for S, statement (B) also holds for S.

This is the soundness theorem

Recall that Δ is satisfiable if there is a truth assignment v such that $v(\delta) = T$ for all $\delta \in \Delta$.

Proof

We shall prove one half of this and leave the other half as an exercise for you. We shall show that if (B) holds then (A) holds.

Suppose that (B) holds – a general principle for all sets Δ. Let ϕ be a formula and Γ a set of formulas. We must show that if $\Gamma \vdash \phi$, then $\Gamma \vDash \phi$. It turns out to be easier to prove the contrapositive of this, namely that

if $\Gamma \nvDash \phi$ then $\Gamma \nvdash \phi$.

Suppose that $\Gamma \nvDash \phi$. Then there is some truth assignment v making all of Γ true and ϕ false. This means that the set $\Gamma \cup \{\neg\phi\}$ is satisfiable (with this v being a suitable truth assignment). It follows from (B) that $\Gamma \cup \{\neg\phi\}$ is consistent. Then by Theorem 3.6(a) we can infer that $\Gamma \nvdash \phi$. ∎

The completeness theorem will show that the converse of (B) above holds, so that a set of formulas Δ turns out to be consistent if and only if it is satisfiable – another way of viewing the remarkable connection between the syntax and semantics of the formal language!

Exercise 3.19 ____

Prove the other half of Theorem 3.7. [*Hints*: Assume (A) – a general principle for all ϕ and Γ. Take a set of formulas Δ and prove (B) by showing that if Δ is inconsistent, then Δ is not satisfiable – assuming that Δ is inconsistent gives you some concrete derivations to exploit.]

We can now show that the set $\{\neg(p \to q), \neg(r \to q)\}$ mentioned earlier is consistent. The set is satisfiable, using the truth assignment v defined by $v(p) = v(r) = T$, $v(q) = F$. So using form (B) in Theorem 3.7, the set is consistent.

As a consequence of the soundness theorem in the form (B) of Theorem 3.7, we can now say something of fundamental importance about the system S, as follows.

Theorem 3.8

The system S is consistent.

Proof

Take the set Δ in (B) of Theorem 3.7 to be empty. Then Δ is satisfied by any truth assignment (in the sense that no such assignment makes any of the formulas in Δ false). Thus Δ is consistent, using form (B) of the soundness theorem, i.e. the system S is consistent. (If you are uneasy about the argument saying that the empty set of formulas is satisfiable, then instead take Δ to be the set $\{(p \to (q \to p))\}$, which is more tangibly satisfiable and therefore consistent! As the one formula in Δ is an axiom of S, any derivation of $\Delta \vdash \phi$ is also one of $\vdash \phi$, and vice versa. So the consistency of Δ entails that the empty set is consistent.) ∎

Given that the point of the system S is to give a starting point for the formalization of more substantial mathematics, had S been inconsistent, all our efforts so far would have been a waste of time! But here's a rhetorical question for you – if we'd not set up the system with its intended meaning firmly in mind, really guaranteeing that the soundness theorem would hold, would it have been obvious from staring at the definition of the formal system that it would be impossible to derive both $\vdash \theta$ and $\vdash \neg\theta$ for some θ?

Mind you, given how tricky it is to derive desired formal theorems, you might have found it entirely believable that one couldn't derive $\vdash \phi$ for all ϕ!

Exercise 3.20 ____

Decide which of the following sets of formulas are consistent. Explain each answer in a way permitted by results established in the book so far. In particular for any inconsistent set Γ, you'll have to show that $\Gamma \vdash \theta$ and $\Gamma \vdash \neg\theta$ for some θ – later on, you'll be allowed to exploit the completeness theorem, which tells you that inconsistency follows if Γ is not satisfiable.

(a) $\{(p \to q), (q \to r), (r \to \neg p)\}$

(b) $\{\neg(p \to q), q\}$

Let us now state and apply the completeness theorem for the system.

Theorem 3.9 Completeness theorem for S

Let ϕ be any formula and Γ a set of formulas. If $\Gamma \vDash \phi$, then $\Gamma \vdash \phi$. (In the case that the set Γ is empty, the result is to be read as saying that if ϕ is a tautology, then it is derivable.)

This is sometimes called the adequacy theorem.

This is a much less trivial result than the soundness theorem – the latter was a straightforward consequence of taking axioms which are tautologies and valid rules of inference for our formal system. In contrast, for the completeness theorem the system has to have enough axioms and rules of inference to derive all logical consequences; and it is in no way obvious that S does have enough. We shall delay our proof of this theorem until the end of the section. For the moment, let us look at some of its applications. First, let us obtain an equivalent formulation of the statement of the theorem in terms of consistency and satisfiability, along the lines of Theorem 3.7 for the soundness theorem.

Theorem 3.10

The following general statements about the system S are equivalent.

(C) For all formulas ϕ and all sets of formulas Γ, if $\Gamma \vDash \phi$, then $\Gamma \vdash \phi$.

(D) For all sets of formulas Δ, if Δ is consistent, then Δ is satisfiable.

When we prove the completeness theorem later, we shall prove it in form (D).

Proof

We shall prove one half of this and leave the other half as an exercise for you. We shall show that if (C) holds then (D) holds.

Suppose that (C) holds – a general principle for all sets Γ and formulas ϕ. Let Δ be a set of formulas. We shall prove the contrapositive of (D), i.e. we shall suppose that Δ is not satisfiable and show that Δ is inconsistent.

Suppose that Δ is not satisfiable. (Take a big breath and read on!) Take any formula θ. Then every truth assignment which satisfies Δ also satisfies both θ and $\neg\theta$!! (Well, you cannot find a truth assignment which gives a counterexample to this last statement, as there aren't any assignments which satisfy Δ!) Thus we have

$$\Delta \vDash \theta \quad \text{and} \quad \Delta \vDash \neg\theta.$$

So by (C) we have

$$\Delta \vdash \theta \quad \text{and} \quad \Delta \vdash \neg\theta,$$

which means that Δ is inconsistent, as required. ∎

Exercise 3.21 ____________________________________

Prove the other half of Theorem 3.10. [*Hint*: You may find Theorem 3.6 useful.]

We can now answer a question raised in Section 2.6 of Chapter 2. Suppose that $\Gamma \vDash \phi$ where the set Γ is infinite. Is there some finite subset Δ of Γ for which $\Delta \vDash \phi$? By the completeness theorem we have $\Gamma \vdash \phi$. A formal derivation of ϕ from assumptions Γ is finitely long so only uses finitely many of these assumptions lying in some finite subset Δ of Γ. This derivation also shows that $\Delta \vdash \phi$. The soundness theorem then gives $\Delta \vDash \phi$, so the answer to the question is yes!

One reason why we are devoting quite a lot of space to investigating the formal system S for the propositional calculus is that this will prepare us well for more mathematically significant work to come on the predicate calculus. We shall prove a completeness theorem for a formal system of predicate calculus and one of its very useful consequences is called the *compactness* theorem. There is a similar result for the propositional calculus and although, like the propositional calculus, it isn't as useful as for the predicate calculus, it is well worth meeting now.

Theorem 3.11 Compactness theorem

Let Γ be an infinite set of formulas in L. If every finite subset of Γ is satisfiable, then so is Γ.

The converse, that if Γ is satisfiable then so is every finite subset, holds trivially.

Proof

The result exploits both the soundness and completeness theorems for S, and hinges on the following observation about consistency:

If every finite subset of Γ is consistent, then so is Γ.

Why does this hold? We shall prove it by showing that if Γ is inconsistent, then some finite subset Δ is also inconsistent. Suppose that Γ is inconsistent. Then for some formula θ there are derivations of both

$$\Gamma \vdash \theta \quad \text{and} \quad \Gamma \vdash \neg\theta.$$

Each of these derivations is finitely long (the key point), so uses only finitely many assumptions from the set Γ. Let Δ be the set of the assumptions used in one or the other of these derivations. Then the latter are also derivations of

$$\Delta \vdash \theta \quad \text{and} \quad \Delta \vdash \neg\theta,$$

so that Δ, which is a finite subset of Γ, is inconsistent.

Now we can prove the compactness theorem. Suppose that every finite subset of Γ is satisfiable. Then by the soundness theorem (Theorem 3.4) every finite subset of Γ is consistent. By the observation above it follows that Γ is consistent. So by the completeness theorem (Theorem 3.9) Γ is satisfiable. ∎

There are alternative proofs of the compactness theorem which make no use of the formal system S and the completeness theorem. For one such, see Exercise 3.28 at the end of this section.

The compactness theorem for propositional calculus can be used to prove results of mathematical significance, usually about infinite sets. We give one example as an appendix to this section. The corresponding result for predicate calculus is much more powerful and most of Chapter 6 is devoted to its applications.

Exercise 3.22

Suppose that $\{\phi_i : i \in \mathbb{N}\}$ is a set of formulas such that every truth assignment makes at least one of the ϕ_i true. Show that there is an $N \in \mathbb{N}$ such that every truth assignment makes at least one of $\phi_0, \phi_1, \ldots, \phi_N$ true. [*Hint*: What can you say about the satisfiability of the set of formulas $\{\neg\phi_i : i \in \mathbb{N}\}$?]

Solution

As every truth assignment satisfies at least one of the ϕ_i, the set $\{\neg\phi_i : i \in \mathbb{N}\}$ is not satisfiable. By the compactness theorem it follows that some finite subset Δ of $\{\neg\phi_i : i \in \mathbb{N}\}$ is not satisfiable. Let N be the largest i for which $\neg\phi_i$ is in Δ. Then

$$\Delta \subseteq \{\neg\phi_0, \neg\phi_1, \ldots, \neg\phi_N\},$$

so that the set $\{\neg\phi_0, \neg\phi_1, \ldots, \neg\phi_N\}$ is not satisfiable. But this means that every truth assignment satisfies at least one of $\phi_0, \phi_1, \ldots, \phi_N$, as required.

Exercise 3.23

Show that the following general statements are equivalent. (Γ and Σ are sets of formulas and ϕ is a formula.)

(E) For all Γ, if every finite subset of Γ is satisfiable, then so is Γ.

(F) For all Σ and ϕ, if $\Sigma \vDash \phi$ then $\Delta \vDash \phi$, for some finite subset Δ of Σ.

There is no need for your solution to use the completeness theorem or any other results about derivations, so see if you can do this purely in terms of satisfiability, logical consequence etc.

Let's now move towards proving the completeness theorem for the system S. We shall prove this in the form

for all sets of formulas Δ, if Δ is consistent, then Δ is satisfiable

of Theorem 3.10. We shall take a consistent set of formulas Δ and show that there is a truth assignment v satisfying it. It will help us first to stand this on its head and ask whether, given a truth assignment v, there is anything interesting about the set of formulas satisfied by v. This is one of the places where we have to pay attention to the formal language L – obviously the formulas available for v to satisfy depend on the propositional variables and connectives from which formulas can be built up. Here we shall continue to take the language L with the variables $\{p_i : i \in \mathbb{N}\}$ and the connectives $\rightarrow, \neg$, and we'll continue to use letters like p to stand for one of the p_is.

Notation

Let v be a truth assignment appropriate for the language L. We shall use Σ_v to denote the set of formulas in L satisfied by v, that is,

$$\Sigma_v = \{\phi : v(\phi) = T\}.$$

What might we say about Σ_v? First, it is consistent – this follows from the soundness theorem, as it is satisfied by v. Next, for every formula ϕ, exactly one of $v(\phi)$ and $v(\neg\phi)$ equals T, so that exactly one of ϕ and $\neg\phi$ belongs to Σ_v. In some sense, as we'll clarify in a moment, Σ_v is as big a consistent set of formulas as we can construct using L. It will help to have some definitions, as follows.

Definitions   *Complete, maximal consistent*

Let Σ be a set of formulas in a language L.

We shall say that Σ is *complete* for L if it is consistent and for each formula ϕ in L, exactly one of ϕ and $\neg\phi$ belongs to Σ.

We shall say that Σ is *maximal consistent* for L if

(i) Σ is consistent;

(ii) for any consistent set of formulas Σ' in L with $\Sigma \subseteq \Sigma'$, we have $\Sigma = \Sigma'$.

Or, equivalently, if ϕ is a formula in L with $\phi \notin \Sigma$, then $\Sigma \cup \{\phi\}$ is inconsistent.

It turns out that for our language L these definitions are equivalent, as we shall now show.

Theorem 3.12

Let Σ be a set of formulas in a language L. Then Σ is complete if and only if it is maximal consistent (for the same language L).

Proof

We shall prove one half of this result and leave the other for you as an exercise.

Suppose that Σ is maximal consistent. Then Σ is consistent, so that for Σ to be complete we need to show that for any formula ϕ, exactly one of ϕ and $\neg\phi$ is in Σ. Clearly we cannot have both ϕ and $\neg\phi$ in Σ, as otherwise Σ would be inconsistent. To show that one of the formulas is in Σ, we'll look at what happens depending on whether or not $\Sigma \cup \{\neg\phi\}$ is consistent.

If both ϕ and $\neg\phi$ are in Σ, then use of the Rule of Assumptions gives both $\Sigma \vdash \phi$ and $\Sigma \vdash \neg\phi$, so that Σ is inconsistent.

If $\Sigma \cup \{\neg\phi\}$ is consistent, then as

$$\Sigma \subseteq \Sigma \cup \{\neg\phi\}$$

and Σ is maximal consistent, we must have

$$\Sigma = \Sigma \cup \{\neg\phi\},$$

so that $\neg\phi \in \Sigma$.

Otherwise, if $\Sigma \cup \{\neg\phi\}$ is inconsistent, then, by Theorem 3.6(a),

$$\Sigma \vdash \phi.$$

But then, by Theorem 3.6(b), $\Sigma \cup \{\phi\}$ is consistent, so using again that Σ is maximal consistent, we have $\phi \in \Sigma$.

Thus if Σ is maximal consistent, then it is complete.   ∎

Exercise 3.24 ⎯⎯⎯⎯⎯⎯⎯⎯⎯⎯⎯⎯⎯⎯⎯⎯⎯⎯⎯⎯⎯⎯⎯⎯

Prove the other half of Theorem 3.12, namely that if Σ is complete, then it is maximal consistent.

⎯⎯⎯⎯⎯⎯⎯⎯⎯⎯⎯⎯⎯⎯⎯⎯⎯⎯⎯⎯⎯⎯⎯⎯⎯⎯⎯⎯⎯⎯⎯⎯⎯⎯

So given a truth assignment v, the set of formulas Σ_v is complete, or equivalently maximal consistent, for L. What we shall use in our proof of the completeness theorem is a complementary result, that every complete set of formulas is satisfied by a unique truth assignment.

Theorem 3.13

Let Σ be a complete set of formulas for the language L. Let v be the truth assignment defined by

$$v(p) = \begin{cases} T, & \text{if } p \in \Sigma, \\ F, & \text{if } \neg p \in \Sigma, \end{cases}$$

for all propositional variables p in L. Then v is the unique truth assignment satisfying Σ.

By unique, we mean relative to the language L. Obviously if we add a new propositional variable q to L, the truth value of q under a truth assignment is irrelevant to whether the assignment satisfies Σ.

Proof

First note that v is well-defined as a function thanks to Σ being complete – for each propositional variable p, exactly one of the formulas p and $\neg p$ is in Σ, so that for each variable p there is a well-defined value of $v(p)$. Should v satisfy Σ, this also forces v to be the unique truth assignment satisfying Σ. This is because if u is any truth assignment satisfying Σ, then it is forced to give the same values as v to each propositional variable, so is the same assignment as v.

By Theorem 2.2 of Section 2.3.

We shall show that, for all formulas ϕ,

$$v(\phi) = T \quad \text{if and only if} \quad \phi \in \Sigma$$

by using induction on the length (number of connectives) of ϕ. As you will see, some of the steps require results about our formal proof system.

The formulas of length 0 are just the propositional variables p, for which the result

$$v(p) = T \quad \text{if and only if} \quad p \in \Sigma$$

holds by the definition of the assignment v.

For the inductive step, suppose that the required result holds for all formulas in L with n or fewer connectives. Let ϕ be a formula with $n+1$ connectives. There are two possibilities for ϕ with which we must cope, depending on whether its principal connective is $\neg$ or $\rightarrow$.

The case when ϕ is of the form $\neg\theta$ is pleasantly straightforward. The inductive hypothesis will apply to θ as it has n connectives. We wish to show that

$$v(\neg\theta) = T \quad \text{if and only if} \quad \neg\theta \in \Sigma.$$

For one way round, suppose that $v(\neg\theta) = T$. Then $v(\theta) = F$, so that by the induction hypothesis we have $\theta \notin \Sigma$. As Σ is complete, we must then have $\neg\theta \in \Sigma$, as required.

For the other way round, suppose that $\neg\theta \in \Sigma$. Then as Σ is complete, we have $\theta \notin \Sigma$. By the inductive hypothesis we then have $v(\theta) = F$. Then $v(\neg\theta) = T$, as required.

The case when ϕ is of the form $(\theta \to \psi)$ is more complicated. As there are n connectives distributed between θ and ψ, the induction hypothesis applies to both these formulas. We shall deal with this case by showing that

(i) if $v((\theta \to \psi)) = T$ then $(\theta \to \psi) \in \Sigma$,

(ii) if $v((\theta \to \psi)) = F$ then $(\theta \to \psi) \notin \Sigma$.

To show (i), note that $v((\theta \to \psi)) = T$ only if $v(\theta) = F$ or $v(\psi) = T$. We shall deal with these two cases separately.

In the case that $v(\theta) = F$, the inductive hypothesis gives us that $\theta \notin \Sigma$, so as Σ is complete we have $\neg\theta \in \Sigma$. By an appropriate substitution, the result of Exercise 3.7(f) of Section 3.2 gives us that

$$\neg\theta \vdash (\theta \to \psi),$$

so that as $\Sigma \vdash \neg\theta$ (by the Rule of Assumptions, as $\neg\theta \in \Sigma$), we have

$$\Sigma \vdash (\theta \to \psi).$$

Then as Σ is consistent we cannot have $\neg(\theta \to \psi) \in \Sigma$. As Σ is complete we then have $(\theta \to \psi) \in \Sigma$, as required.

In the case that $v(\psi) = T$, the inductive hypothesis gives us that $\psi \in \Sigma$. Then by the Rule of Assumptions we have $\Sigma \vdash \psi$, so that using the axiom $(\psi \to (\theta \to \psi))$ and Modus Ponens we obtain $\Sigma \vdash (\theta \to \psi)$. As before, the completeness of Σ gives us that $(\theta \to \psi) \in \Sigma$, as required.

To show (ii), note that $v((\theta \to \psi)) = F$ exactly when $v(\theta) = T$ and $v(\psi) = F$. By the inductive hypothesis this means that $\theta \in \Sigma$ and $\psi \notin \Sigma$, which as Σ is complete means that $\neg\psi \in \Sigma$. Could we have $(\theta \to \psi) \in \Sigma$? If we did, the derivation

$$
\begin{array}{lll}
(1) & \theta & \text{Ass} \\
(2) & (\theta \to \psi) & \text{Ass} \\
(3) & \psi & \text{MP}, 1, 2
\end{array}
$$

would show that

$$\Sigma \vdash \psi,$$

while, as $\neg\psi \in \Sigma$, one use of the Rule of Assumptions gives

$$\Sigma \vdash \neg\psi,$$

contradicting that Σ is consistent. This means that $(\theta \to \psi) \notin \Sigma$, as required.

We have dealt with all possible formulas of length $n + 1$ and the result, for all ϕ, follows by induction. $\blacksquare$

Thanks to this theorem, the mapping

$$v \longmapsto \Sigma_v$$

is a one–one correspondence between the set of all truth assignments on the language L and the set of all complete sets of formulas in L – the inverse map sends a complete set Σ to the unique v satisfying it and clearly the set Σ_v for this v is the original Σ.

We shall at last prove the completeness theorem! As you will see, the extra ingredient we need is that a consistent set of formulas Δ can be shown to be a subset of some complete set Σ. It then follows from Theorem 3.13 that there is a truth assignment satisfying Σ and thus also Δ.

One might be tempted to think that it is obvious that a consistent set is a subset of some maximal consistent set – the latter being equivalent to a complete set. But this is non-trivial!

Theorem 3.9   *Completeness theorem for S*

Let ϕ be any formula and Γ a set of formulas. If $\Gamma \vDash \phi$, then $\Gamma \vdash \phi$.

Proof

Using the result of Theorem 3.10, we shall prove this in the form

for all sets of formulas Δ, if Δ is consistent, then Δ is satisfiable.

Suppose that the set of formulas Δ is consistent. We shall show that there is a complete set of formulas Σ in the same language as Δ with $\Delta \subseteq \Sigma$. A key point is that our language L contains the countably many propositional variables $p_0, p_1, p_2, \ldots$ along with the connectives $\neg, \to$, so that the set of all formulas in L is also countable. This means that we can list all the formulas (without repetitions) as

$$\phi_0, \phi_1, \phi_2, \ldots, \phi_n, \ldots$$

with each formula appearing as ϕ_n for some $n \in \mathbb{N}$. We shall exploit this list to define a chain of sets of formulas Σ_n, for $n \in \mathbb{N}$, recursively, as follows:

$$\Sigma_0 = \Delta,$$
$$\Sigma_{n+1} = \begin{cases} \Sigma_n \cup \{\phi_n\}, & \text{if } \Sigma_n \vdash \phi_n, \\ \Sigma_n \cup \{\neg\phi_n\}, & \text{if } \Sigma_n \nvdash \phi_n, \end{cases} \quad \text{for } n \geq 0.$$

Now we define Σ by

$$\Sigma = \bigcup_{n \in \mathbb{N}} \Sigma_n.$$

Clearly

$$\Delta = \Sigma_0 \subseteq \ldots \subseteq \Sigma_n \subseteq \Sigma_{n+1} \subseteq \ldots \subseteq \Sigma$$

for all $n \in \mathbb{N}$.

We shall show by induction that Σ_n is consistent for all n.

By definition $\Sigma_0 = \Delta$ and we are supposing that Δ is consistent, so the result holds for $n = 0$. For the inductive step, we'll suppose that Σ_n is consistent and show that Σ_{n+1} is consistent. There are two cases, depending on whether or not $\Sigma_n \vdash \phi_n$. In the first case, when $\Sigma_n \vdash \phi_n$, Theorem 3.6(b) gives us that $\Sigma_n \cup \{\phi_n\}$ is consistent, as Σ_n is consistent. In the second case, when

$\Sigma_n \nvdash \phi_n$, Theorem 3.6(a) gives us that $\Sigma_n \cup \{\neg\phi_n\}$ is consistent. In both cases we get that Σ_{n+1} is consistent.

It follows by induction that Σ_n is consistent for all $n \in \mathbb{N}$. We can use this to show that Σ is consistent, as follows.

Suppose that Σ is inconsistent. Then for some formula θ we can derive both $\Sigma \vdash \theta$ and $\Sigma \vdash \neg\theta$. Each of these derivations is finitely long and thus uses only finitely many assumptions out of Σ. Each of these assumptions is a ϕ_n for some n, so appears in Σ by the stage of constructing Σ_{n+1}. As there are only finitely many assumptions involved in these two derivations, this means that they are all included in the same Σ_N, where N is the largest of these ns. But then we have $\Sigma_N \vdash \theta$ and $\Sigma_N \vdash \neg\theta$, which contradicts Σ_N being consistent. Thus Σ must in fact be consistent.

Remember, every formula in the language appears in the list as ϕ_n for some $n \in \mathbb{N}$.

All that remains for Σ to be complete is to show that for all formulas ϕ, exactly one of ϕ and $\neg\phi$ is in Σ. So take any formula ϕ. Then ϕ must appear in the list of all formulas of L as ϕ_n for some $n \in \mathbb{N}$. Then either ϕ_n or $\neg\phi_n$ is inserted into Σ_{n+1} and thus into Σ, i.e. either ϕ or $\neg\phi$ is in Σ. Of course, as Σ is consistent, it cannot contain both ϕ and $\neg\phi$, so we have the required result.

Thus we have a complete set Σ such that $\Delta \subseteq \Sigma$. By Theorem 3.13 there is a truth assignment satisfying Σ and thus satisfying our original consistent set Δ. ∎

This is of course a very indirect way of proving that given Γ and ϕ such that $\Gamma \vDash \phi$, there is a derivation of $\Gamma \vdash \phi$. Indeed the proof gives no clue what such a derivation might look like. However, it gives a very helpful foretaste of the proof of the corresponding result for predicate calculus later on! We've included a constructive proof of a weak form of the completeness theorem as Exercise 3.26 below.

Exercise 3.25 ____

In our proof of the completeness theorem above, we extend the given consistent set Δ to a complete set Σ. The truth assignment satisfying Σ guaranteed by Theorem 3.13 is in fact unique (relative to the language L). Does this mean that there is only one truth assignment satisfying Δ? If you think that this is not necessarily the case, where in our proof do we narrow ourselves down to a single truth assignment?

____

Now that we have proved the soundness and completeness theorems for the system S, we have an alternative way of phrasing the questions we asked about decidability at the end of Section 2.6 of Chapter 2. We can now ask for an algorithmic procedure to decide whether $\Gamma \vdash \phi$ rather than whether $\Gamma \vDash \phi$. To test whether $\Gamma \vDash \phi$ superficially requires us to check each truth assignment satisfying Γ to see whether it also satisfies ϕ. The mechanical nature of formal proof suggests a different sort of procedure to check whether $\Gamma \vdash \phi$.

Consider the problem of deciding whether a single propositional formula with variables in the set $\{p_1, p_2, \ldots, p_n\}$ is derivable from the empty set of assumptions. It can be shown that there is a systematic process which generates a list of all the formal theorems using these variables, with each theorem $\vdash \phi$ appearing in the list after finitely many steps. But this alone does not provide an algorithmic decision procedure – in the case that $\nvdash \phi$, this might only show up once the infinite list of all derivations has been generated. It's not at all obvious from inspecting just the formal system S that there is such an algorithmic procedure. However, there is an easy algorithmic procedure for deciding whether a formula is a tautology – just work out its truth table!

This seems to suggest an advantage to testing logical consequence rather than derivability. But the boot is to some extent on the other foot when we are trying to decide whether $\Gamma \vDash \phi$ in the case when Γ is infinite and is even more in favour of derivation when it comes to predicate logic, regardless of the size of Γ.

In the next section we shall look at some further issues about propositional calculus. We shall look at some alternatives to the formal system S and at whether some of the axioms or rules for a given system depend on each other, or are independent of each other.

Given that the system S is consistent, for those ϕ for which $\vdash \neg\phi$, the latter appears in the list of all formal theorems at some finite stage and the consistency of S then tells us that $\nvdash \phi$. But there are plenty of formulas ϕ for which neither $\vdash \phi$ nor $\vdash \neg\phi$.

An application of the compactness theorem

Here is an application of the compactness theorem which foreshadows more significant applications of the equivalent theorem for predicate calculus later in the book. It concerns partially ordered sets. If you haven't met them before, it might be better to skip the next bit of reading and go straight on to Exercise 3.22. We shall discuss partially ordered sets in somewhat more detail and from scratch in Section 4.4 of Chapter 4. If you have met partial orders before, here's a reminder of some definitions of and about them.

Definitions   Strict partial order

Let A be a set and R a subset of $A \times A$ (so that R is a *binary relation* on A). We shall write aRb as a shorthand for $(a, b) \in R$. R is a *strict partial order* on A, or A is *strictly partially ordered* by R, if it has the following properties:

irreflexive   for all a in A, it is not the case that aRa;

transitive   for all a, b and c in A, if aRb and bRc then aRc.

If a strict partial order R has the following additional property:

linear   for all a and b in A, aRb or $a = b$ or bRa,

it is a *strict linear order* on A, and A is *strictly linearly ordered* by R.

Familiar examples are the usual $<$ on sets like $\mathbb{N}$, $\mathbb{Q}$ and $\mathbb{R}$, which are all strict linear orders. One example of strict partial order which is not linear is 'a is a proper divisor of b' on the set of positive integers. Another is 'U is a proper subset of V' on $\mathscr{P}(\mathbb{N})$, the set of all subsets of $\mathbb{N}$.

It can be shown that for any strict partial order R on a finite set A, one can find a linear order R' on the same set which extends R, that is, such that if aRb then $aR'b$. For instance, if $A = \{a, b, c\}$ and $R = \{(a, b)\}$, which is a strict partial order, then adding the pairs $(b, c), (a, c)$ to R gives a strict linear order R'. The compactness theorem can be used to show that the same holds for a strict partial order R on a countably infinite set A, as follows.

For each ordered pair (a, b) of elements of A, introduce a propositional variable $p_{a,b}$ and let L be the language consisting of these variables along with the connectives $\neg$ and $\rightarrow$. As A is countably infinite, there are countably many of these variables, so that we can regard L as a disguised version of the language to which the compactness theorem applies. We now define a set of formulas Γ in this language as follows, using $\wedge$ and $\vee$ as abbreviations, rather than their equivalent representations using $\neg$ and $\rightarrow$:

So for each $a \neq b$ in A, we have both symbols $p_{a,b}$ and $p_{b,a}$.

$$\Gamma = \{p_{a,b} : a, b \in A, \ aRb\} \cup \{\neg p_{a,a} : a \in A\}$$
$$\cup \{((p_{a,b} \wedge p_{b,c}) \rightarrow p_{a,c}) : a, b, c \in A\}$$
$$\cup \{(p_{a,b} \vee p_{b,a}) : a, b, \in A, \ a \neq b\}.$$

What does this Γ have to do with our problem? We shall show that Γ is satisfiable – this is where the compactness theorem will come in. Then we take a truth assignment v satisfying Γ and from v we can define a binary relation R' on A by

$$aR'b \quad \text{if} \quad v(p_{a,b}) = T.$$

This will be a strict linear order on A extending the partial order R, as required. Why? First, as v satisfies the set $\{p_{a,b} : a, b \in A, \ aRb\}$, we have that for all $a, b \in A$, if

$$aRb,$$

then

$$v(p_{a,b}) = T,$$

so that by the definition of the relation R',

$$aR'b.$$

Thus the relation R' extends the relation R. Next, v satisfies $\{\neg p_{a,a} : a \in a\}$, so that for all $a \in A$ it is not the case that $aR'a$, so that the relation R' is irreflexive. In a similar way, as v satisfies the rest of Γ, the relation R' is transitive and linear on A, so that R' is indeed a strict linear order on A extending R.

So how do we know that Γ is satisfiable? Take any finite subset Δ of Γ. Once we have shown that Δ is satisfiable, it will follow from the compactness theorem that Γ is satisfiable. As Δ is finite, it involves only finitely many of the variables $p_{a,b}$, so refers only to elements in a finite subset B of A. Then Δ must be a subset of the set Δ' defined by

$$\Delta' = \{p_{a,b} : a, b \in B, \ aRb\} \cup \{\neg p_{a,a} : a \in B\}$$
$$\cup \{((p_{a,b} \wedge p_{b,c}) \rightarrow p_{a,c}) : a, b, c \in B\}$$
$$\cup \{(p_{a,b} \vee p_{b,a}) : a, b, \in B, \ a \neq b\},$$

and if Δ' can be satisfied, then so can Δ. Define the subset S of $A \times A$ by

$$S = \{(a, b) : a, b \in B, \ aRb\} \ (\text{or equivalently } R \cap (B \times B)).$$

This set of pairs is the restriction of the strict partial order R to the finite subset B of A, so is itself a strict partial order. It can be shown that a strict partial order on a finite set can be extended to a linear order on the same set, so that S can be extended to a linear order S' on the finite set B. We then use S' to define a truth assignment u on the propositional variables occurring in Δ and Δ' by

$$u(p_{a,b}) = T \quad \text{if} \quad aS'b,$$

for all $a, b \in B$. By similar reasoning to that we used above for v and Γ, u must satisfy Δ' and thus also Δ. This is the last detail required to complete our argument that R can be extended to a linear order R' on A.

We shall prove the same result later in the book using the compactness theorem for the predicate calculus, which is a much more natural environment for axiomatizing the theory of partial and linear orders. But it is interesting to see that the more humble propositional calculus can be used to give the same result!

Further exercises

Exercise 3.26

In this exercise we ask you to prove a weak version of the completeness theorem, namely

This proof is based on one given by the Hungarian logician and computer scientist Laszlo Kalmar (1905–1976).

$$\text{for all formulas } \phi, \text{ if } \vDash \phi \text{ then } \vdash \phi,$$

in a way which is essentially constructive, i.e. lurking within the details of the proof is a recipe for creating a derivation within S of a given tautology ϕ.

(a) Let ϕ be a formula involving variables in the list $p_1, p_2, \ldots, p_n$ and let v be a truth assignment. Define formulas $P_1, P_2, \ldots, P_n$ by

$$P_i = \begin{cases} p_i, & \text{if } v(p_i) = T, \\ \neg p_i, & \text{if } v(p_i) = F, \end{cases} \quad \text{for } i = 1, 2, \ldots, n.$$

Show that if $v(\phi) = T$, then

$$P_1, P_2, \ldots, P_n \vdash \phi,$$

and if $v(\phi) = F$, then

$$P_1, P_2, \ldots, P_n \vdash \neg\phi.$$

[*Hints:* Use induction on the length of ϕ. Bear in mind that for each ϕ you will have to deal with what happens both when $v(\phi) = T$ and when $v(\phi) = F$. And, for the inductive step, when dealing with ϕ of the form $(\theta \to \psi)$, the case $v(\phi) = T$ splits into the subcases $v(\theta) = F$ or $v(\psi) = T$. Obviously you'll need various formal theorems of the system S, but you've met them all before in one guise or other in this section and in Section 3.2. Offhand, the following will probably help, though there are doubtless plenty of good alternatives:

$$\vdash (\phi \to \phi), \quad \vdash (\theta \to \neg\neg\theta), \quad \neg\theta \vdash (\theta \to \psi),$$
$$\psi \vdash (\theta \to \psi), \quad \theta, \neg\psi \vdash \neg(\theta \to \psi);$$

and metatheorems like the Deduction Theorem will come in useful.]

(b) Now suppose that ϕ is a tautology, so that $v(\phi) = T$ for truth assignments on the variables $p_1, p_2, \ldots, p_n$. Show that $\vdash \phi$ as follows.

(i) Let v and v' be truth assignments such that

$$v(p_i) = v'(p_i) \quad \text{for all } i = 1, 2, \ldots, n-1,$$

and

$$v(p_n) = T \quad \text{and} \quad v'(p_n) = F.$$

Define P_i for $i = 1, 2, \ldots, n-1$ as in part (a) using the values both v and v' give to $p_1, p_2, \ldots, p_{n-1}$. Then using the result of part (a) for v and v', we have derivations

$$P_1, P_2, \ldots, P_{n-1}, p_n \vdash \phi$$

and

$$P_1, P_2, \ldots, P_{n-1}, \neg p_n \vdash \phi.$$

Show how to deduce that

$$P_1, P_2, \ldots, P_{n-1} \vdash \phi.$$

The result of Exercise 3.10(f) of Section 3.2 might be useful.

(ii) Explain how to repeat the process above to show that $\vdash \phi$.

Exercise 3.27

Suppose that, for each $i \in \mathbb{N}$, p_i is a propositional variable. Let Σ be a set of sentences of the propositional calculus. Suppose that all truth assignments which satisfy Σ make at least one p_i true. Show that for some $n \in \mathbb{N}$,

$$\Sigma \vDash p_1 \vee p_2 \vee \ldots \vee p_n.$$

Exercise 3.28

In this exercise we ask you to prove the compactness theorem within the realm of truth assignments and satisfiability, without going via the formal proof system.

Suppose that Γ is a set of formulas using the propositional variables p_i for $i \in \mathbb{N}$ such that every finite subset Δ is satisfiable. This exercise will lead you through the construction of a truth assignment v satisfying Γ. We shall define a sequence of truth assignments $\{v_n\}_{n \in \mathbb{N}}$ where the domain of each v_n is the set $\{p_0, p_1, \ldots, p_n\}$. It might make the exercise more readable to say that a truth assignment u *agrees* with v_n on $\{p_0, p_1, \ldots, p_n\}$ when

$$u(p_i) = v_n(p_i) \quad \text{for } i = 0, 1, \ldots, n.$$

We shall ensure that for all $n \geq 0$, v_{n+1} agrees with v_n on $\{p_0, p_1, \ldots, p_n\}$, and we shall use these v_n to define our assignment v satisfying Γ.

(a) Show that there is a truth assignment v_0 defined just on the propositional variable p_0 such that any finite subset of Γ can be satisfied by a truth assignment u agreeing with v_0 on $\{p_0\}$. [*Hints*: If taking $v_0(p_0) = T$ gives v_0 the required property, then we are done. If taking $v_0(p_0) = T$ does not give v_0 the required property, this means that there is some finite subset Δ_0 of Γ which cannot be satisfied by any truth assignment u agreeing with v_0 on p_0. In this case, set $v_0(p_0) = F$ and show that for any finite subset Δ of Γ there is a truth assignment u agreeing with v_0 on p_0 which satisfies the finite set $\Delta \cup \Delta_0$.]

(b) Suppose that for $n \geq 0$ the truth assignment v_n has the property that any finite subset of Γ can be satisfied by a truth assignment u agreeing with v_n on $\{p_0, p_1, \ldots, p_n\}$. Show that there is a truth assignment v_{n+1} agreeing with v_n on $\{p_0, p_1, \ldots, p_n\}$ with the property that any finite subset of Γ can be satisfied by a truth assignment u agreeing with v_{n+1} on $\{p_0, p_1, \ldots, p_{n+1}\}$. [*Hints*: If taking $v_{n+1}(p_{n+1}) = T$ gives v_{n+1} the required property, then we are done. What can be said when taking $v_{n+1}(p_{n+1}) = T$ does not give v_{n+1} the required property? Why does taking $v_{n+1}(p_{n+1}) = F$ in this case give v_{n+1} the required property?]

(c) By parts (a) and (b) we can define a sequence of truth assignments $\{v_n\}_{n \in \mathbb{N}}$ such that v_{n+1} agrees with v_n on $\{p_0, p_1, \ldots, p_n\}$. Define a truth assignment v on the set of all propositional variables by

$$v(p_n) = v_n(p_n) \quad \text{for each } n.$$

Note that v then agrees with each v_n on $\{p_0, p_1, \ldots, p_n\}$. Show that v satisfies Γ. [*Hints*: Take any formula $\phi \in \Gamma$. Then there is a largest n for which the variable p_n appears in ϕ. What can be said about $v_n(\phi)$?]

Exercise 3.29 ____

Let R be a subset of $A \times B$, where A and B are countably infinite sets, such that

1. for each $a \in A$, the set $R_a = \{b \in B : (a, b) \in R\}$ is non-empty and finite;

2. for each finite subset C of A, there are at least as many elements in the set $\bigcup\{R_a : a \in C\}$ as in C.

It can be shown that, for each finite subset C of A, there is a one–one function $f : C \longrightarrow B$ such that $f(a) \in R_a$, for all $a \in C$. Assuming this result, use the compactness theorem to show that there is a one–one function $f : A \longrightarrow B$ such that $f(a) \in R_a$, for all $a \in A$.

[*Hints*: Let L be the language with a propositional variable $p_{a,b}$ for each pair (a, b) in R. Note that as A, B are countable, so is the set of propositional variables of L. For each $a \in A$, devise a formula θ_a which is satisfied by a truth assignment v when $v(p_{a,b}) = T$ for exactly one of the bs in the set R_a. Then define a set Γ consisting of all these formulas θ_a along with appropriate formulas that preclude, for each $b \in B$, $p_{a,b}$ and $p_{a',b}$ being satisfied for distinct $a, a' \in A$. Use the compactness theorem to show that Γ is satisfiable.]

Regard A as a set of women and, for each $a \in A$, R_a as the set of men that a knows. Then the function f gives a way for each woman to marry one of the men she knows and the one–one condition on f means that this can be done monogamously. For a proof of the result, called the *marriage theorem*, when A is a finite set, see Wilson [30].

Can you see what a truth assignment v satisfying Γ has to do with the desired function f?

3.4 Independence of axioms and alternative systems

In this section we shall look at some alternatives to the proof system S discussed so far. We have never claimed that the system S is a particularly obvious formalization of 'natural' argument. It does however do at least one particular job well, namely matching the intended interpretation of formulas well enough to yield both a soundness and completeness theorem. One could ask whether there are formal systems which would achieve the same, but which are in some way 'nicer'. Perhaps their axioms and rules of inference might more obviously capture the essence of 'implies' and 'not' than Ax 1, Ax 2 and Ax 3. Perhaps one could do with fewer axioms and rules. This last remark also prompts a different sort of question about the system S. Are any parts of it redundant in the sense that one of the axioms or rules can in fact be derived from the remaining axioms and rules of the system? Or is each of them independent of the remainder?

But does fewer mean better?

This section includes several results of the sort which cover all derivations of a given formal system. So we will make considerable use of proof by mathematical induction on the length of a derivation, which is the key way of capturing such generality. For instance, we shall give two particular systems and use this method to show that if $\Gamma \vdash \phi$ is derivable within one of the systems, then it is derivable in the other. It will also be used in proving analogues of the soundness theorem aimed at showing when something is *not* derivable, as at the beginning of Section 3.3.

Note that we shall only look at formal systems of the same sort as S, in the sense that their derivations are finite sequences in which each step exploits given axioms and rules of the system. As a reminder, here is a summary of the axioms and rules for S.

There are other useful and important ways of treating formal proof, for instance in the form of a tree.

Axioms: all instances of

(Ax 1) $(\phi \to (\psi \to \phi))$,

(Ax 2) $((\phi \to (\psi \to \theta)) \to ((\phi \to \psi) \to (\phi \to \theta)))$

(Ax 3) $((\neg\phi \to \neg\psi) \to (\psi \to \phi))$,

    where ϕ, ψ, θ are any formulas of L.

Rules: Rule of Assumptions and Modus Ponens.

As you'd expect, by the Rule of Assumptions we mean that within a derivation of $\Gamma \vdash \phi$, a formula from Γ can be used on a line, and by Modus Ponens we mean that if θ and $(\theta \to \psi)$ appear on earlier lines of a derivation, then ψ can be derived.

As our first example of an alternative system, let's take the system S' with the same rules as S, the axioms Ax 1 and Ax 2, but, instead of Ax 3, the axiom scheme Ax 3' defined as follows:

(Ax 3') $((\neg\phi \to \psi) \to ((\neg\phi \to \neg\psi) \to \phi))$,

    for all formulas ϕ, ψ.

We shall use the notation $\Gamma \vdash_S \phi$ for derivations in the system S and the notation $\Gamma \vdash_{S'} \phi$ for derivations in the system S'. First observe that the alternative axiom Ax 3' of S' is a tautology, so that by the completeness theorem for the original system S, any instance of this axiom is a theorem of S. You ought to expect (quite correctly!) that this means that any formal theorem of S' is also a formal theorem of S. Actually the converse also holds and the system S' is fully equivalent to S in the sense that

$$\Gamma \vdash_S \phi \quad \text{if and only if} \quad \Gamma \vdash_{S'} \phi,$$

for all Γ, ϕ. Rather than use the sledgehammer of the completeness theorem for the system S to justify one way round of this, we shall use a method more intrinsic to the nature of derivations, namely mathematical induction on the length of derivation in S, to show that

$$\text{if } \Gamma \vdash_S \phi \text{ then } \Gamma \vdash_{S'} \phi,$$

and mathematical induction on the length of derivation in S' to show that

$$\text{if } \Gamma \vdash_{S'} \phi \text{ then } \Gamma \vdash_S \phi.$$

We shall prove the second of these in grizzly (but, we hope, revealing) detail and leave the first as an exercise for you. We shall assume that $\Gamma \vdash_{S'} \phi$ and show that $\Gamma \vdash_S \phi$. As $\Gamma \vdash_{S'} \phi$ there is a derivation

$$\phi_1, \phi_2, \ldots, \phi_n$$

in S', where ϕ_n is ϕ and all the lines ϕ_i arise using one of the rules or axioms of S', including use of the Rule of Assumptions (which is one of the rules of S') with formulas out of the set Γ. We shall use mathematical induction to show that

$$\Gamma \vdash_S \phi_i$$

for each $i = 1, 2, \ldots, n$.

Basis of induction

The first line (ϕ_1) of the derivation of $\Gamma \vdash_{S'} \phi$ could arise in two different ways: by a use of the Rule of Assumptions; and as an axiom of S', i.e. an instance of one of Ax 1, Ax 2 and Ax 3'. We need to deal with each of these possibilities.

As S also has the Rule of Assumptions and the axiom schemes Ax 1 and Ax 2, if any of these were used to derive ϕ_1 in the S' derivation, the same justification produces a corresponding line ϕ_1 in a derivation in S.

The other possibility is that ϕ_1 is an instance of Ax 3', of the form

$$((\neg\phi \rightarrow \psi) \rightarrow ((\neg\phi \rightarrow \neg\psi) \rightarrow \phi)),$$

and our task must be to show in S that we can derive

$$\vdash_S ((\neg\phi \rightarrow \psi) \rightarrow ((\neg\phi \rightarrow \neg\psi) \rightarrow \phi)).$$

This turns out to be quite reasonable, not least because we can exploit the deduction theorem for S. First we show that from the set of assumptions $\{(\neg\phi \rightarrow \psi), (\neg\phi \rightarrow \neg\psi), \neg\phi\}$ we can derive a contradiction – this is pretty clear, as simple uses of MP give derivations in S of

$$\{(\neg\phi \rightarrow \psi), (\neg\phi \rightarrow \neg\psi), \neg\phi\} \vdash_S \psi,$$

Later we shall consider systems for which the completeness theorem might not hold, in which case the method of mathematical induction on the length of derivation is the only available option.

If you have leapt to the conclusion that all we really need to show is that the only feature of S' which is different from those of S, namely the axiom scheme Ax 3', can be derived in S, then well done!

The ϕ in this formula is not necessarily the same ϕ as derived in the $\Gamma \vdash_{S'} \phi$. It's just that I can't be bothered to use another Greek letter!

and

$$\{(\neg\phi \to \psi),\ (\neg\phi \to \neg\psi),\ \neg\phi\} \vdash_S \neg\psi.$$

Then by Theorem 3.2 (proof by contradiction) of Section 3, which is a result about the system S, we have

$$\{(\neg\phi \to \psi),\ (\neg\phi \to \neg\psi)\} \vdash_S \phi.$$

A couple of applications of the deduction theorem, again a result for S, gives the required result:

$$\vdash_S ((\neg\phi \to \psi) \to ((\neg\phi \to \neg\psi) \to \phi)).$$

So in all cases we have a derivation in S of $\Gamma \vdash_S \phi_1$.

Inductive step

We suppose that, for $k > 1$, we have shown $\Gamma \vdash_S \phi_i$ for each i with $1 \leq i < k$. We shall show that $\Gamma \vdash_S \phi_k$. The kth line of the derivation could have arisen from a use of the Rule of Assumptions or as an instance of an axiom. In these cases the same argument as used for the first line shows that $\Gamma \vdash_S \phi_k$. The one further possibility is that Modus Ponens was used to get the kth line. That means that the kth line is ϕ_k where there are earlier lines ϕ_i, ϕ_j in the original derivation in S' of $\Gamma \vdash_{S'} \phi$ and ϕ_j is the formula $(\phi_i \to \phi_k)$. By the induction hypothesis

$$\Gamma \vdash_S \phi_i \quad \text{and} \quad \Gamma \vdash_S (\phi_i \to \phi_k),$$

so that as Modus Ponens is a rule of the system S, we have a derivation in S of $\Gamma \vdash_S \phi_k$, as required.

This deals with the inductive step, so that the result follows by mathematical induction, that is, if $\Gamma \vdash_{S'} \phi$, then $\Gamma \vdash_S \phi$.

A few observations on this argument. First, once one has seen the full argument, one becomes much more confident that what is required for a convincing argument is to show that any axiom or rule of the system S' is a formal theorem or metatheorem for the system S. Of course if an axiom or rule is in common to both systems, then this doesn't take much doing! Second, a cautionary note. To show that instances of the axiom Ax 3$'$ of S' can be derived in S, we used several results about S (the deduction theorem and proof by contradiction) that actually required quite a lot of justification in a previous section. So when attempting to show that formal theorems of one system are formal theorems of another, be prepared to establish, perhaps with a lot of effort, helpful results and metatheorems of these systems along the way. Some of the next exercises could well give you a taste of this sort of effort (if not for it!).

We are going to ask you to show that if $\Gamma \vdash_S \phi$, then $\Gamma \vdash_{S'} \phi$. Your argument would be greatly assisted if the deduction theorem, which we have proved as a metatheorem for the system S, also holds for the system S'. It does, so our first exercise below asks you to prove this.

Note that Theorem 3.2 is a metatheorem specifically of the system S. Its use here is fine, as we are trying to find a derivation in S. But be careful when trying to construct derivations in other systems, e.g. the system S', not to assume that the same metatheorem necessarily holds for this other system. It may well hold (and does for S'), but the proof of the metatheorem will almost certainly be different than it was for our original system S, as it has to exploit the features of S' rather than S.

Exercise 3.30 ______________________________

Prove the deduction theorem for the system S', namely that if $\Gamma, \phi \vdash_{S'} \psi$, then $\Gamma \vdash_{S'} (\phi \to \psi)$.

Solution

Our full solution would look exactly the same as the proof given for Theorem 3.3 in Section 3.2, except that the one reference to Ax 3 would be replaced by reference to Ax 3′! This reference occurs when we are covering the case that a line $\Gamma, \phi \vdash \psi_i$ arises because ψ_i is an instance of an axiom – and in our proof it turned out not to matter what the axiom looked like. Strictly speaking, we left the details to you as an Exercise 3.11, so there remain some loose ends for you to satisfy yourself about. The real point is that the machinery of the system S needed to turn each line in a derivation of $\Gamma, \phi \vdash \psi$ into a corresponding line contributing to a derivation of $\Gamma \vdash (\phi \to \psi)$ consists of the axioms Ax 1, Ax 2 and the rule Modus Ponens – Ax 3 isn't needed as part of this machinery. The proof only needs to account for a line on which Ax 3, or indeed any other axiom, is used. So, our proof in Section 3 will actually work for any system which includes Ax 1, Ax 2, the rule Modus Ponens and any other axioms.

> There's a subtle but important point here that our proof of the deduction theorem would *not* work as it stands for a system which had extra rules of inference besides Modus Ponens. The use of one of these extra rules to convert a line $\Gamma, \phi \vdash \psi_i$ into a derivation of $\Gamma \vdash (\phi \to \psi_i)$ would require an extra stage of argument compared to our proof in Section 3.2.

Exercise 3.31

Show that if $\Gamma \vdash_S \phi$, then $\Gamma \vdash_{S'} \phi$. [*Hints*: The previous exercise shows that the deduction theorem is also a metatheorem of the system S', which will be helpful. But if you want to use other metatheorems similar to those which hold for S, don't forget that you'll have to show they hold for S'. For instance, the details of the proof of Theorem 3.2 for S, proof by contradiction, make crucial use of Ax 3 – this isn't surprising, as Ax 3 feels like the only part of the system S which gives interesting properties of the $\neg$ symbol. So you cannot use this, or most other results for S involving the $\neg$ symbol without proving them anew in S'. With luck you won't need all such results!]

> At the time of writing we had a proof that needed none of these fancy results about $\neg$! It only needed the deduction theorem and uses of all the rules and axioms of S' except Ax 2.

Exercise 3.32

As the systems S and S' are equivalent in the sense that $\Gamma \vdash_S \phi$ if and only if $\Gamma \vdash_{S'} \phi$ for all Γ, ϕ, the soundness and completeness theorems of Section 3.3 which hold for S must also hold for S'. Suppose that we were trying to prove these metatheorems from scratch using the system S'. Where in the proofs we have given of Theorems 3.4 and 3.9, and any subsidiary results that they use, would the details of the proofs differ, to take account of the difference between S and S'?

Solution

The only difference between the systems is the form of the axiom specifically mentioning the negation symbol $\neg$, Ax 3 in S and Ax 3′ in S'.

In the proof of the soundness theorem, Theorem 3.4, this means that we have to show that whenever an instance ϕ_k of Ax 3′ is used in a derivation of $\Gamma \vdash_{S'} \phi$ and v is a truth assignment satisfying Γ, then $v(\phi_k) = T$. This is very easy to show, simply by checking the truth table of ϕ_k, which is of course a tautology.

Adjusting the proof of the completeness theorem, Theorem 3.9, requires much more thought and effort! The first line of our proof of Theorem 3.9 refers to Theorem 3.10, which is about the equivalence of two formulations of the completeness theorem, labelled (C) and (D). Looking at our proof of Theorem 3.10, there is nothing in the displayed proof of (C) $\implies$ (D) that depends on Ax 3, but the proof of (D) $\implies$ (C), which is left as Exercise 3.21, requires use of a result in Theorem 3.6, an important technical result about consistent sets of formulas. Rather a lot of the proof of Theorem 3.6 is left to you as an exercise! But we do show that if $\Gamma \cup \{\neg\phi\}$ is inconsistent, then $\Gamma \vdash \phi$, applying Theorem 3.2 which shows that proof by contradiction holds for the system S: if $\Gamma, \neg\phi \vdash \psi$ and $\Gamma, \neg\phi \vdash \neg\psi$, then $\Gamma \vdash \phi$. If you persevere and inspect the proof of this last metatheorem in Section 3.2, you will finally see places where we actually use Ax 3 in the derivation of $\Gamma \vdash \phi$. If we replace Ax 3 by Ax 3$'$ within our proof system, the shape of the derivation will almost certainly be different, but at least we already know that there must be a derivation.

You might like to try to prove Theorem 3.2 directly for S' as Exercise 3.39.

It turns out that this is the only part of the proof of Theorem 3.6 which requires adjustment. The proof of the rest of Theorem 3.6(i) (which was left as an exercise) need not involve Ax 3, while the proof of Theorem 3.6(ii) is easy using the deduction theorem. Our proof of the deduction theorem for the system S doesn't use Ax 3, other than as one of the axioms whose use within a formal derivation of $\Gamma, \phi \vdash \psi$ has to be accounted for. So a proof of the deduction theorem as a metatheorem about the system S' would look essentially the same as the one we gave for our original system S.

Furthermore, it turns out that within the rest of our proof of the completeness theorem very few prior results are mentioned. Buried in the proof of Theorem 3.13, we use the result $\neg\theta \vdash (\theta \to \psi)$ of Exercise 3.7(f) of Section 3.2, which has to exploit a negation axiom. So it is only here and in Theorem 3.6 in which there could be some dependence on Ax 3. So by showing that this result holds for S', our proof of the completeness theorem will suffice exactly as it stands for S'.

A moral of the last exercise is that if you are trying to invent an alternative to the system S intended to be capable of deriving $\Gamma \vdash \phi$ precisely when $\Gamma \vDash \phi$, then it's easy to ensure that the soundness theorem will hold. Just make sure than any axiom is a tautology and that any new rule of inference is valid. But ensuring that the completeness theorem holds is somewhat harder. The axioms and rules have to be capable of deriving various formal theorems to underpin the construction of the chain of sets in our proof of Theorem 3.9 and the proof that the union of the sets in the chain is a consistent set of formulas.

We shall look later at a system S'' which at first sight seems to be equivalent to S and S', but turns out to lack the deductive power needed to prove the crucial part of Theorem 3.6(a) which is needed for our proof of the completeness theorem to work. It will transpire that this is not just a failure of our method of proof – the completeness theorem simply doesn't hold for S''. We shall demonstrate this by something similar to the soundness theorem, showing that all formal theorems of S'' have some interesting property testable from 'outside' the formal proof system; and then we show that there's a tautology which doesn't have this special property.

The soundness theorem for S says that whenever $\Gamma \vdash_S \phi$, then $\Gamma \vDash \phi$. The latter property can be tested using truth assignments, which are 'outside' the formal system.

Before we look at the system S'', we shall illustrate this 'soundness theorem' technique in the context of a different sort of question about formal proof systems, as follows. Given a formal system and one of its axioms, is this axiom redundant? That is to say, can it be derived from the remaining rules and axioms of the system? For instance, if our original system S had included as an extra axiom scheme all instances of the formula $(\phi \to \phi)$, then this would

A similar question is whether one of the rules of a system can be inferred from the axioms and remaining rules. In such a case we describe the rule as a *derived rule* of the rest of the system.

have been redundant in this way – we already know that this formula (or any other tautology) is derivable from the original rules and axioms of S. If an axiom or rule of a system cannot be derived from the rest of the system, we say that it is *independent* of the remainder of the system.

In the case of our original system S, each of its axioms and rules turns out to be independent of the rest of the system. How does one demonstrate this? As an example, we shall show that the axiom scheme $\mathrm{Ax}\,2$ is independent of the rest of the system. Our method is to find some property possessed by those formal theorems not involving this axiom, but which is not possessed by all theorems of the full system. In this case the property is rather devious! We shall define some special functions from the set of all formulas to the set $\{0, 1, 2\}$ which are rather like truth assignments, so we shall call each one a *quasi-truth assignment*. For a normal truth assignment v, we can assign a truth value in the set $\{T, F\}$ to each propositional variable and then use the standard truth table rules for the connectives $\neg, \rightarrow$ to assign a truth value $v(\phi)$ to each formula ϕ in the language. Here we shall build up a quasi-truth assignment f from given values $f(p)$ in the set $\{0, 1, 2\}$ for each propositional variable p to assign a value $f(\phi)$ in $\{0, 1, 2\}$ for each formula ϕ in a similar way, except that instead of using standard truth tables for the connectives, we shall use the following tables for $\neg, \rightarrow$:

ϕ	ψ	$(\phi \rightarrow \psi)$
0	0	0
1	0	0
2	0	0
0	1	2
1	1	2
2	1	0
0	2	1
1	2	0
2	2	0

ϕ	$\neg\phi$
0	1
1	0
2	1

This definition of quasi-truth assignment is temporary, applicable only to this one application!

For instance, if f is a quasi-truth assignment such that $f(p) = 2$ and $f(q) = 1$, then

$$f(\neg p) = 1$$
$$f((p \rightarrow q)) = 0$$
$$f((\neg p \rightarrow (p \rightarrow q))) = 0.$$

Let us not pretend for one moment that a quasi-truth assignment has anything to do with truth! Its interpretation of the connectives $\neg$ and $\rightarrow$ has nothing to do with their usual interpretations by 'not' and 'if … then'. It is similar to a truth assignment only in that it is a function assigning values to formulas in some set following the equivalent of truth table rules as with a truth assignment, but not necessarily using $\{T, F\}$ or the usual truth tables.

The point of these particular quasi-truth assignments is that any formula ϕ which is derivable using instances of $\mathrm{Ax}\,1$ and $\mathrm{Ax}\,3$ and the rule Modus Ponens (but no uses of the Rule of Assumptions) has the property that $f(\phi) = 0$ for any quasi-truth assignment f. But some instances of $\mathrm{Ax}\,2$ do not have this property, so that instances of $\mathrm{Ax}\,2$ cannot in general be derived from the rest of the system.

Exercise 3.33

(a) Show that for any formula ϕ which is derivable using instances of $\mathrm{Ax}\,1$ and $\mathrm{Ax}\,3$ and the rule Modus Ponens, $f(\phi) = 0$ for any quasi-truth assignment f.

(b) Let ϕ be the formula

$$((p \to (q \to r)) \to ((p \to q) \to (p \to r))),$$

where p, q, r are propositional variables, which is an instance of $\mathrm{Ax}\,2$. Find a quasi-truth assignment f for which $f(\phi) \neq 0$.

Solution

(a) Use mathematical induction on the length of any derivation of $\vdash \phi$ which makes no use of $\mathrm{Ax}\,2$, just as in a proof of the soundness theorem – indeed, this is a sort of soundness theorem using quasi-truth assignments where 0 is like T. The details are more complicated because there are three 'truth' values rather than two, so that to check whether any quasi-truth assignment f always gives $f((\phi \to (\psi \to \phi))) = 0$ requires checking all $3^2 = 9$ different quasi-truth assignments on ϕ and ψ. Likewise you have to check whether $f(((\neg\phi \to \neg\psi) \to (\psi \to \phi))) = 0$ for all 9 different quasi-truth assignments on ϕ and ψ.

(b) One suitable quasi-truth assignment f is determined by $f(p) = f(q) = 0$, $f(r) = 1$. You can check that $f(\phi) = 2 \neq 0$.

A slightly more roundabout way of showing that $\mathrm{Ax}\,2$ cannot be derived from the rest of the system is to find a formula ϕ derivable using $\mathrm{Ax}\,2$ and a quasi-truth assignment f for which $f(\phi) \neq 0$. If the formula ϕ is very simple, this roundabout method might involve less effort than finding a suitable instance of $\mathrm{Ax}\,2$ and a suitable f, as you can see from the following exercise.

Exercise 3.34

(a) Find a quasi-truth assignment f for which $f((p \to p)) \neq 0$.

(b) Explain why this shows that $\mathrm{Ax}\,2$ is independent of the rest of the system.

Solution

(a) For the quasi-truth assignment f defined by $f(p) = 1$, we have

$$f((p \to p)) = 2 \neq 0.$$

(b) We have $\vdash_S (p \to p)$ for a propositional variable p.   This was derived in Exercise 3.7(c).

Suppose that the axiom scheme $\mathrm{Ax}\,2$ is derivable from the rest of the system. Then any derivation of S which uses an instance of this axiom can be turned into a derivation of the same formula which only uses the axioms and rule of the rest of the system. In particular there is a derivation of $\vdash_S (p \to p)$ using the rest of the system. Then by the first part of Exercise 3.33 we must have $f((p \to p)) = 0$ for the f in part (a). However, $f((p \to p)) = 2 \neq 0$, which gives a contradiction.

We can use the same sort of method to show that some rule or axiom of any system is independent of the rest of the system. We look for a magic property held by all formal theorems of the rest of the system, but not by all theorems of the system using the rule or axiom in question. The magic property will almost inevitably have little to do with the intended interpretations of the symbols of the language. It might be something combinatorial, like 'any formal theorem avoiding use of the axiom contains an odd number of $\to$ symbols' while the full system includes formal theorems with an even number of these symbols – one's mind might have to wander away from things like the normal truth tables.

In the next exercise, we ask you to show that Ax 3 cannot be derived from the rest of the system. As before, the idea is to dream up some property of all formal theorems of the rest of the system S which is not shared by some instance of axiom Ax 3. Of course, this axiom is the only one which does anything interesting involving the $\neg$ symbol, for which the intended interpretation is as negation.

Exercise 3.35

Show that the axiom scheme Ax 3 cannot be derived from the rest of our original system S by each of the following methods.

(a) Find a new sort of quasi-truth assignment on formulas taking the values $\{T, F\}$ with the usual interpretation of $\to$, but an unusual one of $\neg$, with the following properties:

> There are only four possible tables for $\neg$. One of these is the standard interpretation of $\neg$ as negation, for which (ii) cannot hold. So experiment with the other three!

    (i) for each of these quasi-truth assignments f and each formula ϕ derivable using axioms Ax 1, Ax 2 and the rule Modus Ponens, $f(\phi) = T$;

    (ii) for some instance ϕ of Ax 3 and some quasi-truth assignment f, $f(\phi) \neq T$.

    Your answer should include a demonstration of (i) and (ii).

(b) For any formula ϕ, let ϕ^* be the formula obtained by deleting all the negation signs from ϕ. Show that for any derivation of $\vdash \phi$ which avoids use of Ax 3, the formula ϕ^* is a tautology. Now use this result to deduce that some well-chosen instance of Ax 3 isn't a formal theorem of the rest of S.

Exercise 3.36

Show that the rule of Modus Ponens cannot be derived from the remainder of the system S.

We shall now look at an interesting system S'' which turns out to be slightly weaker than the system S' we met earlier by replacing Ax 3$'$ with the seemingly very similar

> By 'slightly weaker' we mean that all formal theorems of S'' are also formal theorems of S', but not vice versa.

$$\text{Ax 3}''\quad ((\phi \to \psi) \to ((\phi \to \neg\psi) \to \neg\phi)),$$

for all ϕ, ψ. Such formulas are all tautologies and as the completeness theorem holds for the system S' (as this derives precisely the same theorems as our original system S, for which we proved completeness), all instances of Ax 3$''$ must be derivable in S'. Thus all formal theorems of S'' are also theorems of S'. The deduction theorem holds for the system S'', for the same reasons as given in our solution to Exercise 3.30, which will help towards deriving the following.

Exercise 3.37 ───────────────────────────────────

Show that the following are derivable in S'' for all formulas ϕ, ψ.

(a) $\vdash_{S''} (\phi \to \neg\neg\phi)$

(b) $\phi, \neg\phi \vdash_{S''} \psi$

Now we'll show that the system S'' is indeed weaker than S.

Exercise 3.38 ───────────────────────────────────

We shall say that a function f from the set of all propositional formulas in L to the set $\{0, 1, 2\}$ is a *3-function* if f obeys the following:

$$f(\neg\phi) = \begin{cases} 0, & \text{if } f(\phi) = 2, \\ 2, & \text{otherwise;} \end{cases}$$

$$f((\theta \to \psi)) = \begin{cases} 0, & \text{if } f(\theta) \geq f(\psi), \\ f(\psi), & \text{if } f(\theta) < f(\psi). \end{cases}$$

We shall write, for all formulas ϕ and sets of formulas Γ,

$$\Gamma \vDash_3 \phi$$

if for all 3-functions f,

$$\max\{f(\gamma) : \gamma \in \Gamma\} \geq f(\phi).$$

In the case when the set Γ is empty, $\vDash_3 \phi$ means that for all 3-functions f we have $f(\phi) = 0$.

(a) Show that if $\Gamma \vdash_{S''} \phi$ then $\Gamma \vDash_3 \phi$.

(b) Hence show that

    (i) $\nvdash_{S''} (\neg\neg p \to p)$, where p is a propositional variable;

    (ii) $\nvdash_{S''} (((p \to q) \to p) \to p)$, where p and q are propositional variables. This formula is called *Pierce's law.*

> Basically an 3-function f is just like a truth assignment: the values of $f(p)$ for all the propositional variables p determine the value of $f(\phi)$ for any formula ϕ. The 0 in $\{0, 1, 2\}$ behaves just like true and 1 and 2 are increasing degrees of falsity. So $f((\theta \to \psi))$ is true when $f(\psi)$ is no falser than $f(\theta)$, and in particular if $f((\theta \to \psi))$ and $f(\theta)$ are true, then so is $f(\psi)$.

The results of part (b) of Exercise 3.38 show that not all tautologies can be derived in the system S'', so that the completeness theorem does not hold for it, and confirm that it is weaker than our standard system. In regard to Pierce's law, one might easily have imagined that the fragment of the system S'' consisting of the axioms Ax 1, Ax 2 and Modus Ponens would suffice for derivations of all tautologies involving only the connective $\to$. But the results of this exercise show that this isn't the case and that some quite strong form of axiom involving negation is required for its derivation.

The result of part (b)(i) also tells us that when p is a propositional variable, $\neg\neg p \nvdash_{S''} p$. This leads us to an example of a set Γ and formula ϕ such that $\Gamma \cup \{\neg\phi\}$ is inconsistent in the system S'' but $\Gamma \nvdash_{S''} \phi$. Just take Γ to be the set $\{\neg\neg p\}$ and ϕ to be p. This means that Theorem 3.6(a) does not hold for the system S'', which is where our method of proof of the completeness theorem for S breaks down for the weaker system S''.

> As the deduction theorem holds for S'', if $\neg\neg p \vdash_{S''} p$, we would have $\vdash_{S''} (\neg\neg p \to p)$.

The real interest in the system S'' and the tautology $(\neg\neg p \to p)$ not being a theorem of it is in connection with the philosophy of mathematics called *Intuitionism*. This philosophy was developed by the Dutch mathematician L.E.J Brouwer (1881–1966) in the early part of the 20th century. Intuitionism

> For a proper explanation of Intuitionism, its logic and view of mathematics, see Dummett [11].

is a possible and coherent response to the philosophical questions raised by use of the infinite in mathematics, for instance on the nature of a proof that a mathematical object exists, whether such a proof must rely on some sort of construction of the object and the nature of such a construction. Intuitionistic logic is based on a particular vision of constructive proof. Most famously, in terms of the difference between Intuitionistic logic and the logic much more commonly used in mathematics (and in this book), a proof of 'ϕ or not ϕ'requires an effective means of constructing a finitely long proof of one of ϕ or 'not ϕ'. So, for instance, an intuitionistic proof that for all real numbers x, 'x is rational or x is not rational' would essentially require a finite procedure which would determine, for any given x, whether it was rational or irrational. As there is no such procedure in standard mathematics, the statement is not an Intuitionistic theorem. Thus $(p \vee \neg p)$ would not in general be a theorem of an Intuitionsitic proof system. From this it follows that being able to prove it is not the case that $\neg p$, what we would regard as proving the statement $\neg\neg p$, is not the same as being able to produce a proof of p, otherwise $(p \vee \neg p)$ would be an Intuitionistic theorem. The system S'' above turns out to be Intuitionistically acceptable, although our arguments about the system almost certainly use modes of reasoning which might not be Intuitionistically acceptable – the Intuitionist view of mathematics is all-encompassing, and statements about Intuitionism have to be made with care by outsiders like the author. Nevertheless, we assert that Exercises 3.46 and 3.47, which are further exercises about the techniques of this section, are also of interest for what they say about other Intuitionistically acceptable proof systems.

Intuitionistic logic takes falsity, for instance a statement like $0 = 1$ about the natural numbers, as a primitive concept and defines 'not ϕ' as meaning that from a proof of ϕ one can construct a proof of falsity. So representing falsity by the propositional constant symbol $\perp$, 'not ϕ' is represented by $(\phi \to \perp)$. Axiom Ax 3$''$ of S'' then stands for

$$((\phi \to \psi) \to$$
$$((\phi \to (\psi \to \perp)) \to (\phi \to \perp))),$$

which is actually derivable from the other axioms and rules of S''.

We shall now say goodbye to the considerable intricacies of formal systems for propositional calculus and start looking at the much richer expressive power of the predicate calculus, in the next chapter.

Further exercises

Exercise 3.39 ________________________________

Show by giving a direct proof in the system S' given at the beginning of this section that proof by contradiction, Theorem 3.2, holds as a metatheorem for S'. That is, show that if $\Gamma, \neg\phi \vdash_{S'} \psi$ and $\Gamma, \neg\phi \vdash_{S'} \neg\psi$, then $\Gamma \vdash_{S'} \phi$. [You may use the deduction theorem for S', which we'll assume that you have shown in Exercise 3.30!]

Exercise 3.40

Suppose L is a propositional language with the single connective $\to$ and that S_1, S_2 are formal systems in L defined as follows:

S_1 **Axioms:**   all instances of

     (Ax 1) $(\phi \to (\psi \to \phi))$,

     (Ax 2) $((\phi \to (\psi \to \theta)) \to ((\phi \to \psi) \to (\phi \to \theta)))$,

        for all formulas ϕ, ψ, θ of L;

   Rules: Rule of Assumptions and Modus Ponens.

S_2 **Axioms:**   none;

   Rules: Rule of Assumptions, Modus Ponens and $\to$Introduction,

where $\to$Introduction is the following rule:

$$\text{if } \Gamma \cup \{\theta\} \vdash \phi \text{ then } \Gamma \vdash (\theta \to \phi).$$

Show that S_1 and S_2 have the same set of theorems, i.e. $\Gamma \vdash_{S_1} \phi$ if and only if $\Gamma \vdash_{S_2} \phi$.

[*Hint*: As S_2 has a rule of inference which modifies the set of assumptions, the method of mathematical induction on the length of a derivation needs modifying for S_2 as follows. For a derivation in S_2, record not only the formulas ϕ_i in the derivation sequence, but also the sets of assumptions Γ_i for each line, so that a derivation of $\Gamma \vdash_{S_2} \phi$ looks like

$$\Gamma_1 \vdash_{S_2} \phi_1, \ \Gamma_2 \vdash_{S_2} \phi_2, \ \ldots, \ \Gamma_n \vdash_{S_2} \phi_n,$$

where $\Gamma_n = \Gamma$ and $\phi_n = \phi$. If the rule $\to$Introduction is used to get line i from line j, the set Γ_i will be different from, but related to, Γ_j according to the rule. To show that $\Gamma \vdash_{S_1} \phi$, use mathematical induction to show that $\Gamma_i \vdash_{S_1} \phi_i$ for $i = 1, 2, \ldots, n$.]

Exercise 3.41

Let S_2' be the system with the rules of S_2 in Exercise 3.40 along with the rule

     if $\Gamma \cup \{\neg\phi\} \vdash \psi$ and $\Gamma \cup \{\neg\phi\} \vdash \neg\psi$, then $\Gamma \vdash \phi$.

Show that our original system S and S_2' have the same set of theorems.

Exercise 3.42

Recall the system S'' which was essentially the system S_1 in the Exercise 3.40 along with the axioms given by all instances of

     Ax 3$''$ $((\phi \to \psi) \to ((\phi \to \neg\psi) \to \neg\phi))$,

for all formulas ϕ, ψ.

Let S^* be the system obtained from S'' by replacing Ax 3$''$ by the axiom

     Ax 3* $((\phi \to \psi) \to (\neg\psi \to \neg\phi))$.

Show that S'' and S^* have the same set of theorems.

The system S_2 has rules for handling the connective $\to$, one (Modus Ponens) saying how to eliminate an $\to$ from the formula being derived and one ($\to$Introduction) showing how to introduce an $\to$ into a formula. These seem much more natural ways of describing formally how to use $\to$ than using axioms like Ax 1 and Ax 2, and systems based on such rules and no axioms are described as *natural deduction.*

This hint applies more generally to systems with rules of inference which result in a change in the set of assumptions.

So S_2' has no axioms, but only rules.

Exercise 3.43

Let T be the formal system defined as follows:

Axioms: all instances of

1. $(\phi \to \phi)$,
2. $(\neg\neg\phi \to \phi)$,
3. $((\neg\phi \to \neg\psi) \to (\psi \to \phi))$,

for all formulas ϕ, ψ of L;

Rule: Modus Ponens.

(a) Show that $\vdash_T (\phi \to \neg\neg\phi)$.

(b) Show that $\nvdash (\neg\phi \to \phi)$.

(c) Show that $\nvdash (p \to (q \to p))$, where p, q are propositional variables. [*Hint:* Our solution uses a property of formal theorems of T that is nothing to do with any truth tables for $\neg, \to$.]

The following definitions will be needed for the next few exercises.

Definitions *n-functions*

A function f from the set of all propositional formulas in some language to the set $\{0, 1, \ldots, n-1\}$, where $n \geq 2$, is said to be an *n-function* if f obeys the following:

$$f((\phi \vee \psi)) = \min\{f(\phi), f(\psi)\}$$
$$f((\phi \wedge \psi)) = \max\{f(\phi), f(\psi)\}$$
$$f(\neg\phi) = \begin{cases} 0, & \text{if } f(\phi) = n-1, \\ n-1, & \text{otherwise,} \end{cases}$$
$$f((\theta \to \psi)) = \begin{cases} 0, & \text{if } f(\theta) \geq f(\psi), \\ f(\psi), & \text{if } f(\theta) < f(\psi). \end{cases}$$

We shall write, for all formulas ϕ and sets of formulas Γ,

$$\Gamma \vDash_n \phi$$

if for all n-functions f,

$$\max\{f(\gamma) : \gamma \in \Gamma\} \geq f(\phi).$$

In the case when the set Γ is empty, $\vDash_n \phi$ means that for all n-functions f we have $f(\phi) = 0$.

These definitions generalize those of a 3-function given earlier in this section.

Basically an n-function f is just like a truth assignment: the values of $f(p)$ for all the propositional variables p determine the value of $f(\phi)$ for any formula ϕ. The 0 in $\{0, 1, \ldots, n-1\}$ behaves just like true and 1, 2 etc. as increasing degrees of falsity. The rules for $f((\phi \vee \psi))$ etc. then give the usual ones when $n = 2$. Likewise when $n = 2$, $\Gamma \vDash_n \phi$ means the same as the usual $\Gamma \vDash \phi$.

One can think of $f((\phi \vee \psi))$ as the 'truest' of $f(\phi)$ and $f(\psi)$.

Exercise 3.44

Suppose that the connective $\leftrightarrow$ is defined in terms of $\wedge$ and $\rightarrow$, so that $(\theta \leftrightarrow \psi)$ is an abbreviation for $((\theta \rightarrow \psi) \wedge (\psi \rightarrow \theta))$. Give the rule for computing the value of $f((\theta \leftrightarrow \psi))$ for any n-function f.

Exercise 3.45

(a) Show that if $\Gamma \vDash_n \phi$ and $n > m$, then $\Gamma \vDash_m \phi$.

(b) Suppose that $m < n$. Give an example of a formula ϕ such that $\vDash_m \phi$ but $\nvDash_n \phi$. [*Hint*: Why is it the case that

$$\vDash_2 ((p_1 \leftrightarrow p_2) \vee (p_1 \leftrightarrow p_3) \vee (p_2 \leftrightarrow p_3))?$$

One explanation uses the fact that there is one more propositional variable than the number of truth values.]

(c) Give an example of a formula ϕ such that $\vDash_2 \phi$ (i.e. $\vDash \phi$) but $\nvDash_n \phi$ for all $n > 2$.

Exercise 3.46

L is a propositional language based on the connectives $\neg, \vee$. A system N for L has the following (natural deduction) rules of inference.

(0) If $\phi \in \Gamma$ then $\Gamma \vdash_N \phi$.

(1) If $\Gamma \vdash_N \phi$ then $\Gamma \vdash_N (\phi \vee \psi)$.

(2) If $\Gamma \vdash_N \psi$ then $\Gamma \vdash_N (\phi \vee \psi)$.

(3) If $\Gamma, \phi \vdash_N \theta$ and $\Gamma, \psi \vdash_N \theta$ then $\Gamma, (\phi \vee \psi) \vdash_N \theta$.

(4) If $\Gamma, \phi \vdash_N \psi$ and $\Gamma, \phi \vdash_N \neg\psi$ then $\Gamma \vdash_N \neg\phi$.

(5) If $\Gamma \vdash_N \neg\neg\phi$ then $\Gamma \vdash_N \phi$.

(a) Show that

  (i) $\vdash_N (\phi \vee \neg\phi)$

  (ii) $\phi, \psi \vdash_N \neg(\neg\phi \vee \neg\psi)$.

(b) Show that for any n-function f, where $n \geq 2$, if $\Gamma \vdash_N \phi$ and the proof uses only rules (0) to (4), then $\max\{f(\gamma) : \gamma \in \Gamma\} \geq f(\phi)$.

(c) Let I be the system with all the rules as above except (5). Use the previous part to show that rule (5) of N is not a derived rule of I and that for any propositional variable p, $(p \vee \neg p)$ is not derivable in I.

(d) Show that for any formula ϕ, $\phi \vdash_I \neg\neg\phi$ and $\neg\neg\neg\phi \vdash_I \neg\phi$.

(e) Show that the soundness and completeness theorems hold for the system N.

(f) Given a set of formulas Γ, define the corresponding set $\neg\neg\Gamma$ to be the set $\{\neg\neg\gamma : \gamma \in \Gamma\}$. Show that $\Gamma \vdash_N \phi$ if and only if $\neg\neg\Gamma \vdash_I \neg\neg\phi$.

(g) Show that $\vDash \neg\phi$ if and only if $\vdash_I \neg\phi$.

The system I with all the rules except (5) is Intuitionistically acceptable.

Some hints! What is the only possibility for the first line of a proof? What rules must be used to obtain a formal theorem with an empty set of assumptions? For a derivation of $\vdash_N \theta$ where θ doesn't have $\neg$ as its principal connective, what rule must have been used? See also the hint for Exercise 3.40 to cope with the rules of inference which alter the set of assumptions.

Exercise 3.47 ___________

A system N' for L has the following (natural deduction) rules of inference.

> (0) If $\phi \in \Gamma$ then $\Gamma \vdash_{N'} \phi$.
>
> (1) If $\Gamma \vdash_{N'} (\phi \wedge \psi)$ then $\Gamma \vdash_{N'} \phi$.
>
> (2) If $\Gamma \vdash_{N'} (\phi \wedge \psi)$ then $\Gamma \vdash_{N'} \psi$.
>
> (3) If $\Gamma \vdash_{N'} \phi$ and $\Gamma \vdash_{N'} \psi$ then $\Gamma \vdash_{N'} (\phi \wedge \psi)$.
>
> (4) If $\Gamma, \phi \vdash_{N'} \psi$ and $\Gamma, \phi \vdash_{N'} \neg\psi$ then $\Gamma \vdash_{N'} \neg\phi$.
>
> (5) If $\Gamma \vdash_{N'} \neg\neg\phi$ then $\Gamma \vdash_{N'} \phi$.

Let the system I' be the system with all the rules of N' except rule (5).

(a) Prove the soundness and completeness theorems for the system N'.

(b) Show that rule (5) is not a derived rule of the system I'.

(c) Show that $\neg\neg\neg\phi \vdash_{I'} \neg\phi$.

(d) Given a set of formulas Γ, define the corresponding set $\neg\neg\Gamma$ to be the set $\{\neg\neg\gamma : \gamma \in \Gamma\}$. Show that $\Gamma \vdash_{N'} \phi$ if and only if $\neg\neg\Gamma \vdash_{I'} \neg\neg\phi$.

(e) Prove that $\vDash \phi$ if and only if $\vdash_{I'} \phi$.

(f) Does this conflict with the results about the system I in Exercise 3.46, e.g. that the tautology $(\phi \vee \neg\phi)$ is not a formal theorem of I?

This is very similar to the previous exercise, but for a system N' using natural deduction rules for $\neg$ and $\wedge$, with the perhaps surprising finale that the Intuitionistically acceptable weaker system obtained by dropping rule (5) can still derive all tautologies expressed using these connectives.

4 PREDICATES AND MODELS

4.1 Introduction: basic ideas

We now come to the main subject of the book, namely a type of formal language and proof system capable of dealing with at least some interesting mathematical statements. Just as we did in Chapter 2 for the propositional calculus, we shall start by describing the formal language within which well-formed formulas are created and the way in which the language is interpreted by the analogue of truth assignments, and formulas end up being true or false under a particular interpretation. We shall then look at the formal proof system in Chapter 5. Before you are hit by several quite complicated-looking formal definitions, you may find it helpful to establish in advance that you already know how to use some of the key underlying ideas.

Suppose that we are given a set A and a binary function $*$ for combining pairs of elements of A to give an element of A. Now, let us ask you the following question:

> That is, $*$ is a function from $A \times A$ to A, sometimes described as a *binary operation* on A.

$$\text{Is it true that for all } x, y \in A, \quad (x * y) = (y * x)?$$

We hope that your reaction is something along the lines of 'this is a daft question, as it all depends on what the set A is and what the function $*$ is'. We would then respond by giving you a specific set A and function $*$ on A; and then, at least in principle, provided that you had the relevant mathematical knowledge about that A and $*$, you would be able to settle the original question. Have a go at this for the sets A and functions $*$ in the exercise below.

Exercise 4.1

For each of the following sets A and functions $*$ on A, decide whether it is true that for all $x, y \in A$, $x * y = y * x$. [You might feel that for most parts, it is enough to give your answer simply as true or false, as appropriate, with no justification.]

(a) $A = \mathbb{N}$, $*$ is $+$.

(b) $A = \mathbb{Z}$, $*$ is $-$.

(c) A is the set of all 2×2 matrices with real coefficients and $*$ is the binary function of matrix multiplication.

(d) A is the set of all 2×2 matrices with real coefficients and $*$ is the binary function of matrix addition.

A relatively minor detail, for our current purposes, of all the functions $*$ in the exercise above is that they are all functions from $A \times A$ to A – they have a value for all pairs of elements of A and that value is in A itself. There are plenty of important functions from sets of the form $A \times B$ to a set C where A, B and C aren't the same set; but in the interests of simplicity, we won't attempt to incorporate them in our model of mathematical language.

The point of Exercise 4.1 is as follows. One formal language which we shall encounter consists of a binary function symbol $*$ and the binary relation symbol $=$, along with propositional connectives like $\wedge$ and $\neg$ which you have already met, and the symbols $\forall$ and $\exists$, which are new. Within this language, one well-formed formula ϕ is given by

$$\forall x \forall y \ (x * y) = (y * x).$$

An interpretation of this language – the counterpart of a truth assignment for interpreting a propositional formula – will be a structure $\mathcal{A}$ which consists of a particular set A and a particular binary function $*_{\mathcal{A}}$ on this A. There'll be specified ways of interpreting the symbols $=$, $\forall$, $\exists$ and the propositional connectives – you may well have seen the symbols $\forall$ and $\exists$ used in everyday maths to represent 'for all' and 'there exists' respectively, and that is indeed how they are going to be interpreted here, referring to elements of the given set A (which will get called the *domain* of the structure $\mathcal{A}$). Also we will choose to interpret the symbol $=$ by equality. A formula like ϕ above can be used to represent statements about such a structure $\mathcal{A}$, and we shall soon define in a formal way what it means for a structure $\mathcal{A}$ to *satisfy* a formula ϕ. We shall also describe this by saying that ϕ is *true in* $\mathcal{A}$. Before we lose you in more technicalities of the formal language and the definition of 'satisfy', please bear in mind that, with any luck, you were effectively using this definition perfectly happily to settle whether

$$\forall x \forall y \ (x * y) = (y * x)$$

was true in various structures $\mathcal{A}$ when you did Exercise 4.1 above. That's the real point of this exercise. You really do know in advance how to use what may appear to be a complicated definition, because this definition models what we do in everyday mathematics!

Another question, looking ahead! Suppose that $\mathcal{A}$ is the structure consisting of the set of all 2×2 matrices with real coefficients with the binary function of matrix multiplication. Is it true in this structure that

$$\forall y \ (x * y) = (y * x)?$$

We hope that your reaction is along the lines of 'well, it depends on which x one takes'. For the question to be well posed, we need to amend it by specifying a value of x. If x is interpreted by the identity matrix

$$\begin{pmatrix} 1 & 0 \\ 0 & 1 \end{pmatrix},$$

then the answer is 'true', while if x is interpreted by the matrix

$$\begin{pmatrix} 0 & 0 \\ 1 & 0 \end{pmatrix},$$

the answer is 'false', as there is some y for which xy does not equal yx. So the answer to the question was sensitive to the value given to x.

Plainly there is something special about the variable x, compared to the variable y, in the formula $\forall y \ (x * y) = (y * x)$. Somehow the variable y is a dummy because of the $\forall y$ – the formula could have been rewritten with the ys replaced by zs, as $\forall z \ (x * z) = (z * x)$, without changing our decisions about its truth or falsity. But the variable x needs to be given a specific value before

To distinguish between a symbol $*$ representing a function in a formal language and its interpretation by an actual function in a structure $\mathcal{A}$, we shall often label the latter with a subscript for the structure, as in $*_{\mathcal{A}}$ (or sometimes a superscript, as in $*^{\mathcal{A}}$).

We shall also explain why interpreting the symbol $=$ as equality is a choice we make, rather than something that is inevitably forced on us.

We hope that even without the formal definitions, you are reading this formula in your head as 'for all y, $(x * y)$ equals $(y * x)$'.

For instance, with the second interpretation of x, interpreting y as $\begin{pmatrix} 1 & 0 \\ 0 & 0 \end{pmatrix}$ gives

$$\begin{pmatrix} 0 & 0 \\ 1 & 0 \end{pmatrix}\begin{pmatrix} 1 & 0 \\ 0 & 0 \end{pmatrix} = \begin{pmatrix} 0 & 0 \\ 1 & 0 \end{pmatrix},$$

and

$$\begin{pmatrix} 1 & 0 \\ 0 & 0 \end{pmatrix}\begin{pmatrix} 0 & 0 \\ 1 & 0 \end{pmatrix} = \begin{pmatrix} 0 & 0 \\ 0 & 0 \end{pmatrix},$$

so that $xy \neq yx$.

we can decide whether or not the formula is true in the structure. There are two points to this discussion. First, we will need to explain what is special about variables like the x here: this is something to do with the shape of the formula, i.e. the syntax of the formal language. In this and other ways, the syntax is going to be more complicated than that for a propositional language. Second, when we define what it means for a structure $\mathcal{A}$ to satisfy a formula ϕ, we will have to build in the particular interpretation of any variables like the x for the question 'is it true in ...?' to make sense.

We shall later describe the y here as a *bound* variable and the x as a *free* variable.

Quite often in everyday maths we introduce a symbol for a special element of a structure, like 0 or 1 or $\mathbf{e}$ for the identity element of a group, when giving axioms for the theory. The symbol for such an element is called a *constant* to distinguish it from the symbols for variables like x, y, z which can represent any of the elements of the structure. Any interpretation of the formal language within a set A has to assign a particular element of A to each constant symbol. Let's look at an example of the use of a constant symbol, adding a constant symbol $\mathbf{e}$ to the language with the binary function symbol $*$ and the equality symbol which we have used above. In this language we can express that a set contains an identity element for the binary function $*$ (one of the axioms for a group) using the statements

Don't worry if you haven't met groups before. We shall give you a brief introduction to them later in this chapter.

$$\forall x((x * \mathbf{e}) = x \wedge (\mathbf{e} * x) = x).$$

Without the constant symbol, we would have to write

Of course, the symbol $\wedge$ will continue to be interpreted by 'and'.

$$\exists y \forall x((x * y) = x \wedge (y * x) = x),$$

There exists y such that for all x...

which is perhaps a bit more cumbersome. A further advantage of the constant symbol when writing down axioms for a group arises when writing down the further axiom for the existence of inverses,

$$\forall x \exists y((x * y) = \mathbf{e} \wedge (y * x) = \mathbf{e}),$$

For all x, there exists y such that...

in that any interpretation has to give the same meaning to the symbol $\mathbf{e}$ in this sentence as in the earlier sentence $\forall x((x * \mathbf{e}) = x \wedge (\mathbf{e} * x) = x)$. Without a constant symbol in the language, the existence of inverses would have to be incorporated into a statement explaining that there was an identity, with something like

$$\exists y(\forall x((x * y) = x \wedge (y * x) = x) \wedge \forall z \exists w((z * w) = y \wedge (w * z) = y)),$$

saying that there is a y which behaves like the identity and then asserting the existence of inverses using this y. This isn't wrong in any way, but again it is a bit cumbersome.

Exercise 4.2

In which of the following structures is the formula

$$\forall x \ (x * \mathbf{e}) = x$$

true? (Our description of each structure gives its domain, i.e. the relevant set A of elements, a description of the binary function interpreting the symbol $*$ and the particular element of A interpreting the constant symbol $\mathbf{e}$.)

(a) $A = \mathbb{Z}$, $*$ is $+$, $\mathbf{e}$ is 0.

(b) $A = \mathbb{Z}$, $*$ is $+$, $\mathbf{e}$ is 1.

(c) $A = \mathbb{Z}$, $*$ is $\times$, $\mathbf{e}$ is 0.

(d) $A = \mathbb{Z}$, $*$ is $\times$, $\mathbf{e}$ is 1.

In the hope of convincing you that you know what you are doing in advance of the proper definitions, here's a further exercise for you. You may well have come across what's called a *binary relation* R on a set A. This just means that R is some set of pairs of elements of A (or equivalently that R is a subset of $A \times A$), though usually this set of pairs is defined in terms of some interesting mathematical property. For instance, R might be the 'less than' relation defined on the set $\mathbb{N}$ of natural numbers, making R the set $\{(a, b) \in \mathbb{N} \times \mathbb{N} : a < b\}$. Alternative ways of saying that the pair (a, b) is in R include the following: a is related to b by R; aRb; and $R(a, b)$. In the following exercises, we will use the $R(a, b)$ notation.

Exercise 4.3

In which of the following structures is the formula

$$\forall x \exists y R(x, y)$$

true? (Our description of each structure gives its domain, i.e. the relevant set A of elements, and a description of binary relation interpreting the symbol R.)

(a) $A = \mathbb{N}$; $<$, i.e. the subset $\{(a, b) \in \mathbb{N} \times \mathbb{N} : a < b\}$

(b) $A = \mathbb{N}$; $>$

(c) $A = \mathbb{N}$; $\leq$

Solution

(a) We hope that you interpreted the formula $\forall x \exists y R(x, y)$ as meaning that for all x in A, there exists a y in A (or, equivalently, there is some y in A) such that x is related to y by R. With $A = \mathbb{N}$ and R interpreted by $<$, this formula is true – for each natural number x, there is a natural number y greater than x.

(b) With this interpretation the formula is false. Although for most natural numbers x, it is the case that there is some natural number y with $x > y$, this does not hold when x is 0. That means it's not true that for all x in $\mathbb{N}$ there exists some y in $\mathbb{N}$ with $x > y$.

(c) With this interpretation the formula is true.

Exercise 4.4

Repeat Exercise 4.3 for each of the following formulas.

(a) $\exists x \forall y R(x, y)$

(b) $\forall x \forall y (\neg R(x, y) \rightarrow R(y, x))$

So far we have given you a foretaste of the ingredients of a formal language, namely how it is interpreted by a structure matching the language and how you work out whether a formula in the language is true in a structure (noting when it is sensible to expect an answer). Our intention has been to suggest that definitions which will appear in the next section and might appear to be quite complicated disguise some fairly natural and straightforward ideas! As one of the aims of our enterprise is to deal with interesting mathematical statements, let's also acquire a feeling that we might indeed be able to use the framework to write formulas which do express interesting mathematical statements. One way in which we can do this is to take a familiar mathematical structure and try to represent some statements about it using a formal language.

Let's take the set $\mathbb{N}$ of natural numbers with its normal arithmetic and order. Suppose that our formal language includes symbols for a binary function $*$, a constant $\mathbf{e}$ and equality $=$. Then let's interpret these symbols in the structure $\mathcal{A}$ with domain $\mathbb{N}$, $*$ as $+$ and $\mathbf{e}$ as 0. Then using $\forall$, $\exists$ and familiar propositional connectives, with their normal interpretation, we can represent at least some statements about natural numbers by formulas. For instance, we can represent 'x is even' by the formula

$$\exists y \; x = (y * y),$$

as the existence of a y in the domain $\mathbb{N}$ such that $x = y + y$ happens precisely when x is even. Similarly we can represent '$x \leq y$' by

$$\exists z \; (x * z) = y,$$

as within $\mathbb{N}$, any z for which $x + z = y$ has to be greater than or equal to 0, forcing y to be greater than or equal to x. In the next exercise, we ask you to try to play the same game. Some ingenuity and knowledge of $\mathbb{N}$ might be required!

Note here the significance of the domain of the structure. If the domain had been the set of rationals, this formula would hold for all x.

When a property of elements in a structure, like 'x is even', can be represented in this way by a formula in the given language, we say that the property is *definable* within the language.

Exercise 4.5

Represent the following statements about $\mathbb{N}$ using the language given above (involving $*$, **e** and $=$, interpreted respectively by $+$, 0 and equality). We reckon that we can solve the parts in the order they are given, so that if you are stumped by one part, we'll allow you to use the result of that part in a later part!

(a) $x < y$

(b) There are at least two even numbers.

(c) There are at most two even numbers. (Ok, this is false! But it is a statement about $\mathbb{N}$ which one can represent using the given language.)

(d) x is divisible by 3.

(e) There is a least natural number.

(f) There is no greatest natural number.

(g) $x = 1$

(h) $x = 2$

Solution

We shall give solutions to (a) and (b), leaving the rest to you.

(a) One solution is to exploit our ability to represent $x \leq y$ and the fact that $x < y$ if and only if $x \leq y$ and $x \neq y$. We can then represent $x < y$ by

$$(\exists z \ (x * z) = y \ \wedge \ \neg \, x = y).$$

There are other solutions, for instance adjusting our earlier answer for $x \leq y$ by making the z non-zero, by

$$\exists z((x * z) = y \ \wedge \ \neg \, z = \mathbf{e}).$$

(b) We know how to represent 'x is even', so we can answer this with a formula saying that there are unequal x and y which are both even. As our representation of 'x is even' happens to use the variable y, we'll avoid confusion by altering our original plan to say that there are unequal x and z which are both even! This gives

$$\exists x \exists z(\neg \, x = z \ \wedge \ (\exists y \ x = (y * y) \wedge \exists y \ z = (y * y))).$$

[If you are worried about our use of $\exists y$ twice in the answer above, you can always change some of the ys to yet another letter, e.g. to give the formula

$$\exists x \exists z(\neg \, x = z \ \wedge \ (\exists y \ x = (y * y) \wedge \exists t \ z = (t * t))).$$

But we reckon that our original answer is fine. Our formula says that there is some y such that $x = y + y$ and that there is some y, which need not be the same as the first y, such that $z = y + y$.]

Exercise 4.6

Show that $x = 0$ can in fact be represented in the language above without using the symbol **e**.

This game is quite challenging! It also raises interesting questions about whether there are limitations to what statements about $\mathbb{N}$ we can represent within a given language. For instance, to what extent can we represent multiplication within our language above? We can represent '$x = yz$' for a specific natural number y like 3, in this case by $x = (z * (z * z))$ – we build the 3-ness into the construction of the formula. But can we represent '$x = yz$' by a single formula, not specifying in advance which specific number y is?

The standard name for the subject being developed in this and the following chapters is the *predicate calculus*. One dictionary definition of a *predicate* is as 'the word or words by which something is said about something'. We tend to use the word 'statement' for 'the word or words by which something is said'. And the 'something' that we are going to make statements 'about' is the elements of a specific set and relationships between them, like whether a set is ordered and has a least element under that order. Such statements are obviously much more complicated than the propositions looked at in Chapter 2 and will be expressed in a much more complicated formal language.

'Predicate' has several other meanings, as can be found in a dictionary. We've given the meaning relevant to this subject.

The use of 'proposition' for the simpler sort of statement is pretty well standard in the subject.

In this chapter we shall begin in Section 4.2 by describing the formal language and the way it is interpreted by corresponding *structures*, which are the analogues of truth assignments for propositional calculus. In Section 4.3 we look at the analogue of tautologies and at logical equivalence between formulas. The main way in which we will apply our framework of formulas in a formal language and their interpretation by structures is to axiomatize mathematical theories. In Section 4.4 we shall look at some examples of these theories and discuss one version of what can be inferred from their axioms, in terms of logical consequence. We conclude the chapter with a look at the ideas of substructures and isomorphisms between structures in Section 4.5.

It is very likely that you have some experience of this from elsewhere in mathematics, perhaps seeing axioms for a theory of order (like $\leq$ or $<$ on $\mathbb{N}$ or $\mathbb{R}$) or group theory.

Further exercises

Exercise 4.7

Take the set $\mathbb{R}$ of real numbers with its normal arithmetic and order. Suppose that our formal language includes symbols for a binary function $*$, a binary function $\circ$ and equality $=$. Then let's interpret these symbols in the structure $\mathcal{A}$ with domain $\mathbb{R}$, $*$ is $+$ and $\circ$ as $\times$. Represent the following statements about $\mathbb{R}$ using this language.

(a) $x = y - z$

(b) $x = 0$

(c) $x \geq 0$

(d) $x > y$

(e) $x = 1$

(f) $x = 1/\sqrt{2}$

In both this and the following exercise, we reckon that we can solve the parts in the order they are given, so that if you are stumped by one part, you can use the result of that part in a later part! But don't feel inhibited from solving the parts in a different order, if you feel that you can do this without a circular argument.

Exercise 4.8 ________________

Take the set $\mathbb{N}$ of natural numbers with its normal arithmetic and order. Suppose that our formal language includes symbols for a binary function $*$, a binary function $\circ$ and equality $=$. Then let's interpret these symbols in the structure $\mathcal{A}$ with domain $\mathbb{N}$, $*$ is $+$ and $\circ$ as $\times$. Represent the following statements about $\mathbb{N}$ using this language.

(a) $x = 0$

(b) $x = 1$

(c) $x \geq y$

(d) x divides y

(e) x is a prime number

(f) There is no greatest prime number.

(g) There are arbitrarily large prime pairs, that is, p and $p + 2$ with both prime.

(h) Every even number greater than 2 is the sum of two primes.

What do you think is normally meant by x divides y in the context of the natural numbers?

Both this and the next part are famous problems of number theory which are unresolved at the time of writing.

4.2 First-order languages and their interpretation

In this section we shall describe first-order languages and their interpretation. This will involve several technical complications, but we hope that you will ride over them by remembering the advice given in the introduction to this chapter. You really do know how to use the definitions that we shall now give – well, most of them, at least when things are kept simple!

The introduction to this chapter should have given you an idea of what sorts of mathematical statement the formal language will represent. Each statement will be about the elements of some set, their images under functions and relationships between them. The most basic sort of statement will be one of two forms: 'this element equals that element' and 'this list of elements is in the relationship R'. In the last section you saw some examples of these basic statements, $(x * \mathbf{e}) = (\mathbf{e} * x)$ and $R(x, y)$. These examples happen to involve a function of two arguments and a binary relation, but in general a function or a relation might involve any finite number of arguments. The first example using $(x * \mathbf{e})$ and $(\mathbf{e} * x)$ illustrates that elements of the set might not only be represented by variables like x, y, z and constant symbols like $\mathbf{e}$, but could be represented by more complicated expressions, which we shall call *terms*. These are built up by applying functions to variables and constants, and indeed by repeatedly applying functions to terms already obtained, as in $((x * \mathbf{e}) * (y * (z * x)))$. The description of the formal language will start by explaining how to construct all these terms. Then we will be able to define the basic statements about terms, which are called *atomic formulas*. Finally we will show how to build up more complicated formulas from these atomic formulas, using the symbols $\forall$ and $\exists$ and combining known formulas with useful propositional connectives like $\neg$ and $\rightarrow$.

The symbols $\exists$ and $\forall$ will always be followed by a symbol for a variable, like x or y or x_i. We call $\exists$ and $\forall$ *quantifiers*. This is because we shall interpret them as expressing a quantity: $\exists x$ as 'there is at least one x' or, more simply, 'there exists an x'; and $\forall x$ as 'for all x'.

First we shall explain what sort of underlying language we shall use.

Definitions   First-order language

A countable *first-order language* L contains countably many variable symbols $x_1, x_2, \ldots, x_n, \ldots$ and some (maybe none) of the following:

1. *function* symbols: $f_{1,1}$, $f_{1,2}$, $f_{1,3}$, $\ldots$, $f_{2,1}$, $f_{2,2}$, $\ldots$, $f_{n,1}$, $\ldots$, $f_{n,m}$, $\ldots$,

2. *relation* symbols: $R_{1,1}$, $R_{1,2}$, $R_{1,3}$, $\ldots$, $R_{2,1}$, $R_{2,2}$, $\ldots$, $R_{n,1}$, $\ldots$, $R_{n,m}$, $\ldots$,

3. *constant* symbols: $\mathbf{c}_1$, $\mathbf{c}_2$, $\mathbf{c}_3$, $\ldots$, $\mathbf{c}_n$, $\ldots$

With the addition of the binary relation symbol $=$, the language is said to be a *language with equality*.

These symbols are called the *non-logical* symbols of the language. The *logical* symbols of the language are the propositional connectives $\wedge, \vee, \neg, \rightarrow, \leftrightarrow$, the quantifiers $\forall, \exists$ and brackets $(,)$.

Having introduced the words 'first-order', we shall promptly drop them for most of the rest of the book and just refer to a 'language'. You should be made aware, however, that there are other formal languages which attempt to model different fragments of natural languages. We shall briefly mention one such alternative, a second-order language, later.

The significance of the first subscript n in the notation $f_{n,m}$ and $R_{n,m}$ is that it gives the number of arguments of the corresponding function or relation, as you will see when we go on to define more complicated expressions within the language. The second subscript m says it is the mth of the symbols requiring n arguments.

In all the examples in the introduction to this chapter, the languages have involved very few symbols, and that is typical of most future examples. It is, however, harmless to look at languages with up to countably infinitely many symbols as allowed for by our definition above. It actually makes sense to look at languages with uncountably many symbols, and we shall come back to this point later in the book. Later on in the book, for the sake of readability, we will normally use a much simpler language, with unsubscripted variables like x, y, z and dropping the subscripts from the function and relation symbols – we will instead specify how many arguments each symbol has, as in the examples in the introduction to this chapter.

Different selections of the function, relation and constant symbols give rise to different languages L.

In many books, the terminology *predicate* symbol is used instead of relation symbol.

Although $=$ is in one sense just another binary symbol, its intended interpretation as equality is so important that it is worth having a special symbol for it and considering it in a special way.

We'll also call $f_{n,m}$ and $R_{n,m}$ *n-place symbols*.

We shall also sometimes use different ways of forming terms and atomic formulas, for instance using *infix* notation for function and relation symbols of two arguments, as in $x + y$ and $x < y$, rather than $+(x, y)$ and $< (x, y)$. Putting the function or relation symbol to the left of the arguments is called *prefix* notation.

Next we give the definition of the set of terms, the expressions that stand for the elements of the domain of an interpretation of the language.

Definition   Term

The set of *terms* of a language L is the set of strings of symbols formed according to the following rules.

1.   All the variable symbols $x_1, x_2, x_3, \ldots$ and all the constant symbols $\mathbf{c}_i$ in L are terms.

2.   If $f_{n,m}$ is a function symbol in L and $\tau_1, \tau_2, \ldots, \tau_n$ are terms, then $f_{n,m}(\tau_1, \tau_2, \ldots, \tau_n)$ is a term.

3.   All terms arise from finitely many applications of 1 and 2.

Here's where we use that $f_{n,m}$ has n arguments.

So all terms are finitely long.

For instance, if the language consists of the function symbols $f_{1,1}, f_{1,2}, f_{3,1}$ and the constant symbols $\mathbf{c}_1, \mathbf{c}_2$, all of the following are terms of the language:

$$x_3, \quad \mathbf{c}_1, \quad f_{1,2}(\mathbf{c}_2), \quad f_{1,2}(f_{1,1}(x_7)), \quad f_{3,1}(x_1, f_{1,1}(\mathbf{c}_2), x_2).$$

At last we can define the basic statements of the language.

Definition   Atomic formula

An *atomic formula* is a string of symbols of the form

$$R_{n,i}(\tau_1, \tau_2, \ldots, \tau_n),$$

where $R_{n,i}$ is a relation symbol of the language and $\tau_1, \tau_2, \ldots, \tau_n$ are terms.

If the language is one with equality, then any string of the form

$$\tau_1 = \tau_2,$$

where τ_1, τ_2 are terms, is also an atomic formula.

Here's where we use that $R_{n,i}$ has n arguments.

If the language includes the function symbols $f_{2,1}, f_{3,4}$, the constant symbol $\mathbf{c}_1$ and the relation symbols $R_{1,1}, R_{2,3}, R_{3,1}$, all of the following are atomic formulas of the language:

$$R_{1,1}(x_5), \quad R_{2,3}(x_4, \mathbf{c}_2), \quad R_{2,3}(f_{2,1}(x_1, \mathbf{c}_1), f_{3,4}(x_7, x_7, x_6)),$$
$$R_{3,1}(x_4, x_5, x_3), \quad R_{3,1}(\mathbf{c}_1, f_{2,1}(f_{2,1}(x_1, x_3), x_4), x_{10});$$

and if the language is also one with equality, then the atomic formulas include:

$$x_3 = \mathbf{c}_1, \quad f_{3,4}(x_1, f_{2,1}(x_1, x_4), \mathbf{c}_1) = x_6.$$

Finally we can define the formulas of the language.

> **Definition   Formula**
>
> The set of *formulas* of a language L is the set of strings of symbols formed according to the following rules.
>
> 1.   All atomic formulas are formulas.
>
> 2.   If ϕ and ψ are formulas and x_i is a variable symbol, then so are
>
> $$\neg\phi \qquad (\phi \wedge \psi) \qquad (\phi \vee \psi) \qquad (\phi \to \psi)$$
> $$(\phi \leftrightarrow \psi) \qquad \forall x_i \phi \qquad \exists x_i \phi$$
>
> 3.   All formulas arise from finitely many applications of 1 and 2.

The x_i in $\forall x_i$ and $\exists x_i$ is called a *quantified variable.*

So all formulas are finitely long.

Examples of formulas in a language including the function symbol $f_{1,6}$, the constant symbol $\mathbf{c}_3$ and the relation symbol $R_{2,4}$ are

$$R_{2,4}(x_3, f_{1,6}(\mathbf{c}_3)) \qquad (R_{2,4}(x_1, x_6) \to \forall x_3 R_{2,4}(f_{1,6}(x_5), x_3))$$
$$\exists x_8 (\forall x_1 (R_{2,4}(x_1, \mathbf{c}_3) \wedge \neg R_{2,4}(f_{1,6}(x_3), x_3)) \leftrightarrow R_{2,4}(x_2, x_7))$$

and if the language is also one with equality, the following are formulas:

$$f_{1,6}(x_2) = \mathbf{c}_3 \qquad (\exists x_2 \ x_2 = f_{1,6}(x_3) \vee \forall x_1 R_{2,4}(x_9, x_1)).$$

Examples of non-formulas include

$$\forall R_{2,4}(x_4, x_1) \quad \text{(there should be a variable after the } \forall)$$
$$\neg \ f_{1,6}(x_5, x_1) = x_2 \quad (f_{1,6} \text{ requires one variable, not two})$$
$$\forall x_1 \ f_{1,6}(x_1) = x_3 = f_{1,6}(x_2) \quad \text{(there's no atomic formula using two } =\text{s).}$$

The third of these examples of non-formulas is the only one which looks like a reasonable mathematical statement, albeit one that fails to conform to our construction rules; and what is doubtless intended could be expressed instead by the legal formula

$$\forall x_1 (\mathbf{f}_{1,6}(x_1) = x_3 \ \wedge \ x_3 = f_{1,6}(x_2)).$$

As our emphasis in this book is on using correct formulas, we won't dwell much on analysing strings with the aim of flushing out non-formulas. Our main aim is to help us read a formula correctly, for instance when trying to establish its truth or falsity under an interpretation – you have already seen this in the context of working out the truth table of a propositional formula.

Exercise 4.9

What can you say about formulas in a language L which has no relation symbols?

As we hope is clear to you from the last exercise, from now on we shall assume that any language L contains at least one relation symbol (which might be the symbol $=$ if L is a language with equality).

The algorithm of Section 2.2 of Chapter 2 can be straightforwardly adapted to check whether a string is a term and then whether a string is a formula. We shall not go into the details here, which are along the same lines, but they are more complicated thanks to needing to check whether certain strings are terms and to allow for the extra construction rules using relation symbols and quantifiers. From now on, with the exception of the next exercise, we shall deal only with strings which *are* formulas. For formulas it will prove to be helpful to produce a tree diagram like the one below, similar to those in Chapter 2, to decompose a string into, and then display how it is built up from, its constituent parts.

We could extend the use of the phrase *principal connective* to include an occurrence of $\forall x_i$ or $\exists x_i$ that is used at the start of a string to create a formula. So testing if a string is a formula would include checking whether it starts with one of $\neg, \forall, \exists$ before we hunt for $\wedge, \vee, \rightarrow, \leftrightarrow$; and if a string does start with $\forall$ or $\exists$, we would check whether what follows is a variable.

$$\forall x_1 (R_{2,1}(x_1, f_{1,4}(x_2)) \rightarrow \exists x_2 R_{2,3}(x_3, x_2))$$

$$(R_{2,1}(x_1, f_{1,4}(x_2)) \rightarrow \exists x_2 R_{2,3}(x_2, x_3))$$

$$R_{2,1}(x_1, f_{1,4}(x_2)) \qquad \exists x_2 R_{2,3}(x_2, x_3)$$

$$R_{2,3}(x_2, x_3)$$

Note that in this sort of diagram the constituent parts are all formulas and the branches all end with atomic formulas – these are the basic building blocks for formulas for this level of analysis and we don't try to analyse the terms inside the atomic formulas as part of the diagram. Following the terminology in the propositional case, the constituent parts are called *subformulas* of the original formula. So the subformulas of $\forall x_1 (R_{2,1}(x_1, f_{1,4}(x_2)) \rightarrow \exists x_2 R_{2,3}(x_3, x_2))$, the formula built up in the diagram above, are all the formulas in the diagram involved at some stage of its construction, namely

Atomic formulas play a similar role in building up formulas to propositional variables.

$$\forall x_1 (R_{2,1}(x_1, f_{1,4}(x_2)) \rightarrow \exists x_2 R_{2,3}(x_3, x_2)),$$
$$(R_{2,1}(x_1, f_{1,4}(x_2)) \rightarrow \exists x_2 R_{2,3}(x_3, x_2)), \quad R_{2,1}(x_1, f_{1,4}(x_2)),$$
$$\exists x_2 R_{2,3}(x_3, x_2), \quad R_{2,3}(x_3, x_2).$$

Definition   Subformula

For all formulas ϕ, their *subformulas* are defined as follows.

1. If ϕ is atomic, then ϕ is the only subformula of itself.

2. If ϕ is one of the forms $\neg\psi$, $\forall x_i\psi$ and $\exists x_i\psi$, then the subformulas of ϕ are ϕ and all subformulas of ψ.

3. If ϕ is one of the forms $(\theta \wedge \psi)$, $(\theta \vee \psi)$, $(\theta \to \psi)$ and $(\theta \leftrightarrow \psi)$, then the subformulas of ϕ are ϕ, all subformulas of θ and all subformulas of ψ.

Exercise 4.10

Which of the following strings of symbols are formulas? For each which is a formula, write down all its subformulas.

(a) $\exists x_1 (\forall x_3 R_{2,2}(x_2, x_3) \vee (x_1 = x_3 \to \forall x_1 R_{1,1}(x_1)))$

(b) $\forall x_3 (R_{1,2}(x_4) \wedge \exists x_1 R_{2,1}(x_1, x_3) \vee R_{1,1}(x_2))$

(c) $(\exists x_1 R_{2,1}(x_1) \leftrightarrow \forall x_5 R_{1,1}(f_{2,1}(x_5, x_5)))$

Solution

(a) For such short strings as in this exercise, we can use some simple rules of thumb, rather than rely on a proper algorithm which will cope with longer strings. For instance, we can attempt to break the string down into its constituent parts, reversing the construction rule 2 for formulas, hoping to arrive at atomic formulas. First one tries to identify the final rule used in the construction of the formula by looking for its principal connective (counting $\forall x_i$ and $\exists x_i$ as connectives for this purpose). For the string

$$\exists x_1 (\forall x_3 R_{2,2}(x_2, x_3) \vee (x_1 = x_3 \to \forall x_1 R_{1,1}(x_1))),$$

the candidate for the principal connective is the $\exists x_1$ at the front. One then removes this and tries the same procedure on the remaining part of the string. The process can be illustrated by the following diagram.

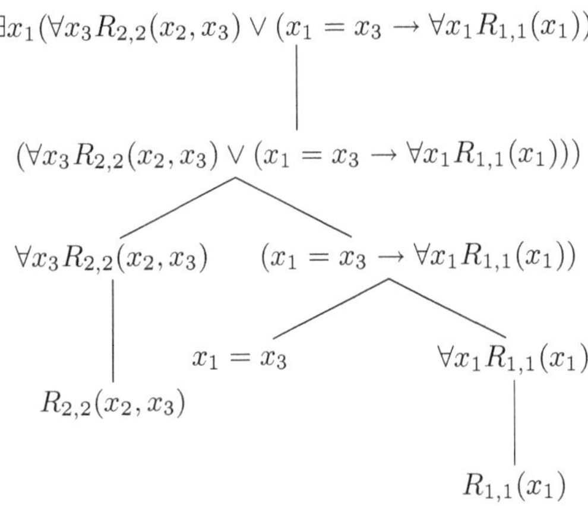

As can be seen, this process breaks down the string into what can be checked to be atomic formulas, with each relation symbol followed by a pair of brackets containing the correct number of what can be identified as terms, and with the $=$ symbol likewise correctly used. Thus the string is a formula. Furthermore, the subformulas are simply all the formulas in the diagram, namely

$$\exists x_1 (\forall x_3 R_{2,2}(x_2, x_3) \vee (x_1 = x_3 \rightarrow \forall x_1 R_{1,1}(x_1))),$$
$$(\forall x_3 R_{2,2}(x_2, x_3) \vee (x_1 = x_3 \rightarrow \forall x_1 R_{1,1}(x_1))),$$
$$\forall x_3 R_{2,2}(x_2, x_3), \quad (x_1 = x_3 \rightarrow \forall x_1 R_{1,1}(x_1)),$$
$$R_{2,2}(x_2, x_3), \quad x_1 = x_3, \quad \forall x_1 R_{1,1}(x_1), \quad R_{1,1}(x_1).$$

(b) After the first obvious stage of breaking down the string by removing the $\forall x_3$, which is the prime candidate to be the principal connective

$$\forall x_3 (R_{1,2}(x_4) \wedge \exists x_1 R_{2,1}(x_1, x_3) \vee R_{1,1}(x_2))$$

$$\big|$$

$$(R_{1,2}(x_4) \wedge \exists x_1 R_{2,1}(x_1, x_3) \vee R_{1,1}(x_2))$$

we arrive at a string $(R_{1,2}(x_4) \wedge \exists x_1 R_{2,1}(x_1, x_3) \vee R_{1,1}(x_2))$ from which there are two possible analyses. One analysis is

$$(R_{1,2}(x_4) \wedge \exists x_1 R_{2,1}(x_1, x_3) \vee R_{1,1}(x_2))$$

$$R_{1,2}(x_4) \qquad \exists x_1 R_{2,1}(x_1, x_3) \vee R_{1,1}(x_2)$$

$$\big|$$
$$?$$

and the other is

$$(R_{1,2}(x_4) \wedge \exists x_1 R_{2,1}(x_1, x_3) \vee R_{1,1}(x_2))$$

$$R_{1,2}(x_4) \wedge \exists x_1 R_{2,1}(x_1, x_3) \qquad R_{1,1}(x_2)$$

$$\big|$$
$$?$$

We know we are heading for trouble at this stage! An algorithm along the lines of that in the appendix to Section 2.2 of Chapter 2 would follow the first of these analyses, treating the connective following the first bracket with count 1 as though it was the principal connective. The possible appearance of brackets in terms and after relation symbols makes the bracket count idea more complicated to use.

In the first case, we arrive at a string $\exists x_1 R_{2,1}(x_1, x_3) \vee R_{1,1}(x_2)$ which is neither atomic nor arises from construction rule 2 for formulas. Note that it lacks the outer brackets (,). In the second case, we can analyse the string $R_{1,2}(x_4) \wedge \exists x_1 R_{2,1}(x_1, x_3)$ no further for similar reasons. So the original string is not a formula.

(c) One of the constituent parts is $R_{2,1}(x_1)$, which isn't an atomic formula – the symbol $R_{2,1}$ requires two arguments, not one. So the original string is not a formula.

In the introduction to this chapter, we gave an indication of how a formal language would be interpreted. The analogue of a truth assignment is a set A of elements with specific functions and relations on A to interpret the function and relation symbols of the language, and with specific elements interpreting the constant symbols. As all these functions etc. on A give the set A some structure, we shall use the word 'structure' to describe the whole specific interpretation of the language.

Definitions Structure

A *structure* $\mathcal{A}$ for a language L is a non-empty set A, called the *domain* of the structure, along with the following:

1. for each function symbol $f_{n,m}$ in L, there is a function $f_{n,m}^{\mathcal{A}} : A^n \longrightarrow A$;

2. for each relation symbol $R_{n,m}$ in L, there is a subset $R_{n,m}^{\mathcal{A}}$ of A^n;

3. for each constant symbol $\mathbf{c}_k$ in L, there is an element $\mathbf{c}_k^{\mathcal{A}}$.

In general, a subset of A^n is called an *n-place relation* on A, so that the subsets $R_{n,m}^{\mathcal{A}}$ are just described as the *relations on* $\mathcal{A}$, and the $\mathbf{c}_k^{\mathcal{A}}$s are called *constants of* $\mathcal{A}$. The functions $f_{n,m}^{\mathcal{A}}$, relations $R_{n,m}^{\mathcal{A}}$ and constants $\mathbf{c}_k^{\mathcal{A}}$ are called the *interpretations in the structure* $\mathcal{A}$ of the corresponding symbols of L.

We shall often write structures using the notation

$$\langle A, \ldots, f_{n,m}^{\mathcal{A}}, \ldots, \ldots R_{n,m}^{\mathcal{A}} \ldots, \ldots \mathbf{c}_k^{\mathcal{A}} \ldots \rangle.$$

For examples where we are using a very limited language and use notation like f, R and $\mathbf{c}$ rather than $f_{n,m}$, $R_{n,m}$ and $\mathbf{c}_k$, we will usually write $f^{\mathcal{A}}$, $R^{\mathcal{A}}$ and $\mathbf{c}^{\mathcal{A}}$ for the interpretations of the symbols in $\mathcal{A}$. If we use infix notation like $+$ or $<$, we may write $+_{\mathcal{A}}$ and $<_{\mathcal{A}}$ for their interpretations.

Note that the interpretation of a relation symbol is given as a set of n-tuples, rather than by a common property of the n-tuples in the set.

Where the structure is mathematically well-known with familiar notations for its functions and relations that obviously match the symbols in L, we'll stick with these familiar notations. For instance, we'll write $\langle \mathbb{N}, +, < \rangle$ for $\mathbb{N}$ with the usual $+$ and $<$ interpreting symbols $f_{2,1}$ and $R_{2,1}$.

If L is a language with equality, how do we interpret the $=$ symbol? The $=$ symbol is really just a 2-place relation used in infix notation, writing $\tau_1 = \tau_2$ rather than $= (\tau_1, \tau_2)$. As such, its interpretation in a structure $\langle A, \ldots \rangle$ could be as any subset of A^2. Of course, our desired interpretation is as actual equality on the set A (or, if you prefer, as the subset $\{(a, a) : a \in A\}$ of A^2). A structure that does this gets a special name, as follows.

Definition Normal structure

Let L be a language with equality. A structure for L is said to be *normal* if the interpretation of $=$ is equality on its domain.

Note that we shall use $=$ both for the relation symbol in the formal language and its interpretation by equality on any set A. Some authors (entirely virtuously) use a different symbol within the formal language to stand for the equality relation. But we shall not, and we hope that the context will make it clear whether we mean the symbol or its interpretation.

In any normal structure $\mathcal{A} = \langle A, \ldots \rangle$, we shall write the interpretation of the symbol $=$ as $=$, standing for the set $\{(a, a) : a \in A\}$, rather than use the notation $=^{\mathcal{A}}$.

Later in the book, when we discuss a formal proof system for predicate calculus which includes axioms for $=$, we shall see that our system cannot completely capture true equality by its axioms and rules. Indeed in Section 5.4 of Chapter 5 we shall look at structures satisfying these axioms in which the symbol $=$ is *not* interpreted by actual equality, and we shall then exploit such structures in the proof of the completeness theorem for predicate calculus in Section 5.5 of Chapter 5. However, in all other parts of the book, we shall assume that *all examples of structures for a language with equality in the book are normal.*

You have essentially seen some examples of structures in the introduction to this chapter and we'll take the risk of not giving you any more right now. Instead we'll move towards explaining when a formula is true in a structure. We hope that the introduction has prepared you for most of the formal definition, but there are some tedious details. For instance, if investigating the truth of the formula $\forall x_1 R_{2,1}(x_1, x_2)$ in a structure $\mathcal{A} = \langle A, R_{2,1}^{\mathcal{A}} \rangle$, you've been alerted to the need to specify which element of A interprets the variable x_2. But in a more complicated formula, like

$$\exists x_2 (R_{2,1}(x_2, x_3) \vee \forall x_1 (R_{2,1}(x_9, x_1) \rightarrow \exists x_9 R_{2,1}(x_9, x_3))),$$

how do we systematically find the variables, which are called *free* variables, for which some element of A needs to be specified? Which appearances of the variables in a formula get interpreted by these specific elements? (Think about the way x_9 occurs in the formula above.) The answers will be forced by the following definitions.

Definitions   Free and bound variables

In any formula of the form $\forall x_i \phi$, all occurrences of the variable x_i within the string $\forall x_i \phi$ are said to be *bound*. Similarly all occurrences of x_i within $\exists x_i \phi$ are said to be bound. Any occurrence of a variable which is not bound is said to be *free*.

In a formula of the form $\forall x_i \phi$ (respectively $\exists x_i \phi$), the subformula ϕ is often said to be the *scope* of the quantifier $\forall x_i$ (respectively $\exists x_i$).

One way to classify the occurrences of variables in a formula as free or bound is to work through the construction of the formula from its subformulas, noting which variables become bound when a quantifier is added to some subformula. The analysis for the formula

$$\exists x_1 (\forall x_3 R_{2,2}(x_2, x_3) \vee (x_1 = x_3 \rightarrow \forall x_1 R_{1,1}(x_1)))$$

is shown below. We work up the construction tree, starting with the atomic subformulas, in which all occurrences of variables are free. Whenever a quantifier is applied to a subformula, we underline all occurrences of the quantified variable within the new subformula that is being constructed – these are now bound occurrences of the variable. Once an occurrence of a variable becomes bound in a subformula, it remains bound wherever this subformula is used in the construction.

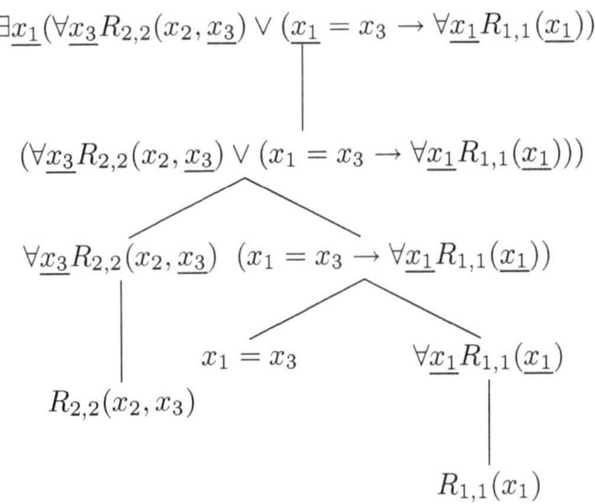

Those occurrences of variables which are not underlined are therefore not bound and hence are free. Note that some occurrences in the above formula of the variable x_3 are bound and some are free – that sort of situation is not unusual. Note also that the occurrence of x_1 in the atomic subformula $R_{1,1}(x_1)$ lies both inside a subformula with principal connective $\forall x_1$ (the subformula $\forall x_1 R_{1,1}(x_1)$ immediately above it in the tree) and inside a larger subformula beginning with $\exists x_1$ (namely the whole formula). To which quantifier is this occurrence of x_1 attached? An answer is given by the following definition.

Definition   *The quantifier which binds a variable*

Suppose that an occurrence of the variable x_i is bound within a formula ϕ. This occurrence must become bound at an earliest stage when working up the construction tree of ϕ, in a subformula ψ of the form either $\forall x_i \theta$ or $\exists x_i \theta$. We then say that the occurrence of x_i is *bound by* the quantifier which is the principal connective of ψ.

So in our example above, the occurrence of x_1 in the atomic subformula $R_{1,1}(x_1)$ is bound by the $\forall x_i$ in the subformula $\forall x_1 R_{1,1}(x_1)$, not the $\exists x_1$ higher up the tree.

Before we ask you to do some exercises identifying free and bound variables in formulas, let's indicate the point of this idea. First of all, when we want to define whether a structure $\mathcal{A} = \langle A, \ldots \rangle$ satisfies a formula ϕ, we now know how to identify which variables are free in ϕ and will thus require specific interpretation by elements of the domain A. Secondly, our explanation of when $\mathcal{A}$ satisfies a formula of the form $\forall x_i \psi$ will reduce this to deciding whether the shorter formula ψ is satisfied in $\mathcal{A}$ for all possible interpretations in A of the free variable x_i in ψ. So, for instance, the truth in $\mathcal{A}$ of the formula which we have just looked at,

$$\exists x_1 (\forall x_3 R_{2,2}(x_2, x_3) \vee (x_1 = x_3 \to \forall x_1 R_{1,1}(x_1))),$$

This answer will be reflected in how we give meaning to the symbols by our definition of when a structure satisfies a formula, as you will see later in Exercise 4.18.

Equivalently, working one's way down the construction tree, the occurrence of x_i is bound by the quantifier lowest down the tree within the scope of which this occurrence of x_i lies.

Of course, to work out the truth of the formula, we also need a specific interpretation of the variable x_2, which is free in the formula.

will be determined by looking at the truth of the shorter formula

$$(\forall x_3 R_{2,2}(x_2, x_3) \lor (x_1 = x_3 \rightarrow \forall x_1 R_{1,1}(x_1)))$$

for interpretations of the variable x_1, which has one free occurrence (appearing in the subformula $x_1 = x_3$).

Indeed, we shall be looking for at least one interpretation of x_1 in A which makes the formula true in $\mathcal{A}$, as we are going to interpret $\exists x_1$ by 'there exists an x_1'.

Exercise 4.11

For each of the following formulas, classify all occurrences of variables as free or bound.

(a) $\exists x_3 (\forall x_2 R_{1,2}(f_{3,1}(x_1, x_3, x_2)) \leftrightarrow (\forall x_1 R_{2,1}(\mathbf{c}_2, x_1) \land \forall x_5 \, f_{2,4}(x_3, x_2) = x_1))$

(b) $(\forall x_5 (f_{1,1}(x_2) = x_5 \lor \exists x_2 R_{2,1}(x_5, x_2)) \rightarrow \exists x_3 (\forall x_1 R_{2,1}(x_1, x_2) \lor x_1 = x_3))$

Solution

(a) The bound occurrences of variables are those underlined in the formula below.

$$\exists \underline{x_3} (\forall \underline{x_2} R_{1,2}(f_{3,1}(x_1, \underline{x_3}, \underline{x_2})) \leftrightarrow (\forall \underline{x_1} R_{2,1}(\mathbf{c}_2, \underline{x_1}) \land \forall \underline{x_5} \, f_{2,4}(\underline{x_3}, x_2) = x_1))$$

Note that the $\forall x_5$ doesn't create any other bound variables. When a formula is constructed as $\forall x_i \phi$, there is no reason why ϕ has to contain free occurrences of x_i.

(b) Not given.

Of course, for most mathematically interesting formulas of the form $\forall x_i \phi$, ϕ would contain free occurrences of x_i. But our set-up doesn't demand that formulas should be interesting!

Exercise 4.12

Write down the scope of each quantifier in the formula below and for each occurrence of a bound variable, indicate the quantifier by which it is bound. (The formula involves a relation symbol R of 2 arguments and variables x, y, z.)

$$\forall y (\exists x (R(y, z) \rightarrow \exists y R(x, y)) \land \neg \forall z R(x, y))$$

We can now explain when a formula is true in a structure. As a formula will in general include variables and more complicated terms, we need to define how terms are interpreted in a structure. Terms are built up from variables and constant symbols. The essence of a constant symbol $\mathbf{c}$ is that a structure $\mathcal{A} = \langle A, \dots \rangle$ includes a specified interpretation of this symbol by an element $\mathbf{c}^A$ of the domain A, and once this is specified, it's fixed forever – that's why the symbol is called a constant! But the variables, as the name suggests, can be interpreted by any elements of A. So we must specify how each one is to be interpreted and then this will then determine the interpretation of every

Some books use the word *valuation* for what we call an interpretation of all the variables.

more complicated term, as per the following definition.

Definition   Interpretation of terms

Let $\mathcal{A} = \langle A, \ldots \rangle$ be a structure for a language L. Suppose that the variables $x_1, x_2, x_3, \ldots$ are interpreted respectively by elements $a_1, a_2, a_3, \ldots$ of A. We shall abbreviate this interpretation by $\vec{x}/\vec{a}$. Then the interpretation in $\mathcal{A}$ of each term τ of the language under this interpretation of the variables, which we shall write as $\tau[\vec{x}/\vec{a}]^{\mathcal{A}}$, is defined recursively as follows.

1a   For each variable x_i, we define $x_i[\vec{x}/\vec{a}]^{\mathcal{A}} = a_i$.

1b   For each constant symbol $\mathbf{c}_k$, we define $\mathbf{c}_k[\vec{x}/\vec{a}]^{\mathcal{A}} = \mathbf{c}_k^{\mathcal{A}}$.

2   If $f_{n,m}$ is a function symbol in L and $\tau_1, \tau_2, \ldots, \tau_n$ are terms, then

$$f_{n,m}(\tau_1, \tau_2, \ldots, \tau_n)[\vec{x}/\vec{a}]^{\mathcal{A}} = f_{n,m}^{\mathcal{A}}\left(\tau_1[\vec{x}/\vec{a}]^{\mathcal{A}}, \tau_2[\vec{x}/\vec{a}]^{\mathcal{A}}, \ldots, \tau_n[\vec{x}/\vec{a}]^{\mathcal{A}}\right).$$

> In practice, a term involves only finitely many variables and we only need the interpretation of these. We shall still use the notation $\vec{x}/\vec{a}$ in this case.

As an example, let the language L consist of the function symbols $f_{1,1}, f_{2,1}$ and the constant symbol $\mathbf{c}_1$, along with equality. Let $\mathcal{A}$ be the structure $\langle \mathbb{N}, f_{1,1}^{\mathcal{A}}, +, 5, = \rangle$, where $f_{1,1}^{\mathcal{A}}(n) = n^2$, for all $n \in \mathbb{N}$. Suppose that the variables $x_1, x_2, x_3, \ldots$, are interpreted respectively by $1, 2, 3, \ldots$, which we shall write as $\vec{a}$. Then

> We hope that it is obvious in the way we have described the structure $\mathcal{A}$ that $+$ stands for the standard function of two variables on $\mathbb{N}$ interpreting the function symbol $f_{2,1}$ of two variables and that 5 is the interpretation of the constant symbol $\mathbf{c}_1$.

$$
\begin{aligned}
f_{2,1}(\mathbf{c}_1, f_{1,1}(x_6))[\vec{x}/\vec{a}]^{\mathcal{A}} &= f_{2,1}^{\mathcal{A}}(\mathbf{c}_1[\vec{x}/\vec{a}]^{\mathcal{A}}, f_{1,1}(x_6)[\vec{x}/\vec{a}]^{\mathcal{A}}) \\
&= f_{2,1}^{\mathcal{A}}(\mathbf{c}_1^{\mathcal{A}}, f_{1,1}^{\mathcal{A}}(x_6[\vec{x}/\vec{a}]^{\mathcal{A}})) \\
&= f_{2,1}^{\mathcal{A}}(5, f_{1,1}(6)) \quad (\text{as } a_6 = 6) \\
&= f_{2,1}^{\mathcal{A}}(5, 6^2)) \\
&= 5 + 36 = 41.
\end{aligned}
$$

Exercise 4.13 ────────────────────────

Take the same set-up as above, so that L is the language consisting of the function symbols $f_{1,1}, f_{2,1}$ and the constant symbol $\mathbf{c}_1$, along with equality; $\mathcal{A}$ is the structure $\langle \mathbb{N}, f_{1,1}^{\mathcal{A}}, +, 5, = \rangle$, where $f_{1,1}^{\mathcal{A}}(n) = n^2$, for all $n \in \mathbb{N}$; and the variables $x_1, x_2, x_3, \ldots$, are interpreted respectively by $1, 2, 3, \ldots$, which we shall write as $\vec{a}$. Find the values of $\tau[\vec{x}/\vec{a}]^{\mathcal{A}}$ for each of the following terms τ.

(a) x_4

(b) $f_{2,1}(x_3, \mathbf{c}_1)$

(c) $f_{2,1}(f_{1,1}(x_2), f_{2,1}(x_8, f_{1,1}(\mathbf{c}_1)))$

For our main definition we shall need a variation of the $\tau[\vec{x}/\vec{a}]^{\mathcal{A}}$ notation to cover the case where we are given an interpretation $\vec{a}$ of the variables and want to alter the interpretation of one of the variables, say x_i, to the element b of A.

Definition   *Changing the interpretation of a term*

We shall write $\vec{x}/\vec{a}[x_i/b]$ to signify that the interpretation of the variable x_i in the interpretation $\vec{a}$ of the variables has been changed (from a_i) to b.

So $\vec{x}/\vec{a}[x_i/b]$ replaces a_i by b in the ith place.

So with the interpretation $\vec{a}$ in Exercise 4.13, $\vec{x}/\vec{a}[x_2/87]$ stands for the interpretation of the variables $x_1, x_2, x_3, x_4, \ldots$, respectively as $1, 87, 3, 4 \ldots$, while repeated use of the notation gives examples like

$$\vec{x}/\vec{a}[x_2/87][x_1/9][x_2/1008]$$

which interprets the variables respectively as $9, 1008, 3, 4, \ldots$. Just in case you are wondering why we have this ghastly piece of notation, bear in mind that we shall have to explain when a formula like

$$(x_2 = f_{1,1}(x_4) \land \forall x_2\, f_{1,1}(x_2) = f_{1,1}(x_8))$$

is true in a structure $\mathcal{A} = \langle A, \ldots \rangle$. To make the question of whether this formula is true or false meaningful, we need to specify an interpretation of the free variables within it. We can take an interpretation $\vec{a}$ of all possible variables, but you can see that here we are only interested in the interpretation of x_2, x_4 and x_8. Inevitably we shall then have to decide whether, with this interpretation of these variables, both of

$$x_2 = f_{1,1}(x_4)$$

and

$$\forall x_2\, f_{1,1}(x_2) = f_{1,1}(x_8)$$

are true in $\mathcal{A}$. The specified interpretation of x_2 matters for the first of these formulas, but is irrelevant for the second. For the second of these formulas we shall need to override the interpretation of x_2 by considering its interpretation of all possible b in A to see whether $f_{1,1}(x_2) = f_{1,1}(x_8)$ is true in all of these cases, with the original specified interpretation of x_8.

Now for the main definition – at last! The definition connects structures, sets with relations and functions on them out in the real mathematical world, with strings of symbols which conform to rules for being a formula. The definition gives meanings to the individual symbols in the language and connects these strings with properties of any structure in which they are then true.

The formulas are the syntax of the language and the structures which interpret them according to this definition give the semantics of the language.

Just as the driver of a truth assignment v satisfying a propositional formula is the values v gives to the propositional variables, the key part of this definition is which atomic formulas are true in a structure for each possible interpretation of the variables $\vec{x}$. The truth of more complicated formulas then follows using our earlier truth table rules for the propositional connectives and the obvious interpretations of the symbols $\forall$ and $\exists$, namely that $\forall x_i \phi$ is true in $\mathcal{A}$ if and only if ϕ is true in $\mathcal{A}$ for all $b \in A$, and a similar interpretation of $\exists$ by 'there exists'.

Definitions Satisfaction

Let $\mathcal{A} = \langle A, \ldots, f_{n,m}^{\mathcal{A}}, \ldots, \ldots R_{n,m}^{\mathcal{A}} \ldots, \ldots \mathbf{c}_k^{\mathcal{A}} \ldots \rangle$ be a structure for a language L (with $\mathcal{A}$ normal if L is a language with equality). Suppose that the variables $x_1, x_2, x_3, \ldots$, are interpreted respectively by $a_1, a_2, a_3, \ldots$, which as before we'll abbreviate as $\vec{x}/\vec{a}$. Let ϕ be a formula of L. The relation

$$\mathcal{A} \vDash_{\vec{x}/\vec{a}} \phi,$$

which we shall read as

> the structure $\mathcal{A}$ *satisfies* the formula ϕ when $x_1, x_2, x_3, \ldots$, are interpreted by $a_1, a_2, a_3, \ldots$,

is defined recursively on the construction of ϕ as follows.

1. Atomic formulas:

   (a) for each relation symbol $R_{n,m}$ in L and terms $\tau_1, \tau_2, \ldots, \tau_n$,
   $\mathcal{A} \vDash_{\vec{x}/\vec{a}} R_{n,m}(\tau_1, \tau_2, \ldots, \tau_n)$ if and only if

   $$(\tau_1[\vec{x}/\vec{a}]^{\mathcal{A}}, \tau_2[\vec{x}/\vec{a}]^{\mathcal{A}}, \ldots, \tau_n[\vec{x}/\vec{a}]^{\mathcal{A}}) \in R_{n,m}^{\mathcal{A}};$$

   (b) if τ_1, τ_2 are terms, then $\mathcal{A} \vDash_{\vec{x}/\vec{a}} \tau_1 = \tau_2$ if and
   only if $\tau_1[\vec{x}/\vec{a}]^{\mathcal{A}} = \tau_2[\vec{x}/\vec{a}]^{\mathcal{A}}$.

2. For any formula of one of the forms $\neg\phi$, $(\phi \wedge \psi)$, $(\phi \vee \psi)$,
   $(\phi \rightarrow \psi)$, $(\phi \leftrightarrow \psi)$, truth tables laws are followed, e.g.

   > $\mathcal{A} \vDash_{\vec{x}/\vec{a}} \neg\phi$ if and only if it is not the case that $\mathcal{A} \vDash_{\vec{x}/\vec{a}} \phi$;
   > $\mathcal{A} \vDash_{\vec{x}/\vec{a}} (\phi \vee \psi)$ if and only if $\mathcal{A} \vDash_{\vec{x}/\vec{a}} \phi$ or $\mathcal{A} \vDash_{\vec{x}/\vec{a}} \psi$.

3. $\mathcal{A} \vDash_{\vec{x}/\vec{a}} \forall x_i \phi$ if and only if for all $b \in A$, $\mathcal{A} \vDash_{\vec{x}/\vec{a}[x_i/b]} \phi$.

4. $\mathcal{A} \vDash_{\vec{x}/\vec{a}} \exists x_i \phi$ if and only if there is some $b \in A$ for
   which $\mathcal{A} \vDash_{\vec{x}/\vec{a}[x_i/b]} \phi$.

When $\mathcal{A} \vDash_{\vec{x}/\vec{a}} \phi$, we shall also use terminology like

> the formula ϕ *is true in*, or *is satisfied by*, the structure $\mathcal{A}$ when $x_1, x_2, x_3, \ldots$ are interpreted by $a_1, a_2, a_3, \ldots$.

These definitions stem from the work of Alfred Tarski (1902–83), the founder of model theory, to which Chapter 6 is a brief introduction.

The symbol $\vDash$ is used both for a structure satisfying a formula and for logical consequence, $\Gamma \vDash \phi$. This may seem potentially confusing, but the context usually makes the intended meaning clear.

1(b) is relevant when L is a language with equality and just says that the interpretations of the terms are equal.

For a given formula ϕ, it will turn out that we only need the interpretation of any free variables in ϕ, rather than all variables in the language. See Theorem 4.1 later in the section.

Although the underlying idea behind this definition is very straightforward, it is quite elaborate when written down in detail. The definition permits a systematic approach to interpreting complicated formulas in a structure and deciding whether or not they are true. When we look at the formal proof system which provides one of our frameworks for establishing when one formula is a consequence of another, the formal system manipulates symbols and formulas mechanically, just as strings of symbols obey rules about allowable shapes, not knowing how the symbols are to be interpreted. Our definition above will prove to be very important in nailing down suitable axioms and rules of inference for this formal system – it turns out that we'll have to be very careful about rules for manipulating the quantifiers.

As an example of using this definition, let's take the set-up of Exercise 4.13 above. So L is a language with equality, the function symbols $f_{1,1}, f_{2,1}$ and the constant symbol $\mathbf{c}_1$; $\mathcal{A}$ is the structure $\langle \mathbb{N}, f_{1,1}^{\mathcal{A}}, +, 5, = \rangle$, where $f_{1,1}^{\mathcal{A}}(n) = n^2$, for all $n \in \mathbb{N}$; and the variables $x_1, x_2, x_3, \ldots$, are interpreted respectively by $1, 2, 3, \ldots$, which we shall write as $\vec{a}$.

Recall our convention that all structures for a language with equality are normal.

First let us investigate whether $\mathcal{A} \vDash_{\vec{x}/\vec{a}} f_{1,1}(x_3) = f_{2,1}(x_4, \mathbf{c}_1)$. This is an atomic formula and thus holds if and only if

$$f_{1,1}(x_3)[\vec{x}/\vec{a}]^{\mathcal{A}} = f_{2,1}(x_4, \mathbf{c}_1)[\vec{x}/\vec{a}]^{\mathcal{A}},$$

that is, if and only if

$$3^2 = 4 + 5,$$

which is true. Thus it is the case that $\mathcal{A} \vDash_{\vec{x}/\vec{a}} f_{1,1}(x_3) = f_{2,1}(x_4, \mathbf{c}_1)$.

If we slightly tweak the formula in the above example and ask whether

$$\mathcal{A} \vDash_{\vec{x}/\vec{a}} f_{1,1}(x_3) = f_{2,1}(x_7, x_8),$$

we hope that this is pretty obviously false (as $3^2 \neq 7 + 8$). We'll describe this by saying that $\mathcal{A}$ doesn't satisfy the given formula with the given interpretation of the variables and also write it using a slash through the $\vDash$ symbol, which here would get written as

$$\mathcal{A} \nvDash_{\vec{x}/\vec{a}} f_{1,1}(x_3) = f_{2,1}(x_7, x_8).$$

Notation: $\nvDash$ stands for 'doesn't satisfy'. It also stands for 'is not a logical consequence' – the context will make the meaning clear!

Let's look at something more complicated than an atomic formula, in particular something involving a quantifier. Is it the case that

$$\mathcal{A} \vDash_{\vec{x}/\vec{a}} \forall x_7 \, f_{1,1}(x_3) = f_{2,1}(x_7, x_8)?$$

As the principal connective is $\forall x_7$, this holds if and only if

$$\text{for all } b \in \mathbb{N}, \mathcal{A} \vDash_{\vec{x}/\vec{a}[x_7/b]} f_{1,1}(x_3) = f_{2,1}(x_7, x_8).$$

The notation $\vec{x}/\vec{a}[x_7/b]$ says that for each i except 7, the variable x_i continues to be interpreted by the element $a_i = i$ in the domain $\mathbb{N}$ of $\mathcal{A}$, while x_i is now interpreted by the natural number b. Plainly it is not the case that $3^2 = b + 8$ holds for all $b \in \mathbb{N}$, so that

$$\mathcal{A} \nvDash_{\vec{x}/\vec{a}} \forall x_7 \, f_{1,1}(x_3) = f_{2,1}(x_7, x_8).$$

You may have noticed in this example that when testing whether

$$\mathcal{A} \vDash_{\vec{x}/\vec{a}} \forall x_i \phi,$$

we have to check whether for all $b \in A$,

$$\mathcal{A} \vDash_{\vec{x}/\vec{a}[x_i/b]} \phi,$$

where the notation $\vec{x}/\vec{a}[x_i/b]$ means that the value b overrides whatever value happened to be given to the variable x_i in $\vec{a}$. The same sort of thing happens when testing whether $\mathcal{A} \vDash_{\vec{x}/\vec{a}} \exists x_i \phi$, as you will see in the next exercise. In both cases, the original value of x_i is irrelevant when the the x_is appearing in ϕ are bound.

Exercise 4.14

Does $\mathcal{A} \vDash_{\vec{x}/\vec{a}} \exists x_7 \; f_{1,1}(x_3) = f_{2,1}(x_7, x_8)$? (As before, the interpretation $\vec{a}$ is $1, 2, 3, \ldots$.)

Solution

This time the principal connective is $\exists x_7$. So $\mathcal{A} \vDash_{\vec{x}/\vec{a}} \exists x_7 \; f_{1,1}(x_3) = f_{2,1}(x_7, x_8)$ holds if and only if

there is some $b \in \mathbb{N}, \mathcal{A} \vDash_{\vec{x}/\vec{a}[x_7/b]} f_{1,1}(x_3) = f_{2,1}(x_7, x_8)$,

that is, there is some $b \in \mathbb{N}$ for which $3^2 = b + 8$. As there is such a b, namely $b = 1$, it is the case that $\mathcal{A} \vDash_{\vec{x}/\vec{a}} \exists x_7 \; f_{1,1}(x_3) = f_{2,1}(x_7, x_8)$.

You may have noticed in the last exercise that although the original question was whether $\exists x_7 \; f_{1,1}(x_3) = f_{2,1}(x_7, x_8)$ is satisfied in $\mathcal{A}$ with a particular interpretation of the variables including interpreting x_7 as 7, the definition of $\vDash$ for a formula beginning with $\exists x_7$ completely disregards the original interpretation of the variable. This was also the case in the preceding worked example when the formula began with $\forall x_7$. This is typical when resolving the truth of formulas beginning with $\forall x$ or $\exists x$. In general the only variables in a formula which need an interpretation by elements of the domain are those appearing in it which are free. Of course, variables can occur free in some subformulas and bound in others, so you have to remain alert.

So the way the definition of $\vDash$ works, if the question is whether $\forall x \phi$, or respectively $\exists x \phi$, is satisfied when x is interpreted by a particular element of the domain, one completely ignores this particular element and looks only at whether ϕ is satisfied by all, or respectively some, interpretations of x by elements of the domain, not necessarily having much to do with the particular one given.

Exercise 4.15

Does $\mathcal{A} \vDash_{\vec{x}/\vec{a}} (\forall x_7 \; f_{1,1}(x_3) = f_{2,1}(x_7, x_8) \vee x_7 = f_{2,1}(x_3, x_4))$, where as before the interpretation $\vec{a}$ is $1, 2, 3, \ldots$?

Solution

As the principal connective of the formula is $\vee$, it is satisfied under the given interpretation if

$\mathcal{A} \vDash_{\vec{x}/\vec{a}} \forall x_7 \; f_{1,1}(x_3) = f_{2,1}(x_7, x_8)$ or $\mathcal{A} \vDash_{\vec{x}/\vec{a}} x_7 = f_{2,1}(x_3, x_4)$.

Note that x_7 is bound in the first of the subformulas and free in the second. We know from earlier that $\mathcal{A} \nvDash_{\vec{x}/\vec{a}} \forall x_7 \; f_{1,1}(x_3) = f_{2,1}(x_7, x_8)$. But as the interpretations of x_7, x_3, x_4 are respectively 7, 3, 4, we do have $\mathcal{A} \vDash_{\vec{x}/\vec{a}} x_7 = f_{2,1}(x_3, x_4)$, so that

$\mathcal{A} \vDash_{\vec{x}/\vec{a}} (\forall x_7 \; f_{1,1}(x_3) = f_{2,1}(x_7, x_8) \vee x_7 = f_{2,1}(x_3, x_4))$.

To work out the truth value of a propositional formula under a truth assignment, one often builds up from the inside, obtaining truth values for successively bigger subformulas. But for first-order formulas, one is likelier to work one's way in from the outside.

Exercise 4.16

Let L and $\mathcal{A}$ be the same language and structure as in the discussion above, but now let $\vec{a}$ be *any* interpretation in the domain $\mathbb{N}$ of the variables $x_1, x_2, x_3, \ldots$. Let b and c be any elements of $\mathbb{N}$. Show that

$\mathcal{A} \vDash_{\vec{x}/\vec{a}[x_1/b]} \forall x_7 \; f_{1,1}(x_3) = f_{2,1}(x_7, x_8)$ if and only if

$\mathcal{A} \vDash_{\vec{x}/\vec{a}[x_1/c]} \forall x_7 \; f_{1,1}(x_3) = f_{2,1}(x_7, x_8)$.

We hope that it is obvious that the same sort of result would have held in Exercise 4.16 for all other variables x_i (as well as the x_1) except for x_3 and x_8 which appear free in the formula. This exercise illustrates the more general result that whether or not $\mathcal{A} \vDash_{\vec{x}/\vec{a}} \phi$ for a given ϕ and sequence $\vec{a}$ of elements in the domain of $\mathcal{A}$ doesn't depend on the interpretation within $\vec{a}$ of variables x_i which don't appear in ϕ as free variables. It may seem a bit grand, but we'll state this as a theorem.

Theorem 4.1

Suppose that $\mathcal{A} = \langle A, \ldots \rangle$ is a structure for a language L and that the sequence $\vec{a}$ of elements of A interprets the variables $x_1, x_2, x_3, \ldots$. Let ϕ be a formula of L in which x_i does not appear as a free variable and let $b, c \in A$. Then

$$\mathcal{A} \vDash_{\vec{x}/\vec{a}[x_i/b]} \phi \text{ if and only if } \mathcal{A} \vDash_{\vec{x}/\vec{a}[x_i/c]} \phi.$$

The proof of this and similar 'obvious' results can require considerable persistence, though not necessarily great inspiration. We shall leave it to the end of the section.

As a consequence of Theorem 4.1, when we want to know whether $\mathcal{A} \vDash_{\vec{x}/\vec{a}} \phi$, the only variables for which we need to specify an interpretation are the variables which appear free in ϕ. Indeed, if ϕ has no free variables, we don't need to specify the interpretation of any of the variables. Such a formula merits a special name. We will thus introduce an extra definition and variants of our notation as follows.

Definition and notation

A formula with no free variables is called a *sentence*.

If the formula ϕ has free variables included in the list $x_1, x_2, \ldots, x_n$, we can indicate this by writing $\phi(x_1, x_2, \ldots, x_n)$. If in addition $a_1, a_2, \ldots, a_n$ are elements of the domain of a structure $\mathcal{A}$, we shall write

$$\mathcal{A} \vDash_{x_1/a_1, x_2/a_2, \ldots, x_n/a_n} \phi$$

and

$$\mathcal{A} \vDash_{x_1/a_1, x_2/a_2, \ldots, x_n/a_n} \phi(x_1, x_2, \ldots, x_n)$$

to indicate that $\mathcal{A}$ satisfies ϕ when $x_1, x_2, \ldots, x_n$ are interpreted by $a_1, a_2, \ldots, a_n$.

This doesn't mean that every one of $x_1, x_2, \ldots, x_n$ appears free in ϕ, only that any free variable is amongst these.

When ϕ is a sentence, so there are no free variables in ϕ, we can write simply $\mathcal{A} \vDash \phi$.

In some books, a sentence is called a *closed formula*.

When we come to axiomatizing mathematical theories, we shall normally use sentences rather than formulas with free variables. This is because the axioms typically express some property of all elements x of a set or assert the existence of an element x with a special property. It's thus natural for the corresponding formula to bind the x with a $\forall$ or $\exists$ as appropriate.

The next exercise asks you to look at some slightly more complicated formulas. So far we have only looked at formulas containing at most one quantifier. When we look at formulas containing more than one, say

$$\forall x_1 \exists x_2 \forall x_3 \phi,$$

we need to take care over the letters we use to represent elements of the structure interpreting x_1, x_2, x_3 as we successively strip off the quantifiers. Let $\mathcal{A} = \langle A, \ldots \rangle$ be a structure for the language and $\vec{a}$ a sequence of elements of the domain A interpreting the variables $\vec{x}$. Then

$$\mathcal{A} \vDash_{\vec{x}/\vec{a}} \forall x_1 \exists x_2 \forall x_3 \phi$$

if and only if for all $b \in A$,

$$\mathcal{A} \vDash_{\vec{x}/\vec{a}[x_1/b]} \exists x_2 \forall x_3 \phi.$$

This holds if and only if for all $b \in A$, there is some $c \in A$ such that

$$\mathcal{A} \vDash_{\vec{x}/\vec{a}[x_1/b][x_2/c]} \forall x_3 \phi.$$

Notice that we have chosen the letter c, different from the letter b, for the element of A that depends on b. This doesn't mean that element c isn't in fact equal to b. It does however allow for the possibility that c might not be equal to b. Strictly speaking, in our definition of when a formula beginning with a quantifier is satisfied in a structure, we should have emphasised that the element of the domain interpreting the quantified variable will not necessarily be one of the elements previously given as interpreting one of the variables. So, taking our example one stage further, we have

$$\mathcal{A} \vDash_{\vec{x}/\vec{a}} \forall x_1 \exists x_2 \forall x_3 \phi$$

if and only if for all $b \in A$, there is some $c \in A$ such that for all $d \in A$,

$$\mathcal{A} \vDash_{\vec{x}/\vec{a}[x_1/b][x_2/c][x_3/d]} \phi.$$

In general, checking this for a typical structure $\mathcal{A}$ could be quite challenging! If you have ever studied real analysis, you might recognize that the complexity of the quantifiers is at the same level as checking whether the limit of an infinite sequence $\{y_n\}$ as $n \to \infty$ equals l, the usual definition of which is

for all $\varepsilon > 0$, there is an N such that for all n, if $n \geq N$ then $|y_n - l| < \varepsilon$.

This complexity might lead one to suspect that deciding whether a given ϕ is satisfied in a structure $\mathcal{A}$ is going to be very difficult, especially if the domain of $\mathcal{A}$ is infinite.

Exercise 4.17 ───

As earlier, let $\mathcal{A}$ be the structure $\langle \mathbb{N}, f_{1,1}^{\mathcal{A}}, +, 5, = \rangle$, where $f_{1,1}^{\mathcal{A}}(n) = n^2$, for all $n \in \mathbb{N}$. For each of the following formulas ϕ, decide whether

$$\mathcal{A} \vDash_{x_1/2, x_2/7, x_3/2, x_4/3} \phi.$$

In each case give your answer in a fairly full way, explaining each use of a part of the definition of $\mathcal{A} \vDash_{\vec{x}/\vec{a}} \phi$.

We won't ask you to give such full explanations in most later exercises! But it is important that you know in principle how to give such explanations.

(a) $\forall x_1 \exists x_2\ f_{1,1}(x_1) = x_2$

(b) $\forall x_2 \exists x_1\ f_{1,1}(x_1) = x_2$

(c) $(\exists x_3\ f_{1,1}(x_3) = \mathbf{c}_1 \to f_{2,1}(x_1, x_4) = x_2)$

(d) $\forall x_2 (f_{2,1}(x_2, x_1) = \mathbf{c}_1 \to x_2 = x_4)$

Solution

We shall do the solution to part (a) and won't give the details of the other solutions, but hope that you obtained the answers as to whether $\mathcal{A} \vDash_{\bar{x}/\bar{a}} \phi$ as (b) no and (c) and (d) yes. For (a) we shall argue as follows. We have

$$\mathcal{A} \vDash_{x_1/2,x_2/7,x_3/2,x_4/3} \forall x_1 \exists x_2 \; f_{1,1}(x_1) = x_2$$

if and only if for all $b \in \mathbb{N}$,

$$\mathcal{A} \vDash_{x_1/2,x_2/7,x_3/2,x_4/3[x_1/b]} \exists x_2 \; f_{1,1}(x_1) = x_2,$$

which is the same as

$$\mathcal{A} \vDash_{x_1/b,x_2/7,x_3/2,x_4/3} \exists x_2 \; f_{1,1}(x_1) = x_2.$$

This holds if and only if for all $b \in \mathbb{N}$ there is some $c \in \mathbb{N}$ such that

$$\mathcal{A} \vDash_{x_1/b,x_2/7,x_3/2,x_4/3[x_2/c]} f_{1,1}(x_1) = x_2,$$

which is the same as

$$\mathcal{A} \vDash_{x_1/b,x_2/c,x_3/2,x_4/3} f_{1,1}(x_1) = x_2.$$

As $f_{1,1}(x_1) = x_2$ is an atomic formula, this holds if and only if for all $b \in \mathbb{N}$ there is some $c \in \mathbb{N}$ such that

$$f_{1,1}^{\mathcal{A}}(b) = c,$$

that is,

$$b^2 = c.$$

This is plainly true, as for each $b \in \mathbb{N}$ we can take c to equal the natural number b^2.

While our main interest is in the interpretation of elegantly constructed formulas, the definition tells us how to interpret inelegant and confusing formulas, like

$$\forall x_1 \exists x_1 R_{1,1}(x_1).$$

Which of the quantifiers really dominates this? The $\forall x_1$ or the $\exists x_1$? There may not be any natural answer to this problem, but the definition forces our hand (and matches our earlier definition of which quantifier binds a bound variable).

Exercise 4.18

Let the language L have only the relation symbol $R_{1,1}$. Does the structure $\langle \mathbb{N}, \{n \in \mathbb{N} : n \text{ is even}\}\rangle$ satisfy the formula $\forall x_1 \exists x_1 R_{1,1}(x_1)$?

We can generalize the result we hope you obtained in the last exercise by use of Theorem 4.1, which we shall now restate and prove.

Theorem 4.1

Suppose that $\mathcal{A} = \langle A, \ldots \rangle$ is a structure for a language L and that the sequence $\vec{a}$ of elements of A interprets the variables $x_1, x_2, x_3, \ldots$. Let ϕ be a formula of L in which x_i does not appear as a free variable and let $b, c \in A$. Then

$$\mathcal{A} \models_{\vec{x}/\vec{a}[x_i/b]} \phi \text{ if and only if } \mathcal{A} \models_{\vec{x}/\vec{a}[x_i/c]} \phi.$$

Proof

This is a general result for all formulas ϕ and the basic method for proving such a result, as with so many earlier results about propositional formulas, is by mathematical induction on the length of ϕ. As ever, there are several sensible measures of the *length* of a formula: we will take the number of connectives and quantifiers. We shall phrase the induction hypothesis as follows. For all formulas ϕ of L with $\leq n$ connectives and quantifiers in which the variable x_i does not appear as a free variable, all structures $\mathcal{A} = \langle A, \ldots \rangle$ for L, and all interpretations $\vec{a}$ and $b, c \in A$,

$$\mathcal{A} \models_{\vec{x}/\vec{a}[x_i/b]} \phi \text{ if and only if } \mathcal{A} \models_{\vec{x}/\vec{a}[x_i/c]} \phi.$$

For $n = 0$, ϕ is an atomic formula, of one of the forms $R_{n,m}(\tau_1, \tau_2, \ldots, \tau_n)$ or $\tau_1 = \tau_2$, where the τ_i are terms, none of which involve the variable x_i. An easy induction on the length of terms not involving x_i (with length measured by e.g. the number of function symbols used in a term) shows that

$$\tau[\vec{x}/\vec{a}[x_i/b]]^{\mathcal{A}} = \tau[\vec{x}/\vec{a}[x_i/c]]^{\mathcal{A}} \text{ (and both equal } \tau[\vec{x}/\vec{a}]^{\mathcal{A}}),$$

so that in the case that ϕ is $R_{n,m}(\tau_1, \tau_2, \ldots, \tau_n)$, we have

$$(\tau_1[\vec{x}/\vec{a}[x_i/b]]^{\mathcal{A}}, \tau_2[\vec{x}/\vec{a}[x_i/b]]^{\mathcal{A}}, \ldots, \tau_n[\vec{x}/\vec{a}[x_i/b]]^{\mathcal{A}}) \in R_{n,m}^{\mathcal{A}}$$

if and only if

$$(\tau_1[\vec{x}/\vec{a}[x_i/c]]^{\mathcal{A}}, \tau_2[\vec{x}/\vec{a}[x_i/c]]^{\mathcal{A}}, \ldots, \tau_n[\vec{x}/\vec{a}[x_i/c]]^{\mathcal{A}}) \in R_{n,m}^{\mathcal{A}},$$

so that

$$\mathcal{A} \models_{\vec{x}/\vec{a}[x_i/b]} R_{n,m}(\tau_1, \tau_2, \ldots, \tau_n) \text{ if and only if}$$
$$\mathcal{A} \models_{\vec{x}/\vec{a}[x_i/c]} R_{n,m}(\tau_1, \tau_2, \ldots, \tau_n).$$

Similarly, if x_i doesn't appear in τ_1 or τ_2, we can show that

$$\mathcal{A} \models_{\vec{x}/\vec{a}[x_i/b]} \tau_1 = \tau_2 \text{ if and only if } \mathcal{A} \models_{\vec{x}/\vec{a}[x_i/c]} \tau_1 = \tau_2$$

(again with both of these holding if and only if $\mathcal{A} \models_{\vec{x}/\vec{a}} \tau_1 = \tau_2$, given that x_i doesn't appear in the terms).

Now assume that the hypothesis holds for all ϕ of length $\leq n$ in which x_i does not appear and that we have such a ϕ of length $n + 1$. We should really discuss all possible forms of ϕ, depending on its principal connective, but we shall only discuss the case when this principal connective is a universal quantifier, when ϕ is of the form $\forall x_j \psi$, with x_i not free in ϕ, and leave other

Which, if any, of these can arise obviously depends on what symbols are in the specific language L.

In an atomic formula, the free variables are simply those variables appearing in the formula. With no quantifiers around, all variables are free!

cases for you as an exercise. There are two cases for $\forall x_j \psi$, one where $j \neq i$ and the other where $j = i$.

In the case $j \neq i$, the variable x_i doesn't appear free in ψ and

$$\mathcal{A} \vDash_{\vec{x}/\vec{a}[x_i/b]} \forall x_j \psi$$

if and only if

$$\text{for all } d \in A, \ \mathcal{A} \vDash_{\vec{x}/\vec{a}[x_j/d][x_i/b]} \psi.$$

By the induction hypothesis (as ψ has length n and the hypothesis applies to *all* sequences $\vec{a}$, thus including the sequence $\vec{a}$ where a_j is replaced by d), this holds if and only if

$$\text{for all } d \in A, \ \mathcal{A} \vDash_{\vec{x}/\vec{a}[x_j/d][x_i/c]} \psi,$$

hence if and only if

$$\mathcal{A} \vDash_{\vec{x}/\vec{a}[x_i/c]} \forall x_j \psi,$$

as required.

In the case $j = i$, when we strip off the $\forall x_i$ the initial interpretation of x_i is overridden by a general $d \in A$, so we have both that

$$\mathcal{A} \vDash_{\vec{x}/\vec{a}[x_i/b]} \forall x_i \psi \text{ if and only if for all } d \in A, \ \mathcal{A} \vDash_{\vec{x}/\vec{a}[x_i/d]} \psi,$$

and

$$\mathcal{A} \vDash_{\vec{x}/\vec{a}[x_i/c]} \forall x_i \psi \text{ if and only if for all } d \in A, \ \mathcal{A} \vDash_{\vec{x}/\vec{a}[x_i/d]} \psi,$$

so that

$$\mathcal{A} \vDash_{\vec{x}/\vec{a}[x_i/b]} \forall x_i \psi \text{ if and only if } \mathcal{A} \vDash_{\vec{x}/\vec{a}[x_i/c]} \forall x_i \psi.$$

∎

Exercise 4.19

Fill in some of the detail in our proof of Theorem 4.1 by justifying the inductive step in the cases when the principal connective of ϕ of length $n + 1$ is one of $\wedge$, $\neg$ and an existential quantifier $\exists x_j$.

As the connectives $\wedge$ and $\neg$ are adequate, we can claim that the theorem has been proved, once one has successfully completed this exercise!

As one application of Theorem 4.1, we can show how the definition of satisfaction copes with a formula which begins with an essentially redundant quantifier, that is, has the form $\forall x \phi$ or $\exists x \phi$, where the variable x does not appear free in ϕ. Let's consider the case of $\forall x_i \phi$, where x_i does not appear free in ϕ, in a language L. We shall show that for any structure $\mathcal{A} = \langle A, \ldots \rangle$ for L and sequence $\vec{a}$ of elements of A interpreting the variables $x_1, x_2, x_3, \ldots$,

Such formulas are perfectly legal, but are a bit silly! It would be poor taste to write down such a formula with a redundant quantifier.

$$\text{if } \mathcal{A} \vDash_{\vec{x}/\vec{a}} \phi \quad \text{then } \mathcal{A} \vDash_{\vec{x}/\vec{a}} \forall x_i \phi.$$

We shall ask you to show the converse in Exercise 4.20.

Let a_i be the element of A interpreting x_i in the given sequence $\vec{a}$ and let c be any element of A. By Theorem 4.1, as x_i is not free in ϕ, we have

$$\mathcal{A} \vDash_{\vec{x}/\vec{a}} \phi \quad \text{if and only if } \mathcal{A} \vDash_{\vec{x}/\vec{a}[x_i/c]} \phi,$$

so that as

$$\mathcal{A} \vDash_{\vec{x}/\vec{a}} \phi,$$

In the statement of Theorem 4.1, if we take b to be the element a_i already interpreting x_i in $\vec{a}$, the sequence $\vec{a}[x_i/b]$ is simply the original sequence $\vec{a}$.

we have, for all $c \in A$, that

$$\mathcal{A} \vDash_{\bar{x}/\bar{a}[x_i/c]} \phi,$$

which gives

$$\mathcal{A} \vDash_{\bar{x}/\bar{a}} \forall x_i \phi,$$

as required.

Exercise 4.20

Suppose that $\mathcal{A} = \langle A, \ldots \rangle$ is a structure for a language L and that the sequence $\bar{a}$ of elements of A interprets the variables $x_1, x_2, x_3, \ldots$. Let ϕ be a formula of L in which x_i does not appear as a free variable. Show that

$$\text{if } \mathcal{A} \vDash_{\bar{x}/\bar{a}} \forall x_i \phi \quad \text{then } \mathcal{A} \vDash_{\bar{x}/\bar{a}} \phi.$$

We now have the basic definition of when a formula is satisfied by a structure. In the next section we shall look at formulas that are always true and some important equivalences between formulas.

Further exercises

Exercise 4.21

Let the language L consist of the one place relation symbol Q and the two place relation symbol R. Which of the following formulas are true in the given structures? Brief explanations will do.

(a) $\forall x \exists y (R(x,y) \wedge Q(y))$

(b) $\exists x \forall y R(x,y)$

(c) $\forall x (R(x,x) \rightarrow Q(x))$

(d) $\forall x \forall y \exists z ((R(x,z) \wedge R(z,y)) \vee \neg R(x,y))$

> Structures (giving the interpretation of Q before that of R):
>
> $$\mathcal{A} = \langle \mathbb{R}, \ \{x : x \text{ is rational}\}, < \rangle,$$
> $$\mathcal{B} = \langle \mathbb{R}, \ \{x : x \text{ is negative}\}, < \rangle,$$
> $$\mathcal{C} = \langle \mathbb{N}, \ \{x : x \text{ is even}\}, \ \{(x,y) : x \text{ divides } y\} \rangle.$$

In the context of natural numbers x, y in $\mathbb{N}$, 'x divides y' means that $y = kx$ for some $k \in \mathbb{N}$.

Exercise 4.22

Formalize the following statements about $\mathbb{N}$ using a language *without* equality with a 1-place relation symbol P, a 3-place relation symbol S and a constant symbol $\mathbf{c}$. The intended interpretation is the structure $\mathcal{A}$ with domain $\mathbb{N}$, $P^{\mathcal{A}} = \{x : x \text{ is prime}\}$, $S^{\mathcal{A}} = \{(x,y,z) : x + y = z\}$, $\mathbf{c}^{\mathcal{A}} = 2$. (You may abbreviate the formula of one part if you use it in a later part.)

(a) x equals 0. (Note that there is no symbol for 0 in the given language.)

(b) x is greater than y.

(c) Every even number greater than two is the sum of two primes.

(d) There are arbitrarily large prime pairs, where $\{a, b\}$ is said to be a *prime pair* if both a and b are prime and a and b differ by 2, e.g. $\{29, 31\}$ and $\{101, 103\}$.

Exercise 4.23

Let τ be a term in a language L involving the variable x_1. Let τ' be a term obtained by replacing some, maybe all, of the occurrences of x_1 in τ by the variable x_2. Let $\mathcal{A}$ be a structure for L and let $\vec{a}$ be an interpretation of the variables $\vec{x}$ by elements of the domain of $\mathcal{A}$. Show that if $a_1 = a_2$, then

$$\tau[\vec{x}/\vec{a}]^{\mathcal{A}} = \tau'[\vec{x}/\vec{a}]^{\mathcal{A}}.$$

For instance, if τ is the term $(f(x_1, \mathbf{c}, g(x_3, x_1), x_2)$, where f and g are respectively 4-place and 2-place function symbols and $\mathbf{c}$ is a constant symbol, then τ' might be $(f(x_1, \mathbf{c}, g(x_3, x_2), x_2)$.

[*Hints*: Use mathematical induction on the number k of function symbols in the term τ. The base case when $k = 0$ is in some sense the most interesting part of the argument, as this is where the replacement of an occurrence of x_1 by x_2 really happens!]

Exercise 4.24

Suppose that $\mathcal{A} = \langle A, \ldots \rangle$ is a structure for a language L and that the sequence $\vec{a}$ of elements of A interprets the variables $x_1, x_2, x_3, \ldots$. Let ϕ be a formula of L in which x_i does not appear as a free variable. Show that

$$\mathcal{A} \vDash_{\vec{x}/\vec{a}} \phi \quad \text{if and only if} \quad \mathcal{A} \vDash_{\vec{x}/\vec{a}} \exists x_i \phi.$$

[*Hint*: Use Theorem 4.1.]

Exercise 4.25

(For those with the background from e.g. set theory!)

(a) Suppose that the only function symbol contained in the first-order language L is $f_{1,1}$. Explain why the set of terms of the language is countable.

(b) Suppose that the first-order language L contains all of the function symbols $f_{1,1}\ f_{1,2},\ f_{1,3},\ \ldots,\ f_{2,1},\ f_{2,2},\ \ldots,\ f_{n,1},\ \ldots,\ f_{n,m}\ \ldots$, all of the relation symbols $R_{1,1},\ R_{1,2},\ R_{1,3},\ \ldots,\ R_{2,1},\ R_{2,2},\ \ldots,\ R_{n,1},\ \ldots,\ R_{n,m}\ \ldots$ and all of the constant symbols $\mathbf{c}_1,\ \mathbf{c}_2,\ \mathbf{c}_3,\ \ldots$.

(i) Explain why the set of terms of this language is countable.

(ii) Is the set of formulas of this language countable or uncountable? Explain your answer.

Exercise 4.26

Suppose that ϕ is a formula in L with free variables contained in the list $x_1, x_2, \ldots, x_n$ and that S is a symbol of L, for either a function or relation or constant, which does not appear in ϕ. Show that for any structure $\mathcal{A}$ for the language L and interpretation of the free variables $\vec{x}$ by elements $\vec{a}$, the truth of $\mathcal{A} \vDash_{[\vec{x}/\vec{a}]} \phi$ is independent of the interpretation $S^{\mathcal{A}}$ of S in $\mathcal{A}$. [*Hints*: Probably the most convincing method is to regard the result as one which holds for all formulas ϕ not involving the symbol S and then prove it by mathematical induction on the length of these formulas, for a fixed structure $\mathcal{A}$. One useful formulation of an induction hypothesis is as follows: the truth of $\mathcal{A} \vDash_{[\vec{x}/\vec{a}]} \phi$ is independent of $S^{\mathcal{A}}$, for all formulas ϕ not involving S of length $\leq n$ and all interpretations $\vec{a}$ of the free variables $\vec{x}$ – the last bit helps because if ϕ has subformulas of the form $\exists x \psi$ or $\forall x \psi$, the truth of $\mathcal{A} \vDash_{[\vec{x}/\vec{a}]} \phi$ usually involves investigating interpretations of $\vec{x}$ by elements besides those in the original sequence $\vec{a}$.]

Such an obvious result! So tedious to demonstrate!

4.3 Universally valid formulas and logical equivalence

In this section, we shall look at the extension of the idea of a tautology to formulas expressed in a first-order language. A tautology is a propositional formula which is true under all truth assignments, i.e. all interpretations of the language. The analogue for formulas written in a first-order language is a universally valid formula, as per the following definition.

Some authors use the phrase *logically valid* instead of universally valid.

Definition   Universally valid

Let $\phi(x_1, x_2, \ldots, x_n)$ be a formula in a language L with free variables contained in the list $x_1, x_2, \ldots, x_n$. We say that ϕ is *universally valid* if for all structures $\mathcal{A}$ for L (normal structures when L includes equality) and all interpretations of the variables $x_1, x_2, x_3, \ldots$, respectively by $a_1, a_2, a_3, \ldots$, in the domain of $\mathcal{A}$,

$$\mathcal{A} \vDash_{\vec{x}/\vec{a}} \phi(x_1, x_2, \ldots, x_n).$$

As before, we'll abbreviate interpreting the variables $x_1, x_2, x_3, \ldots$ respectively by $a_1, a_2, a_3, \ldots$ as $\vec{x}/\vec{a}$.

We hope that the importance of universally valid formulas is obvious – they correspond to our preconceptions of statements which are always true because of their shape, regardless of their interpretation.

A simple example of a universally valid formula (in a language including the 2-place relation symbol R) is

$$(\forall x_1 R(x_1, x_3) \rightarrow \forall x_1 R(x_1, x_3)).$$

This is essentially a disguised version of the tautology $(p \rightarrow p)$, with the formula $\forall x_1 R(x_1, x_3)$ substituted for the propositional variable p. It is universally valid because of the definition of $\vDash$ saying that $\mathcal{A} \vDash_{\vec{x}/\vec{a}} (\phi \rightarrow \psi)$ when it is the case that

if $\mathcal{A} \vDash_{\vec{x}/\vec{a}} \phi$ then $\mathcal{A} \vDash_{\vec{x}/\vec{a}} \psi$,

using the usual truth table laws for $\rightarrow$. In this case

if $\mathcal{A} \vDash_{\vec{x}/\vec{a}} \forall x_1 R(x_1, x_3)$ then $\mathcal{A} \vDash_{\vec{x}/\vec{a}} \forall x_1 R(x_1, x_3)$

is always true, precisely because the shape of this is 'if p then p' which is a tautology. The formula is satisfied regardless of how $\mathcal{A}$ actually interprets R or the value a_3 given to x_3.

This situation, where a formula is really a tautology involving various propositional variables disguised by substituting first-order formulas in a consistent manner, is worth a definition.

Definition   Substitution instance of tautology

Suppose that ϕ is a tautology which is built up from propositional variables in the list $p_1, p_2, \ldots, p_n$ and that $\theta_1, \theta_2, \ldots, \theta_n$ are formulas of a first-order language L. Then the formula ψ obtained by replacing each occurrence of p_i by θ_i, for all $i = 1, 2, \ldots, n$, is called a *substitution instance of the tautology* ϕ.

Theorem 4.2

Every substitution instance of a tautology is universally valid.

There are many universally valid formulas besides those which are substitution instances of tautologies. Look at

$$(\forall x_1 R(x_1, x_3) \rightarrow \forall x_2 R(x_2, x_3)).$$

This is not a substitution instance of a tautology. However, with our experience of dummy variables in mathematics, like the x_1 and x_2 in this formula, we do expect it to be true in all interpretations.

Exercise 4.27 ___________________________

Show that $(\forall x_1 R(x_1, x_3) \rightarrow \forall x_2 R(x_2, x_3))$ is universally valid.

Solution

We need to show that something holds for all structures for the language. So let $\mathcal{A} = \langle A, R^{\mathcal{A}} \rangle$ be any structure for the language used in this formula. Then A is any non-empty set and $R^{\mathcal{A}}$ is any 2-place relation on A (meaning a subset of $A \times A$). The only free variable in the formula is x_3. So we need to show that for all interpretations of x_3, i.e. for all $a_3 \in A$,

$$\mathcal{A} \vDash_{x_3/a_3} (\forall x_1 R(x_1, x_3) \rightarrow \forall x_2 R(x_2, x_3)).$$

> To test universal validity of a formula ϕ, we need only look at structures for the language used in the construction of ϕ. The irrelevance of other symbols was shown in Exercise 4.26 in the previous section.

The argument depends on whether or not $\mathcal{A} \vDash_{x_3/a_3} \forall x_1 R(x_1, x_3)$, so we split it into two cases.

If $\mathcal{A} \nvDash_{x_3/a_3} \forall x_1 R(x_1, x_3)$, then truth table laws for interpreting $\rightarrow$ vacuously give the required result.

If $\mathcal{A} \vDash_{x_3/a_3} \forall x_1 R(x_1, x_3)$, then for all $a \in A$,

$$\mathcal{A} \vDash_{x_1/a, x_3/a_3} R(x_1, x_3),$$

> The interesting case!

which means that for all $a \in A$,

$$(a, a_3) \in R^{\mathcal{A}}.$$

We can now build up a new formula satisfied by $\mathcal{A}$ with the given interpretation of x_3. Keeping an eye on the required answer, we deduce that for all $a \in A$,

$$\mathcal{A} \vDash_{x_2/a, x_3/a_3} R(x_2, x_3),$$

so that

$$\mathcal{A} \vDash_{x_3/a_3} \forall x_2 R(x_2, x_3).$$

Thanks to the truth table for $\rightarrow$, we can then infer that in this case we also have

$$\mathcal{A} \vDash_{x_3/a_3} (\forall x_1 R(x_1, x_3) \rightarrow \forall x_2 R(x_2, x_3)).$$

Thus the given formula is universally valid.

The cunning part of the last solution was to eliminate reference to the formal language and obtain the information that, for all $a \in A$, the pair (a, a_3) was in $R^{\mathcal{A}}$. We then had considerable freedom about what new formula would be true in $\mathcal{A}$. We chose to obtain that for all $a \in A$,

$$\mathcal{A} \vDash_{x_2/a, x_3/a_3} R(x_2, x_3),$$

because that's what led to the answer required by the question. But we could have equally well deduced that for all $a \in A$,

$$\mathcal{A} \vDash_{x_{11}/a, x_{97}/a_3} R(x_{11}, x_{97}),$$

so that

$$\mathcal{A} \vDash_{x_{97}/a_3} \forall x_{11} R(x_{11}, x_{97}),$$

had we so wanted.

Exercise 4.28

Show that each of the following is universally valid. (g is a 1-place function symbol, f is a 3-place function symbol and R a 2-place relation symbol R.)

(a) $(x_1 = x_2 \rightarrow f(x_1, x_2, x_3) = f(x_2, x_1, x_3))$
(b) $(\forall x_1 R(x_1, x_3) \rightarrow \forall x_4 R(x_4, x_3))$
(c) $(\exists x_1 R(x_1, x_3) \rightarrow \exists x_2 R(x_2, x_3))$
(d) $(x_1 = x_2 \rightarrow (R(x_1, g(x_1)) \rightarrow R(x_1, g(x_2))))$
(e) $(\forall x R(g(x), x) \rightarrow \forall x \exists y R(y, x))$

Solution

We give a solution only for part (a) and leave the others to you. As the formula includes the $=$ symbol, we need to consider normal structures for the language of the form $\mathcal{A} = \langle A, f^{\mathcal{A}}, = \rangle$. We need to show that for all such $\mathcal{A}$ and all $a_1, a_2, a_3 \in A$,

$$\mathcal{A} \vDash_{x_1/a_1, x_2/a_2, x_3/a_3} (x_1 = x_2 \rightarrow f(x_1, x_2, x_3) = f(x_2, x_1, x_3)).$$

If

$$\mathcal{A} \nvDash_{x_1/a_1, x_2/a_2, x_3/a_3} x_1 = x_2,$$

this result holds vacuously thanks to the truth table for $\rightarrow$.

If

$$\mathcal{A} \vDash_{x_1/a_1, x_2/a_2, x_3/a_3} x_1 = x_2,$$

then as $\mathcal{A}$ is a normal structure, a_1 must equal a_2. This means that the interpretations of the terms $f(x_1, x_2, x_3)$ and $f(x_2, x_1, x_3)$ are connected by

$$\begin{aligned} f(x_1, x_2, x_3)[\vec{x}/\vec{a}]^{\mathcal{A}} &= f^{\mathcal{A}}(a_1, a_2, a_3) \\ &= f^{\mathcal{A}}(a_2, a_1, a_3) \quad (\text{as } a_1 = a_2) \\ &= f(x_2, x_1, x_3)[\vec{x}/\vec{a}]^{\mathcal{A}}, \end{aligned}$$

so that

$$\mathcal{A} \vDash_{x_1/a_1, x_2/a_2, x_3/a_3} f(x_1, x_2, x_3) = f(x_2, x_1, x_3).$$

Thus in this case we also have

$$\mathcal{A} \vDash_{x_1/a_1, x_2/a_2, x_3/a_3} (x_1 = x_2 \to f(x_1, x_2, x_3) = f(x_2, x_1, x_3)),$$

showing that $(x_1 = x_2 \to f(x_1, x_2, x_3) = f(x_2, x_1, x_3))$ is universally valid.

> In a normal structure, the interpretation $=^{\mathcal{A}}$ is actual equality on the domain A.

Alongside the idea of a formula being true in all interpretations, we have the more modest idea of a formula being true in *some* interpretation, given by the following definition.

Definition   Satisfiable formula

Let $\phi(x_1, x_2, \ldots, x_n)$ be a formula in a language L with free variables contained in the list $x_1, x_2, \ldots, x_n$. We say that ϕ is *satisfiable* if there is some structure $\mathcal{A}$ for L and some interpretation of the variables $x_1, x_2, x_3, \ldots$ respectively by $a_1, a_2, a_3, \ldots$ in the domain of $\mathcal{A}$ such that

$$\mathcal{A} \vDash_{\vec{x}/\vec{a}} \phi(x_1, x_2, \ldots, x_n).$$

Likewise a set Γ of formulas is *satisfiable* if there is a structure and interpretation within it of the free variables which simultaneously satisfies all the formulas in Γ.

Exercise 4.29 ———————————————————————

Show that the following formulas are satisfiable (where f is a 2-place function symbol and R a 2-place relation symbol).

(a) $f(x_1, f(x_2, x_3)) = f(f(x_1, x_2), x_3)$

(b) $(\forall x \exists y R(x, y) \wedge \forall x (\exists y R(x, y) \to R(x, f(x, x))))$

(c) $\neg \forall x (\exists y R(x, y) \to R(x, f(x, x)))$

Solution

(a) One of many structures and interpretations satisfying the formula is $\langle \mathbb{N}, +, = \rangle$ with x_1, x_2, x_3 interpreted by $12, 4, 9$ respectively.

(b) Not given.

(c) Not given.

Exercise 4.30

(a) Show that the formula $\exists y \, \neg y = y$ in a language L with equality is not satisfiable (in any normal structure).

(b) Show that the formula $(\forall x \exists y \, \neg x = y \rightarrow \exists y \, \neg y = y)$ is satisfiable (in some normal structure).

Exercise 4.31

Let ϕ be a formula in a language L. Show that ϕ is universally valid if and only if $\neg \phi$ is not satisfiable.

It will be valuable to compile – and for you to remember and recognize! – a list of helpful and relatively simple universally valid formulas. Perhaps the most important of these is one which corresponds to how we intend to use the universal quantifier $\forall$. We are after all aiming to use the predicate language as a framework for deriving mathematical theorems and so far we've not gone much beyond exploiting the propositional connectives. One of the most fundamental ways in which $\forall$ is used in reasoning was analyzed over 2000 years ago by the Greek philosopher Aristotle, one of the founders of the study of logic, in his description of certain forms of argument called *syllogisms*. The shape of probably the most important of these syllogisms is as follows.

all A's are B's: X is an A: therefore X is a B,

that is, if all objects of a certain sort have a property, then any particular one of these objects has this property. Examples of this style of argument include:

all elephants have a trunk: Nellie is an elephant: therefore Nellie has a trunk;

and, slightly disguised,

all real numbers have a cube root: therefore for any real number x, the number x^2 has a cube root.

In the framework of our formal language and structures for it, this becomes

if $\phi(x)$ holds for all x in a structure, then $\phi(x)$ holds for any particular value of x.

A simple, but often useful, special case of this is represented by the formula

$$(\forall x \phi(x) \rightarrow \phi(x)),$$

and we would like you to show that this is universally valid.

Exercise 4.32

Suppose that the only free variable in the formula $\phi(x)$ is x. Show that the formula $(\forall x \phi(x) \rightarrow \phi(x))$ is universally valid.

Solution

Let $\mathcal{A} = \langle A, \ldots \rangle$ be any structure for the language. Note that there are some free occurrences of x in $(\forall x \phi(x) \to \phi(x))$, namely the x in the second copy of the subformula $\phi(x)$, so that we have to show that for all $a \in A$

$$\mathcal{A} \vDash_{x/a} (\forall x \phi(x) \to \phi(x)).$$

If

$$\mathcal{A} \nvDash_{x/a} \forall x \phi(x),$$

then vacuously

$$\mathcal{A} \vDash_{x/a} (\forall x \phi(x) \to \phi(x)). \cdot$$

The interesting case is when

$$\mathcal{A} \vDash_{x/a} \forall x \phi(x).$$

As x is not free in $\forall x \phi(x)$, this means that

$$\mathcal{A} \vDash \forall x \phi(x).$$

so that for the given $a \in A$,

$$\mathcal{A} \vDash_{x/a} \phi(x).$$

Thus

$$\mathcal{A} \vDash_{x/a} (\forall x \phi(x) \to \phi(x)),$$

as required.

> Thanks to the truth table for $\to$!

> This is a consequence of Theorem 4.1 of Section 4.2.

Exercise 4.33

Show that if x is not free in ϕ, then $(\forall x \phi \to \phi)$ is universally valid. [*Hint*: Use Theorem 4.1 of Section 4.2.]

We shall be ambitious and aim to extend 'if $\phi(x)$ holds for all x in a structure, then $\phi(x)$ holds for any particular value of x' to allow the 'particular value of x' to be any term τ of the language. This key principle governing the use of $\forall$ then becomes

$$(\forall x \phi(x) \to \phi(\tau)),$$

for any term τ, where by $\phi(\tau)$ we mean the formula which results from $\phi(x)$ by replacing all free occurrences of x by the term τ. This formula tells us how to 'eliminate' the $\forall x$ from the formula $\forall x \phi$.

> We shall also allow ϕ to contain free variables besides x. This will cause no difficulties.

Alas, there's an irritating complication, which we shall have to deal with at some length. The complication is illustrated by the following example. Let $\phi(x)$ be the innocent-looking formula

$$\exists y \, \neg x = y,$$

for which the sentence $\forall x \phi(x)$ is satisfied in any structure with a domain containing at least 2 elements; and for the term τ, take simply the variable y. Then $(\forall x \phi(x) \to \phi(\tau))$ becomes

$$(\forall x \exists y \, \neg x = y \to \exists y \, \neg y = y),$$

which, as we hope you discovered in Exercise 4.30, is only satisfiable in a very limited range of structures – it is certainly not universally valid. The problem is something to do with the term τ being the variable y which becomes bound by the quantifier $\exists y$ lurking within the formula $\phi(x)$. Just in case you fear that there are no terms τ safe to use when eliminating the $\forall x$ in this particular example, let us assure you that any term τ not involving y will do fine. It will be instructive for you to work through an example with a suitable τ in the next exercise.

Exercise 4.34

Let f be a function symbol of two arguments. Show that the formula

$$(\forall x \exists y \,\neg\, x = y \rightarrow \exists y \,\neg\, f(z, x) = y)$$

So $\tau = f(z, x)$.

is universally valid.

Solution

Let $\mathcal{A} = \langle A, f^*, = \rangle$ be any structure for the language involved in the formula, where f^* is a function from A^2 to A. Note that the formula involves both z and x as free variables (the occurrence of x in the subformula $\exists y \,\neg\, f(z, x) = y$ is free), so we need to show the formula is satisfied by $\mathcal{A}$ for all interpretations of these variables, say a of x and c of z, where $a, c \in A$.

If

$$\mathcal{A} \not\models_{x/a, z/c} \forall x \exists y \,\neg\, x = y,$$

then vacuously

Thanks to the truth table for $\rightarrow$!

$$\mathcal{A} \models_{x/a, z/c} (\forall x \exists y \,\neg\, x = y \rightarrow \exists y \,\neg\, f(z, x) = y).$$

The interesting case is when

$$\mathcal{A} \models_{x/a, z/c} \forall x \exists y \,\neg\, x = y.$$

We must show that

$$\mathcal{A} \models_{x/a, z/c} \exists y \,\neg\, f(z, x) = y.$$

As

$$\mathcal{A} \models_{x/a, z/c} \forall x \exists y \,\neg\, x = y$$

and the variable x is not free in $\forall x \exists y \,\neg\, x = y$, this means that

This is a consequence of Theorem 4.1 of Section 4.2.

$$\mathcal{A} \models_{z/c} \forall x \exists y \,\neg\, x = y.$$

So for the given $a, c \in A$,

$$\mathcal{A} \models_{x/f^*(a,c), z/c} \exists y \,\neg\, x = y,$$

which means that there is some $b \in A$ for which

$$\mathcal{A} \models_{x/f^*(a,c), y/b, z/c} \neg\, x = y,$$

and that, accounting for the $\neg$ and dispensing with the formal language,

$$f^*(a, c) \neq b$$

in $\mathcal{A}$. Re-introducing the formal language, we can now say that for the given

$a, c \in A$ and the b above,

$$A \nvDash_{x/a, y/b, z/c} f(x, z) = y,$$

so that

$$A \vDash_{x/a, y/b, z/c} \neg f(x, z) = y,$$

and for the given $a, c \in A$,

$$A \vDash_{x/a, z/c} \exists y \neg f(x, z) = y,$$

as required.

In general, to rescue $(\forall x \phi(x) \rightarrow \phi(\tau))$ as a logically valid formula, we shall add the stipulation that the term τ doesn't involve variables that would become bound by quantifiers hidden inside $\phi(x)$. We shall introduce some terminology for this situation.

Definition Freely substitutable

Let ϕ be a formula, x a variable and τ a term. Let $\phi(\tau)$ be the formula obtained by replacing all the free occurrences of x in $\phi(x)$ by τ. Then τ is said to be *freely substitutable for x in ϕ* if none of the variables in τ becomes bound in the places where τ is substituted for the free x in ϕ.

In the case when x is not free in ϕ, the formula $\phi(\tau)$ is simply the original ϕ and we adopt the convention that in this case τ is said to be freely substitutable for x in ϕ.

> We might say that τ has been *substituted for the free x in $\phi(x)$*.

> This convention for x not free in ϕ seems to make posh arguments involving 'freely substitutable' more straightforward. Plainly the interesting use of the definition is when x *is* free in ϕ.

As an example of the use of this definition, take a language with equality and including a function symbol f and relation symbol R, both of two arguments, and let ϕ be the formula

$$(\exists w \exists x (R(x, y) \lor \forall t \, w = t) \rightarrow \forall z R(z, x)).$$

Note that ϕ has x and y as free variables. Let τ be the term $f(t, w)$. Then the result of substituting τ for the free x in ϕ is

$$(\exists w \exists x (R(x, y) \lor \forall t \, w = t) \rightarrow \forall z R(z, f(t, w))).$$

> The earlier occurrences of x in ϕ are bound, so are not available to be substituted by the term.

The variables in τ, namely t and w, do not fall within the scope of any quantifier that binds them – the free x for which τ has been substituted lies within the scope of a $\forall z$, not within the scope of the $\exists w$, $\exists x$ and $\forall t$ earlier in ϕ which might potentially bind the variables in τ – so τ is freely substitutable for x in ϕ.

If, however, we substitute τ for the free y in ϕ, we obtain the formula

$$(\exists w \exists x (R(x, f(t, w)) \lor \forall t \, w = t) \rightarrow \forall z R(z, x)),$$

and the w in $f(t, w)$ now falls within the scope of the $\exists w$. So τ is not freely substitutable for y in ϕ.

Exercise 4.35

Let ϕ be the formula

$$\exists y(f(x,y) = t \leftrightarrow (\exists z \forall x R(z,x) \vee \forall t R(x,t))).$$

Which of the following terms is freely substitutable for x in ϕ? And which of them is freely substitutable for z in ϕ? [R is a 2-place relation symbol, f is a 2-place function symbol and $\mathbf{c}$ is a constant symbol.]

(a) $f(\mathbf{c},\mathbf{c})$ (b) z (c) $f(y,x)$ (d) $f(x,x)$

Exercise 4.36

Our definition of 'τ is freely substitutable for x in ϕ' is relatively informal. Give a more formal definition in terms of the number of connectives and quantifiers in ϕ.

Solution

We shall work with a fixed term τ and treat all formulas as built up from the adequate set of connectives $\{\neg, \wedge\}$ and $\forall, \exists$.

The term τ is freely substitutable for x in any atomic formula. (Our definition stipulates that this happens regardless of whether x appears in the atomic formula, although the more interesting case is when x does appear.)

If ϕ is of the form $\neg\psi$, τ is freely substitutable for x in ϕ exactly when τ is freely substitutable for x in ψ. If ϕ is of the form $(\psi \wedge \theta)$, τ is freely substitutable for x in ϕ exactly when τ is freely substitutable for x in both ψ and θ.

If ϕ is one of the forms $\forall y\psi$ and $\exists y\psi$, where y is a variable other than x, τ is freely substitutable for x in ϕ exactly when τ is freely substitutable for x in ψ and y does not appear in τ.   *This is of course the most interesting case.*

If ϕ is one of the forms $\forall x\psi$ and $\exists x\psi$, τ is freely substitutable for x in ϕ.   *This uses our convention for when x is not free in ϕ.*

We can now state the first, and perhaps the most fundamental, universally valid formula involving $\forall$ using this new terminology.

Theorem 4.3

Let ϕ be a formula and let the term τ be freely substitutable for x in ϕ. Then

$$(\forall x\phi(x) \rightarrow \phi(\tau))$$

is universally valid.

We have already proved the special case of this theorem when the term τ is simply the variable x in Exercise 4.32. We shall now work towards the proof of the more general result, which we shall leave as Exercise 4.37. The interesting application of the result, and the non-trivial aspect of its proof, arises when x is free in ϕ. If you inspect our solution to Exercise 4.34, where $\phi(x)$ was the formula $\exists y \neg y = x$ and τ was $f(x,z)$, you might spot that the argument hinges on showing that for all $a, c \in A$,

$$\mathcal{A} \vDash_{x/f^*(a,c),z/c} \exists y \neg x = y,$$

if and only if

$$\mathcal{A} \vDash_{x/a,z/c} \exists y \, \neg f(x,z) = y,$$

that is,

$$\mathcal{A} \vDash_{x/f^*(a,c),z/c} \phi(x),$$

if and only if

$$\mathcal{A} \vDash_{x/a,z/c} \phi(f(x,z)).$$

The proof of Theorem 4.3 depends on showing the more general form of this, which we will state as a theorem. When reading the theorem, recall that $\tau[\vec{x}/\vec{a}]^{\mathcal{A}}$ stands for the interpretation in $\mathcal{A}$ of the term τ when the variables in τ are interpreted by the corresponding elements in $\vec{a}$. Recall that we use the notation $\vec{x}/\vec{a}[x_1/\tau[\vec{x}/\vec{a}]^{\mathcal{A}}]$ to stand for the interpretation of the variables in which the interpretation of x_1 is altered to $\tau[\vec{x}/\vec{a}]^{\mathcal{A}}$.

Theorem 4.4

Let ϕ be a formula, x_1 a variable and τ a term which is freely substitutable for x_1 in ϕ. Let $\mathcal{A} = \langle A, \ldots \rangle$ be a structure for the underlying language L and let $\vec{a}$ be an interpretation of the variables $\vec{x}$ of L. Then

$$\mathcal{A} \vDash_{\vec{x}/\vec{a}[x_1/\tau[\vec{x}/\vec{a}]^{\mathcal{A}}]} \phi(x_1)$$

if and only if

$$\mathcal{A} \vDash_{\vec{x}/\vec{a}} \phi(\tau).$$

The statement of the theorem is something of a mouthful; but we hope that its meaning is clear. We state it for the particular variable x_1 for a small bit of clarity! Note that when x_1 is not free in ϕ, the result is a simple consequence of Theorem 4.1 of Section 4.2, as the interpretation of x_1 is irrelevant to whether $\mathcal{A} \vDash_{\vec{x}/\vec{a}} \phi$.

We shall leave the proof of this theorem as an optional exercise for you.

Exercise 4.56.

Exercise 4.37 ————————————————————

Use the result of Theorem 4.4 to prove Theorem 4.3.

Exercise 4.38 ————————————————————

Let ϕ be a formula and let the term τ be freely substitutable for x in ϕ. Show that

$$(\phi(\tau) \rightarrow \exists x \phi(x))$$

is universally valid.

———————————————————————————————————

We have dealt at length with one aspect of handling the universal quantifier $\forall$, namely how to eliminate it from a formula. We shall leave the reverse process, namely how to introduce a universal quantifier in front of a formula, till Chapter 5. For the moment, we shall ask you to do an exercise which illustrates what *won't* work!

Exercise 4.39 ────────────────────────────────

Show that the formula $(\phi(x) \to \forall x \phi(x))$ is not in general universally valid. [*Hints*: Take a language with equality including the constant symbol $\mathbf{c}$ and let $\phi(x)$ be the formula $x = \mathbf{c}$. Now find a normal structure and an interpretation of the free variable x in which $(\phi(x) \to \forall x \phi(x))$ is false.]

────────────────────────────────

We shall now ask you to check some further useful universally valid formulas.

Exercise 4.40 ────────────────────────────────

Show that each of the following formulas is universally valid.

(a) $(\exists y \forall x \phi \to \forall x \exists y \phi)$

(b) $(\forall x(\phi \to \psi) \to (\forall x \phi \to \forall x \psi))$

(c) $(\forall x \phi(x) \to \phi(\mathbf{c}))$, where $\mathbf{c}$ is a constant symbol.

(d) $(\forall x R(x, f(x)) \to \forall x \exists y R(x, y))$, where f is a 1-place function symbol and R is a 2-place relation symbol.

────────────────────────────────

Many universally valid formulas are of the form $(\phi \leftrightarrow \psi)$ and as with propositional formulas, it is useful to regard the ϕ and ψ in such cases as equivalent in some way. The definition below encompasses the one already encountered for propositional formulas in Chapter 2.

> **Definition   Logical equivalence**
>
> The formulas ϕ and ψ are *logically equivalent*, written as $\phi \equiv \psi$, if every structure and interpretation satisfying ϕ also satisfies ψ and vice versa.

Exercise 4.41 ────────────────────────────────

Show that $\phi \equiv \psi$ if and only if $\vDash (\phi \leftrightarrow \psi)$.

────────────────────────────────

There are many very useful logical equivalences, some of which are in the following exercise.

Exercise 4.42 ────────────────────────────────

Demonstrate each of the following logical equivalences.

(a) $\neg \exists x \phi \equiv \forall x \neg \phi$

(b) $\neg \forall x \phi \equiv \exists x \neg \phi$

Solution

(a) Let $\mathcal{A} = \langle A, \ldots \rangle$ be a structure for the language and let $\vec{a}$ be any interpretation of the variables $\vec{x}$ in L by elements of A.

Suppose first that

$$\mathcal{A} \vDash_{\vec{x}/\vec{a}} \neg \exists x \phi.$$

Then it is not the case that

$$\mathcal{A} \vDash_{\vec{x}/\vec{a}} \exists x \phi,$$

so that it is not the case that there is some $b \in A$ such that

$$\mathcal{A} \vDash_{\vec{x}/\vec{a}[x/b]} \phi.$$

That means whatever $b \in A$ one takes,

$$\mathcal{A} \vDash_{\vec{x}/\vec{a}[x/b]} \neg \phi,$$

so that

$$\mathcal{A} \vDash_{\vec{x}/\vec{a}} \forall x \neg \phi.$$

For the converse, the argument above essentially reverses. Suppose that

$$\mathcal{A} \vDash_{\vec{x}/\vec{a}} \forall x \neg \phi.$$

Then for all $b \in A$,

$$\mathcal{A} \vDash_{\vec{x}/\vec{a}[x/b]} \neg \phi,$$

so that for all $b \in A$ it is not the case that

$$\mathcal{A} \vDash_{\vec{x}/\vec{a}[x/b]} \phi.$$

Thus it is not the case that for some $b \in A$,

$$\mathcal{A} \vDash_{\vec{x}/\vec{a}[x/b]} \phi,$$

so it is not the case that

$$\mathcal{A} \vDash_{\vec{x}/\vec{a}} \exists x \phi,$$

which means that

$$\mathcal{A} \vDash_{\vec{x}/\vec{a}} \neg \exists x \phi.$$

(b) Not given.

> Observe that we have to exploit our understanding of how to use 'for all' and 'there exists' in natural language, as well as of 'not'.

As $\neg\neg\psi \equiv \psi$, it follows from this exercise that

$$\neg \forall x \neg \phi \equiv \neg\neg \exists x \phi$$
$$\equiv \exists x \phi,$$

and similarly that

$$\neg \exists x \neg \phi \equiv \forall x \phi.$$

This means that we could in principle represent all formulas, up to logical equivalence, using a limited set of connectives like $\{\neg, \rightarrow\}$ and just one of the quantifiers $\forall$ and $\exists$, as from one of these and $\neg$ we can represent the other.

Exercise 4.43

Suppose that x is not free in ϕ. Show that $(\phi \wedge \forall x\psi) \equiv \forall x(\phi \wedge \psi)$ and
$(\phi \wedge \exists x\psi) \equiv \exists x(\phi \wedge \psi)$.

Solution

Let $\mathcal{A} = \langle A, \ldots \rangle$ be a structure for the language and let $\vec{a}$ be any interpretation
of the variables $\vec{x}$ in L by elements of A. We shall show that

$$\text{if } \mathcal{A} \vDash_{\vec{x}/\vec{a}} (\phi \wedge \forall x\psi) \quad \text{then} \quad \mathcal{A} \vDash_{\vec{x}/\vec{a}} \forall x(\phi \wedge \psi),$$

and will leave it for you to show that

$$\text{if } \mathcal{A} \vDash_{\vec{x}/\vec{a}} \forall x(\phi \wedge \psi) \quad \text{then} \quad \mathcal{A} \vDash_{\vec{x}/\vec{a}} (\phi \wedge \forall x\psi)$$

and that

$$\mathcal{A} \vDash_{\vec{x}/\vec{a}} (\phi \wedge \exists x\psi) \quad \text{if and only if} \quad \mathcal{A} \vDash_{\vec{x}/\vec{a}} \exists x(\phi \wedge \psi).$$

Suppose that

$$\mathcal{A} \vDash_{\vec{x}/\vec{a}} (\phi \wedge \forall x\psi).$$

Then

$$\mathcal{A} \vDash_{\vec{x}/\vec{a}} \phi \quad \text{and} \quad \mathcal{A} \vDash_{\vec{x}/\vec{a}} \forall x\psi.$$

Looking at the second of these, we have that for all $b \in A$,

$$\mathcal{A} \vDash_{\vec{x}/\vec{a}[x/b]} \psi.$$

Looking at the first, as x is not free in ϕ, by Theorem 4.1 we have that for all
$b \in A$,

$$\mathcal{A} \vDash_{\vec{x}/\vec{a}[x/b]} \phi.$$

As the interpretations of the variables are the same, we can glue these together
to obtain, for all $b \in A$,

$$\mathcal{A} \vDash_{\vec{x}/\vec{a}[x/b]} (\phi \wedge \psi),$$

so that

$$\mathcal{A} \vDash_{\vec{x}/\vec{a}[x/b]} \forall x(\phi \wedge \psi),$$

as required.

As with formulas of propositional calculus, the relation of logical equivalence
is an equivalence relation on the set of formulas of a first-order language.

Exercise 4.44

Show each of the following, for all formulas ϕ, ψ, θ.

(a) $\phi \equiv \phi$

(b) If $\phi \equiv \psi$ then $\psi \equiv \phi$.

(c) If $\phi \equiv \psi$ and $\psi \equiv \theta$, then $\phi \equiv \theta$.

This is the analogue of
Exercise 2.35 in Section 2.4 of
Chapter 2 for propositional
languages. The solution here needs
to account for the slightly more
complicated definition of $\equiv$ for a
first-order language.

Just as for propositional calculus, we often have a formula θ and want to replace all occurrences of the subformula ϕ by a formula ϕ'. In Chapter 3, we introduced the notation $\theta[\phi'/\phi]$ for the resulting formula. The main interest here is that if ϕ and ϕ' are logically equivalent, then so are θ and $\theta[\phi'/\phi]$.

For instance, if θ is the formula

$$\forall x \exists y R(x, y)$$

ϕ is the formula $\exists y R(x, y)$ and ϕ' is the formula $\neg \forall y \neg R(x, y)$, then $\theta[\phi'/\phi]$ is the formula

$$\forall x \neg \forall y \neg R(x, y).$$

Exercise 4.45 ⎯⎯⎯⎯⎯⎯⎯⎯⎯⎯⎯⎯⎯⎯⎯⎯⎯⎯⎯⎯

Let θ be a formula with subformula ϕ and ϕ' any formula, where all these formulas are built up using the connectives $\neg, \wedge$ and the quantifiers $\forall, \exists$. Show that if $\phi \equiv \phi'$, then $\theta \equiv \theta[\phi'/\phi]$. [*Hint*: Adapt the hints given for Exercise 2.46 of Section 2.4 of Chapter 2.]

⎯⎯⎯⎯⎯⎯⎯⎯⎯⎯⎯⎯⎯⎯⎯⎯⎯⎯⎯⎯⎯⎯⎯⎯⎯⎯⎯⎯⎯⎯⎯⎯⎯⎯⎯⎯⎯⎯

As for propositional formulas, we can ask the question whether the equivalence class of formulas logically equivalent to a given formula ϕ contains a 'nice' formula. One answer is given in terms of prenex normal form, defined as follows.

Definitions   Prenex normal form

A formula ϕ is in *prenex normal form* if it has the form

$$Q_1 y_1 Q_2 y_2 \ldots Q_n y_n \theta,$$

where each Q_i is a quantifier $\forall$ or $\exists$, the y_is are variables and the formula θ involves no quantifiers (often described as being *quantifier-free*). The string of quantifiers $[Q_1 y_1 Q_2 y_2 \ldots Q_n y_n$ is called the *prefix* of the formula ϕ.

For instance, $\exists x \forall y (P(x) \rightarrow \neg Q(y))$ and $\forall z \exists x \neg (Q(z) \wedge P(x))$ are both in prenex normal form. These are in fact logically equivalent formulas, so that prenex normal forms for formulas won't be unique.

Theorem 4.5

Any formula is logically equivalent to a formula in prenex normal form.

Proof

We shall show the result for any first-order formula ϕ built up using the connectives $\wedge$ and $\neg$ along with the quantifiers $\forall$ and $\exists$. As $\{\wedge, \neg\}$ is an adequate set of connectives, the result then holds for all first-order formulas. The proof will show how to construct a ϕ^* in prenex normal form logically equivalent to ϕ by exploiting the following logical equivalences involving quantifiers, as well as standard ones essentially involving $\neg$ and $\wedge$, like $(\phi \wedge \forall x \psi) \equiv (\forall x \psi \wedge \phi)$.

1. $\forall x \phi(x) \equiv \forall y \phi(y)$, if x is free in $\phi(x)$ and y doesn't appear in $\phi(x)$.

2. $\exists x \phi(x) \equiv \exists y \phi(y)$, if x is free in $\phi(x)$ and y doesn't appear in $\phi(x)$.

3. $\neg \exists x \phi \equiv \forall x \neg \phi$

4. $\neg \forall x \phi \equiv \exists x \neg \phi$

5. $(\phi \wedge \forall x \psi) \equiv \forall x (\phi \wedge \psi)$, if x is not free in ϕ.

6. $(\phi \wedge \exists x \psi) \equiv \exists x (\phi \wedge \psi)$, if x is not free in ϕ.

> If x isn't free in ϕ and y doesn't appear in ϕ, then fairly trivially $\forall x \phi \equiv \forall y \phi$ and $\exists x \phi \equiv \exists y \phi$.

> Similarly $(\forall x \psi \wedge \phi) \equiv \forall x (\psi \wedge \phi)$ and $(\exists x \psi \wedge \phi) \equiv \exists x (\psi \wedge \phi)$, if x is not free in ϕ.

The method, as you should by now expect for a result which holds for all formulas, is mathematical induction on the length (number of connectives and quantifiers) in ϕ.

The result is trivial for formulas of length 0, as these are atomic formulas, thus involving no quantifiers and already in prenex normal form.

For the inductive step, assume that the result holds for all formulas with $\leq n$ connectives and quantifiers and take any formula of length $n + 1$. This formula must have one of the forms $\forall x \phi$, $\exists x \phi$, $\neg \phi$ and $(\phi \wedge \psi)$. We shall deal with the last of these cases and leave the rest for you as an exercise. In all cases, our aim is to 'factor' quantifiers to the outside.

For the case $(\phi \wedge \psi)$, by the inductive hypothesis the subformulas ϕ and ψ are logically equivalent to formulas in prenex normal form, say

$$\phi \equiv Q_1 y_1 Q_2 y_2 \ldots Q_r y_r \theta \quad \text{and} \quad \psi \equiv Q_1' y_1' Q_2' y_2' \ldots Q_s' y_s' \theta',$$

where each Q_i and Q_j' is a quantifier $\forall$ or $\exists$, and θ and θ' are quantifier-free. By Exercise 4.45, the formula $(\phi \wedge \psi)$ is logically equivalent to the conjunction of these formulas in prenex normal form. We are itching to exploit equivalences 5 and 6, but have to be careful. If there is no overlap between the variables appearing in ϕ and those in ψ, there is no problem. But if a quantified variable in one appears as a variable in the other we have to be careful. For instance if we are dealing with $(\forall x \theta \wedge \forall x \theta')$, x is not free in $\forall x \theta$, so equivalence 5 can be used to move the $\forall x$ from the $\forall x \theta'$ to give

$$(\forall x \theta \wedge \forall x \theta') \equiv \forall x (\forall x \theta \wedge \theta');$$

but as x might be free in θ', we can't simply use equivalence 5 to manipulate the subformula $(\forall x \theta \wedge \theta')$ any further. The trick here is to use equivalence 1 to change the xs in $\forall x \theta$ into a variable appearing nowhere else in θ or θ', and then use equivalence 5.

More generally, if the quantified variable in one of the $Q_j' y_j'$s is the same as in one of the $Q_i y_i$s, then we essentially use the appropriate one of equivalences 1 and 2 in the list above to replace $Q_j' y_j'$ by a similar quantifier using a variable not yet used anywhere in the formulas involved. For instance, suppose that ψ is equivalent to $\forall x \exists y \forall z \theta'$ and the variable y is one of the variables quantified in a $Q_i y_i$ in ϕ. We take a new variable, w say, not used in the prenex normal forms for ϕ or ψ. Then by equivalence 2,

$$\exists y \forall z \theta' \equiv \exists w \forall z \theta'',$$

where the free occurrences of y in θ' are replaced by ws to give θ''; and then

$$\forall x \exists y \forall z \theta' \equiv \forall x \exists w \forall z \theta''.$$

We can thus modify the prenex normal form for ψ so that there is no overlap between its quantified variables and those in the prenex normal form for ϕ.

We still have the case where there is some overlap between a quantified variable in the prenex normal form of one of ϕ and ψ with a free variable in the other, which would again be an obstacle to using equivalences 5 and 6, which is our ultimate aim for factoring quantifiers to the outside. But we can do the same trick as earlier and replace the offending quantified variables by ones which don't appear anywhere else. Finally we can now make repeated use of equivalences 5 and 6 to obtain a formula logically equivalent to $(\phi \wedge \psi)$ of the form

$$Q_1 y_1 Q_2 y_2 \ldots Q_r y_r Q_1' y_1' Q_2' y_2' \ldots Q_s' y_s' (\theta \wedge \theta').$$

$\blacksquare$

Exercise 4.46 ____

Fill the gaps left in the inductive step of the proof above.

____

We can use the details of this proof to find a formula in prenex normal form equivalent to

$$\exists x \neg (\forall t S(t, x, y) \wedge \neg \forall y \exists x S(x, y, z)),$$

where S is a 3-place relation symbol. As the proof explains how to get the prenex normal form of a formula from the prenex normal form of certain of its subformulas, a practical algorithm can start from its atomic subformulas and work through its tree diagram. Here the subformulas $S(x, y, z)$, $\exists x S(x, y, z)$ and $\forall y \exists x S(x, y, z)$ are already in prenex normal form, but the next stage, the subformula $\neg \forall y \exists x S(x, y, z)$, is not. But this is logically equivalent to

$$\exists y \forall x \neg S(x, y, z),$$

which is in prenex normal form.

The next stage is to find the prenex normal form of the subformula

$$(\forall t S(t, x, y) \wedge \neg \forall y \exists x S(x, y, z))$$

from those of $\forall t S(t, x, y)$ (which is already in prenex normal form) and of $\neg \forall y \exists x S(x, y, z)$, which we're taking as $\exists y \forall x \neg S(x, y, z)$. Annoyingly the quantified variables in the latter formula overlap with free variables in the former. But following the method of the proof, we note that the variables v, w appear nowhere in $\forall t S(t, x, y)$ or $\exists y \forall x \neg S(x, y, z)$, so as

$$\exists y \forall x \neg S(x, y, z) \equiv \exists v \forall w \neg S(w, v, z),$$

we can use equivalences 5 and 6 in the proof to obtain

$$(\forall t S(t, x, y) \wedge \exists y \forall x \neg S(x, y, z) \equiv \forall t \exists v \forall w (S(t, x, y) \wedge \neg S(w, v, z)).$$

Using equivalences 3 and 4 in the proof then gives

$$\neg(\forall t S(t, x, y) \wedge \neg \forall y \exists x S(x, y, z)) \equiv \exists t \forall v \exists w \neg (S(t, x, y) \wedge \neg S(w, v, z)),$$

so that a logically equivalent formula to

$$\exists x \neg (\forall t S(t, x, y) \wedge \neg \forall y \exists x S(x, y, z))$$

We hope that your filling in the gaps in our proof explains how we have obtained this using equivalences 3 and 4 in the proof.

in prenex normal form is

$$\exists x \exists t \forall v \exists w \neg (S(t, x, y) \wedge \neg S(w, v, z)).$$

Exercise 4.47

Establish the following logical equivalences which will be useful for more directly obtaining a prenex normal form for formulas using $\vee$, $\rightarrow$ and $\leftrightarrow$.

(a) $(\phi \vee \forall x \psi) \equiv \forall x (\phi \vee \psi)$, if x is not free in ϕ.

(b) $(\phi \vee \exists x \psi) \equiv \exists x (\phi \vee \psi)$, if x is not free in ϕ.

(c) $(\phi \rightarrow \forall x \psi) \equiv \forall x (\phi \rightarrow \psi)$, if x is not free in ϕ.

(d) $(\phi \rightarrow \exists x \psi) \equiv \exists x (\phi \rightarrow \psi)$, if x is not free in ϕ.

(e) $(\forall x \psi \rightarrow \phi) \equiv \exists x (\psi \rightarrow \phi)$, if x is not free in ϕ.

(f) $(\exists x \psi \rightarrow \phi) \equiv \forall x (\psi \rightarrow \phi)$, if x is not free in ϕ.

Exercise 4.48

For each of the following formulas using a 2-place relation symbol R, find an equivalent formula in prenex normal form.

(a) $\forall t (\exists x R(x, y) \wedge \forall y (R(t, y) \wedge \neg R(z, x)))$

(b) $\exists y ((\neg \exists x R(x, t) \rightarrow \forall z R(x, w)) \vee \exists x R(x, w))$

Exercise 4.49

Show that any formula is logically equivalent to a formula in prenex normal form

$$Q_1 y_1 Q_2 y_2 \ldots Q_n y_n \theta,$$

where the variables y_is are distinct. [*Hint*: Look at Exercise 4.20 and the preamble to it in Section 4.2.]

Prenex normal form leads to another interesting sort of formula said to be in *Skolem form*. Although we shall only make one use of this sort of formula in the book, discussion of it gives some useful practice with the ideas of universally valid formula and satisfiability. The key idea is introduced in the following exercise.

The rest of this section might be omitted on a first reading of the book.

Exercise 4.50

Let $\phi(x, y)$ be a quantifier-free formula with free variables x, y and let f be a 1-place function symbol which does not appear in ϕ.

(a) Show that $(\forall x \phi(x, f(x)) \rightarrow \forall x \exists y \phi(x, y))$ is universally valid.

(b) Is $(\forall x \exists y \phi(x, y) \rightarrow \forall x \phi(x, f(x)))$ universally valid? Explain your answer.

(c) Show that $\forall x \exists y \phi(x, y)$ is satisfiable if and only if $\forall x \phi(x, f(x))$ is satisfiable.

Solution

(a) Suppose that $\mathcal{A}$ is a structure for the language involved in the formula. If $\mathcal{A} \nvDash \forall x \phi(x, f(x))$, then trivially

$$\mathcal{A} \vDash (\forall x \phi(x, f(x)) \rightarrow \forall x \exists y \phi(x, y)).$$

If $\mathcal{A} \vDash \forall x \phi(x, f(x))$, then for all $a \in A$,

$$\mathcal{A} \vDash_{x/a} \phi(x, f(x)),$$

so by Theorem 4.4,

$$\mathcal{A} \vDash_{x/a, \, y/f^{\mathcal{A}}(a)} \phi(x, y).$$

Then for all $a \in A$, there is some $b \in A$ (namely $f^{\mathcal{A}}(a)$) such that

$$\mathcal{A} \vDash_{x/a, \, y/b} \phi(x, y),$$

so that for all $a \in A$,

$$\mathcal{A} \vDash_{x/a} \exists y \phi(x, y),$$

so that

$$\mathcal{A} \vDash \forall x \exists y \phi(x, y).$$

(b) This is not in general universally valid. For a counterexample, suppose that the formula $\phi(x, y)$ is $R(x, y)$, where R is a 2-place relation symbol. Now let $\mathcal{A}$ be the structure $\langle \mathbb{N}, <, f^{\mathcal{A}} \rangle$ for the language consisting of R and f, where $<$ is the usual strict order on $\mathbb{N}$ and $f^{\mathcal{A}}(n) = n$ for all $n \in \mathbb{N}$. Then

$$\mathcal{A} \vDash \forall x \exists y R(x, y)$$

but

$$\mathcal{A} \nvDash \forall x R(x, f(x)),$$

so that

$$\mathcal{A} \nvDash (\forall x \exists y R(x, y) \rightarrow \forall x R(x, f(x))).$$

(c) The argument in one direction is easy. If some structure $\mathcal{A}$ satisfies $\forall x \phi(x, f(x))$, then as $(\forall x \phi(x, f(x)) \rightarrow \forall x \exists y \phi(x, y))$ is universally valid, $\mathcal{A}$ also satisfies $\forall x \exists y \phi(x, y)$.

For the converse, suppose that $\mathcal{A} = \langle A, \ldots \rangle$ is a structure for the language L used in the formula ϕ which satisfies $\forall x \exists y \phi(x, y)$. The function symbol f does not appear in L, so we shall expand the structure $\mathcal{A}$ to provide an interpretation f^* of f. As f is a 1-place symbol, we need f^* to be a function from A^1 to A, that is, from A to itself. Of course we want the expanded structure $\mathcal{A}^*$ to satisfy $\forall x \exists y \phi(x, y)$. So how do we define $f^*(a)$ for each $a \in A$? As $\mathcal{A} \vDash \forall x \exists y \phi(x, y)$, we have for all $a \in A$

$$\mathcal{A} \vDash_{x/a} \exists y \phi(x, y),$$

so that for each $a \in A$ there is some $b \in A$ such that

> Suppose L and L^* are languages with $L \subseteq L^*$. If $\mathcal{A}$ and $\mathcal{A}^*$ are structures for L and L^*, respectively, with the same domain A and the same interpretations of each symbol in L, then $\mathcal{A}^*$ is called an *expansion* of $\mathcal{A}$.

$$\mathcal{A} \vDash_{x/a, y/b} \phi(x, y).$$

For each $a \in A$ we shall define $f^*(a)$ to be one of these corresponding bs. Expanding the structure $\mathcal{A}$ by adding f^* defined in this way to interpret the symbol f, giving the structure $\mathcal{A}^* = \langle A, \dots, f^* \rangle$, we have

$$\mathcal{A}^* \vDash \forall x \phi(x, f(x)),$$

so that $\forall x \phi(x, f(x))$ is satisfiable.

Note that the detail in our argument in (c) shows not only that $\forall x \exists y \phi(x, y)$ is satisfiable if and only if $\forall x \phi(x, f(x))$ is satisfiable, but shows that a structure satisfying one of them essentially satisfies the other. One way round, a structure satisfying $\forall x \phi(x, f(x))$ also satisfies $\forall x \exists y \phi(x, y)$. The other way round, a structure $\mathcal{A}$ satisfying $\forall x \exists y \phi(x, y)$ can be expanded to a structure $\mathcal{A}^*$ which interprets the extra function symbol in the right sort of way to satisfy $\forall x \phi(x, f(x))$. We shall want this level of detail later in the book.

Now is the time to own up to a subtle and controversial step in our argument in part (c) above! This step is the one from

for each $a \in A$ there is some $b \in A$ such that $\mathcal{A} \vDash_{x/a, y/b} \phi(x, y)$

to saying that

for each $a \in A$ we define $f^*(a)$ to be one of these corresponding bs.

If there are several suitable bs for a given a, we've not given a rule for choosing a particular one of these bs and this means one can dispute whether we have really defined the function f^*. If the domain A has some useful in-built structure, we could perhaps do better and be more specific about how to define each $f^*(a)$. For instance, if $A = \mathbb{N}$, so that each b is also in $\mathbb{N}$, we can define $f^*(a)$ as the least b such that $\mathcal{A} \vDash_{x/a, y/b} \phi(x, y)$. But if A hasn't got the right sort of structure, there might be no nice way of tying down a special b for each a – for instance, if $A = \mathbb{R}$, subsets of $\mathbb{R}$ do not in general contain a least element, so we'd need to look for a different sort of property of the bs for a given a.

In the study of the foundations of mathematics, this is a very important issue, bound up with the definition of infinite sets and their properties. The step in our argument where for each $a \in A$ we choose one of the corresponding bs to be $f^*(a)$ depends on what is called the *axiom of choice*, which we discuss further in Section 6.4 of Chapter 6 when we give an outline of some of the theory of infinite sets. Do note that many of the key results in the rest of the book depend on the axiom of choice, hence our relaxed use of it in the solution to Exercise 4.50(c).

The function f^* created from the structure $\mathcal{A}$ and the formula $\forall x \exists y \phi(x, y)$ in Exercise 4.50(c) is called a *Skolem function* for this formula in $\mathcal{A}$. The significance of this exercise is in the way that adding the function symbol f to the language used in ϕ can be used to dispose of the $\exists y$ and y in the sentence $\forall x \exists y \phi(x, y)$ and obtain a simpler sentence which is satisfiable precisely when the original formula is satisfiable. The significant aspect of the 'simpler' is that it is what is called a *universal* formula, namely one of

These functions were introduced and exploited by the Norwegian mathematician Thoralf Skolem (1887–1963).

the form $\forall x_1 \forall x_2 \ldots \forall x_k \phi(x_1, x_2, \ldots, x_k, z_1, \ldots, z_m)$, where ϕ is quantifier-free. We shall investigate a nice property of universal formulas later in this chapter.   In Theorem 4.7 of Section 4.5.

The idea of introducing the function symbol f to get rid of a $\exists$ can be extended to more complicated sentences, in particular ones in prenex normal form. We illustrate how in the following exercise.

Exercise 4.51

Let $\phi(x_1, x_2, x_3, x_4, x_5)$ be a quantifier-free formula with free variables x_1, x_2, x_3, x_4, x_5. Let f be a 2-place function symbol and g a 3-place function symbol not used in ϕ.

(a) Show that the sentence

$$(\forall x_1 \forall x_2 \forall x_3 \phi(x_1, x_2, x_3, f(x_1, x_2), g(x_1, x_2, x_3))$$
$$\to \forall x_1 \forall x_2 \exists y_1 \forall x_3 \exists y_2 \phi(x_1, x_2, x_3, y_1, y_2))$$

is universally valid.

(b) Show that

$$\forall x_1 \forall x_2 \exists y_1 \forall x_3 \exists y_2 \phi(x_1, x_2, x_3, y_1, y_2)$$

is satisfiable if and only if

$$\forall x_1 \forall x_2 \forall x_3 \phi(x_1, x_2, x_3, f(x_1, x_2), g(x_1, x_2, x_3))$$

is satisfiable.

We hope that your solution to part (b) took a structure $\mathcal{A} = \langle A, \ldots \rangle$ satisfying

$$\forall x_1 \forall x_2 \exists y_1 \forall x_3 \exists y_2 \phi(x_1, x_2, x_3, y_1, y_2)$$

and expanded it to a structure $\mathcal{A}^* = \langle A, \ldots, f^*, g^* \rangle$ with f^* and g^* defined to ensure that for all $a_1, a_2, a_3 \in A$,   For those worried about this sort of thing, the axiom of choice is needed to define these and all other Skolem functions.

$$\mathcal{A}^* \vDash_{x_1/a_1, x_2/a_2, x_3/a_3, y_1/f^*(a_1,a_2), y_2/g^*(a_1,a_2,a_3)} \phi(x_1, x_2, x_3, y_1, y_2).$$

The functions f^* and g^* are further examples of Skolem functions for a formula in a structure, here for $\forall x_1 \forall x_2 \exists y_1 \forall x_3 \exists y_2 \phi(x_1, x_2, x_3, y_1, y_2)$ in the structure $\mathcal{A}$. We introduced the function symbol f to correspond to the $\exists y_1$ in the formula

$$\forall x_1 \forall x_2 \exists y_1 \forall x_3 \exists y_2 \phi(x_1, x_2, x_3, y_1, y_2)$$

and made f a 2-place symbol because the $\exists y_1$ was preceded by two universal quantifiers, $\forall x_1$ and $\forall x_2$. The interpretation of the y_1 depends on the interpretation of the two variable x_1 and x_2. We introduced the function symbol g to correspond to the $\exists y_2$ and made it a 3-place symbol because the $\exists y_2$ was preceded by three universal quantifiers, with the interpretation of the y_2 depending on the interpretation of the three variables x_1, x_2 and x_3.

We shall outline how to extend these ideas to arbitrarily complex sentences. In the one application of the construction later in the book, we shall start with sentences not necessarily in prenex normal form. Of course, given any sentence ψ, we can first construct a logically equivalent sentence in prenex normal form, very roughly of the form

$$\forall x_1 \ldots \exists y_1 \ldots \forall x_i \ldots \exists y_j \ldots \phi(x_1, \ldots, x_n, y_1, \ldots, y_m),$$

where ϕ is quantifier-free, the x_is are the universally quantified variables and the y_js are the existentially quantified variables. For the sake of notational simplicity, we assume that the universal quantifiers in the prefix appear in the order $\forall x_1, \forall x_2, \ldots, \forall x_n$ and the existential quantifiers in the order $\exists y_1, \exists y_2, \ldots, \exists y_m$, doubtless jumbled up with the $\forall x_i$s, just as in the formula used in Exercise 4.51 where $n = 3$ and $m = 2$. By Exercise 4.49 we can ensure that the quantified variables are distinct, to avoid any redundancy in this notation. To simplify this prenex normal form into a universal sentence, for each existential quantifier $\exists y_j$ we introduce a new function symbol f_j of as many arguments as there are universal quantifiers preceding the $\exists y_j$ in the prefix. So if $\exists y_j$ is preceded by $\forall x_1, \forall x_2, \ldots, \forall x_k$ (and earlier $\exists y_i$s), we make f_j a k-place function symbol. We then replace the y_j in ϕ by $f_j(x_1, x_2, \ldots, x_k)$.

Check that all this accords with what we did in Exercise 4.51.

There's a technical detail we've avoided so far! If the first quantifiers in the prenex normal form are not universal but existential, so that the prenex normal form looks like

$$\exists y_1 \exists y_2 \ldots \exists y_r \forall x_1 \ldots \phi(x_1, \ldots, y_1, \ldots),$$

rather than introduce new function symbols for $y_1, y_2, \ldots, y_r$, we replace these variables by new constant symbols $\mathbf{c}_1, \mathbf{c}_2, \ldots, \mathbf{c}_r$.

Some books regard constant symbols as 0-place function symbols, which might help to simplify our description here. Given a structure $\mathcal{A}$ satisfying the original sentence ψ, we might call the interpretations of the new constant symbols *Skolem constants* for the formula in $\mathcal{A}$.

The point of all this is that by using these new function and constant symbols we can construct from the original sentence ψ, via its prenex normal form, a universal sentence ψ^{Sk} of the form

$$\forall x_1 \ldots \forall x_n \phi(x_1, \ldots, x_n, \ldots, f_j(x_1, x_2, \ldots, x_k), \ldots)$$

using extra function symbols not originally in ψ, called a *Skolem form* for ψ, which is satisfiable if and only if ψ is satisfiable, and such that $(\psi^{\mathrm{Sk}} \to \psi)$ is universally valid. A proof of this result requires a straightforward extension of the arguments in our solution to Exercise 4.50. These arguments actually prove something a bit stronger than that ψ is satisfiable if and only if ψ^{Sk} is satisfiable. We shall state this stronger result as a theorem, as we shall use it in Section 6.4 of Chapter 6 as one way of proving a significant result of the subject, called the *downward Löwenheim–Skolem theorem*.

Theorem 4.6

Let ψ be a sentence with Skolem form ψ^{Sk}, using extra function and constant symbols not originally in ψ. Then $(\psi^{\mathrm{Sk}} \to \psi)$ is logically valid. If $\mathcal{A}$ is a structure for the language of ψ which satisfies ψ, there is an expansion $\mathcal{A}^*$ of $\mathcal{A}$, adding interpretations of the extra function and constant symbols (by Skolem functions and constants), which satisfies ψ^{Sk}, while any structure satisfying ψ^{Sk} also satisfies ψ.

In this section we have started to look at formulas which are 'always true'. These are very important in their own right and entirely non-trivial by comparison with propositional tautologies. The question arises of whether there is an algorithmic procedure for deciding whether a predicate formula is universally valid. For propositional formulas there's a straightforward procedure, just by using truth tables. A given propositional formula ϕ involves only finitely many propositional variables. So all we need to do is test its truth value under each of the finitely many different truth assignments on these variables – if they all come out as true, the formula is a tautology, so is universally valid. But with a first-order formula ϕ, although it involves only finitely many symbols in a language L, there will usually be infinitely many possible domains for a structure for L. Also for each domain, there could many different ways of interpreting the relation, function and constant symbols appearing in ϕ. If our only procedure for testing the universal validity of ϕ involves checking infinitely many structures, it's unlikely that this will give a real decision procedure. Even checking the truth of ϕ of the shape $\forall x \psi(x)$ in a single structure $\mathcal{A}$ with an infinite domain could be unfeasible, should it be the case that our procedure relies on testing whether $\mathcal{A} \vDash_{x/a} \psi(x)$ for all a in this infinite domain. But don't despair just yet! Even if there is no decision procedure, it's still useful and interesting to establish that particular individual formulas are universally valid! Perhaps the formal proof system which we shall look at in the next chapter genuinely gives a better way of finding universally valid formulas (assuming that we have a soundness and completeness theorem connecting formal theorems and universally valid formulas).

> This is just leading you on! There is *no* algorithmic procedure which will decide whether a predicate formula is universally valid.

We shall turn now to the main point of our predicate language, which is to provide a framework for expressing interesting mathematical theories and finding their logical consequences. This is the subject of our next section.

Further exercises

Exercise 4.52 ______________________________

Determine whether each of the following formulas is universally valid. (P, Q are 1-place relation symbols and R is a 2-place relation symbol.)

(a) $(\forall x(P(x) \to Q(x)) \to (\forall x P(x) \to \forall x Q(x)))$

(b) $((\forall x P(x) \to \forall x Q(x)) \to \forall x(P(x) \to Q(x)))$

(c) $(\forall x \exists y R(x, y) \to \exists y \forall x R(x, y))$

(d) $(\forall y \forall x R(x, y) \to \forall x \exists y R(x, y))$

Exercise 4.53 ______________________________

For each of the following formulas, what is the smallest possible (non-empty) domain of any structure in which it is satisfiable? In each case give a specific structure of this minimal size. (R is a 2-place relation symbol, f is a 1-place function symbol and $\mathbf{c}$ is a constant symbol.)

(a) $\exists x \exists y(R(x, y) \to \neg R(y, y))$

(b) $\forall x \exists y \neg x = y$

(c) $(\forall x \neg f(x) = \mathbf{c} \wedge \forall x \forall y(f(x) = f(y) \to x = y))$

Exercise 4.54 ______________________________

Let f be a function symbol of two arguments. Show that the formula

$$(\forall x \exists y \neg x = y \to \exists y \neg f(y, x) = y)$$

is not universally valid.

Exercise 4.55 ___________

In a language for first-order predicate calculus with one 2-place relation symbol R, let ϕ denote the sentence

$$(\forall x \exists y R(x,y) \wedge (\forall x \forall y \forall z((R(x,y) \wedge R(y,z)) \rightarrow R(x,z))$$
$$\wedge (\forall x \neg R(x,x) \wedge \exists x \forall y \neg R(y,x))))$$

and ψ the sentence

$$\forall x \forall y (\neg R(x,y) \vee \neg R(y,x)).$$

For each of the four combinations of one of the set $\{\phi, \neg\phi\}$ with one of the set $\{\psi, \neg\psi\}$, decide whether the combination is satisfiable. If it is satisfiable, then give a structure satisfying it; if not, explain why not.

Exercise 4.56 ___________

Prove Theorem 4.4. [*Hints*: Make use of the more formal definition of 'τ is freely substitutable for x in ϕ' in Exercise 4.36 and argue in terms of the number of connectives and quantifiers in ϕ. Theorem 4.1 of Section 4.2 will help deal with cases where x_1 is not free in a formula.]

4.4 Some axiom systems and their consequences

In this section we shall look at some of the interesting mathematical theories that can be represented within the framework of the predicate calculus dealt with in this book. In practice in mathematics, the word 'theory' tends to be used to mean either all the mathematical consequences of a set of axioms or all the properties shared by some class of structures. For instance, the theory of groups can either be regarded as all consequences of a certain set of axioms or as the properties of those structures called groups.

The proper name for what we are dealing with is *first-order* predicate calculus. We will mention second-order languages later in the section.

Our preliminary definition of what we shall mean by a theory for the rest of this book is that it is the set of all consequences of a set of sentences in a first-order language which we shall call the *axioms* of the theory. We shall be more precise about what we mean by 'consequences' quite soon. We shall call any structure which satisfies these sentences a *model* of the theory.

Don't worry if you don't know much about groups. We shall discuss them later in this section.

In this section we shall describe several standard mathematical theories by giving axioms for them. In each case, the axioms completely determine what mathematicians mean informally by the theory. Later in the book we shall look at some mathematical theories described by means other than first-order axioms for which the question arises whether there is a set of first-order axioms that captures the essence of the theory. In such a case we shall describe the theory as *first-order axiomatizable*, or *axiomatizable* for short.

Strictly speaking we should call these sentences *non-logical axioms* to distinguish them from formulas which are instances of axioms in the formal proof system, like the instance $(\phi \rightarrow (\psi \rightarrow \phi))$ of Ax 1. The latter is often called a *logical axiom*.

For this section, we shall look only at some mathematical theories normally described by giving axioms for them. We start with the simple, but we hope well-known, example of equivalence relations.

We'll see later that several very important mathematical theories are not first-order axiomatizable. Note that some authors use 'axiomatizable' only when there is an algorithmic procedure for deciding whether a sentence is one of the axioms.

Equivalence relations

> *Definitions   Equivalence relations*
>
> Let L be a language with a binary relation symbol R. The theory of *equivalence relations* has the following axioms.
>
> 1. $\forall x R(x, x)$ (R is reflexive)
> 2. $\forall x \forall y (R(x, y) \to R(y, x))$ (R is symmetric)
> 3. $\forall x \forall y \forall z ((R(x, y) \land R(y, z)) \to R(x, z))$ (R is transitive)
>
> Let $\mathcal{A} = \langle A, R^* \rangle$, where the domain A is a non-empty set and R^* is a subset of $A \times A$, be a structure satisfying the axioms – what we shall call a *model* of them. Then R^* is said to be an *equivalence relation* on A, and the model $\mathcal{A}$ is often described (confusingly) also as an equivalence relation.

In everyday mathematics, the relation is normally written in infix notation, using xRy instead of $R(x, y)$.

Translating the axioms, if $\mathcal{A}$ satisfies the axioms, then R^* has the following properties:

for all $a \in A$, $(a, a) \in R^*$;

for all $a, b \in A$, if $(a, b) \in R^*$, then $(b, a) \in R^*$;

for all $a, b, c \in A$, if $(a, b) \in R^*$ and $(b, c) \in R^*$, then $(a, c) \in R^*$.

We hope that you have already encountered equivalence relations in your mathematical studies. If not, then basically $(a, b) \in R^*$ is a way of saying that 'a is the same as b in a certain way'. We saw examples of equivalence relations earlier in the book, namely the relation of logical equivalence $\equiv$ on the set of formulas in a propositional language and on the set of formulas of a first-order language. Here are some further simple examples of equivalence relations.

See Exercise 2.35 in Section 2.4 of Chapter 2 and Exercise 4.44 in Section 4.3 of this chapter.

Example 4.1

(a) $\langle A, = \rangle$ for any non-empty set A. The three axioms are plainly satisfied by equality on the set A.

(b) $\langle A, R^* \rangle$, where A is the set of all lines in the plane and $(a, b) \in R^*$ if the line a is parallel to the line b. The axioms are plainly satisfied. You might, however, like to pause and reflect what full proofs, particularly of transitivity, might look like, depending on how 'parallel' is defined.

(c) $\langle \{0, 1, 2, 3\}, \{(0, 0), (1, 1), (2, 2), (3, 3), (0, 1), (1, 0)\} \rangle$. The axioms can be checked slightly laboriously, e.g. explaining why transitivity holds for each of the 4^3 different ways of interpreting the x, y, z.

(d) $\langle \mathbb{Z}, R^* \rangle$, where $(m, n) \in R^*$ if $m - n = 6k$ for some $k \in \mathbb{Z}$. (The relation can also be expressed as '6 divides $m - n$' and as 'm is congruent to n modulo 6'.) Verifying that this is an equivalence relation takes somewhat more effort, and perhaps requires more ingenuity, than the previous examples. We have to check that each of the three axioms is satisfied.

For all $n \in \mathbb{Z}$, we have $n - n = 0 = 6 \times 0$, so that R^* is reflexive.

Throughout this section, structures are normal, that is, $=$ is interpreted as equality on the domains of the structures.

The effort is worthwhile as this is a mathematically important equivalence relation.

For all $m, n \in \mathbb{Z}$, if $(m, n) \in R^*$, then $m - n = 6k$ for some integer k. Then

$$n - m = -(m - n) = -6k = 6(-k),$$

so that, as $-k$ is an integer, $(n, m) \in R^*$. Thus R^* is symmetric.

For all $m, n, p \in \mathbb{Z}$, if $(m, n) \in R^*$ and $(n, p) \in R^*$, then $m - n = 6k$ and $n - p = 6j$ for some integers k, j. Then

$$m - p = (m - n) + (n - p) = 6k + 6j = 6(k + j),$$

so that, as $k + j$ is an integer, $(m, p) \in R^*$. Thus R^* is transitive. $\quad\blacklozenge$

Example 4.2

$\langle Form(P, S), \equiv \rangle$, where $Form(P)$ is the set of propositional formulas built up from the set P of propositional variables using connectives in the set S and $\equiv$ is logical equivalence. We asked you to show that this is an equivalence relation in Exercise 2.35 of Section 2.4 of Chapter 2. $\quad\blacklozenge$

Exercise 4.57

Let r be a positive integer. Show that the structure $\langle \mathbb{Z}, R^* \rangle$, where $(m, n) \in R^*$ if $m - n = rk$ for some $k \in \mathbb{Z}$, is an equivalence relation.

For the exercises in this section, take a reasonably relaxed attitude to the level of detail within your solutions!

Exercise 4.58

Decide which of the following structures are equivalence relations.

(a) $\langle \mathbb{Z}, R^* \rangle$, where $(m, n) \in R^*$ if $m + n = 6k$ for some $k \in \mathbb{Z}$.

(b) $\langle \mathbb{Z}, R^* \rangle$, where $(m, n) \in R^*$ if $m + n = 2k$ for some $k \in \mathbb{Z}$.

(c) $\langle \mathbb{R}, R^* \rangle$, where $(a, b) \in R^*$ if $a - b$ is rational.

(d) $\langle \mathbb{R}, R^* \rangle$, where $(a, b) \in R^*$ if $a - b$ is irrational.

Now that we have one example of a standard mathematical theory given by axioms, we should give a precise definition of 'theory'. In everyday mathematics, a theory is more than a set of axioms: it is really the set of all consequences of the axioms. In a book on logic it is rather important to make more precise what is meant by 'consequence'! We have already used the word in the context of the propositional calculus when saying that a formula ϕ is a logical consequence of a set Γ, written as $\Gamma \vDash \phi$, and we shall extend this terminology to predicate calculus. Our definition of a theory, given a set of axioms, will then be the set of all logical consequences of the axioms.

Definitions   *Logical consequence and theory*

Let Γ be a set of formulas and ϕ a formula in a language L. Then ϕ is a *logical consequence* of Γ, or equivalently Γ *logically implies* ϕ, if for every structure $\mathcal{A}$ for L (normal structure when L includes equality) and interpretation $\vec{a}$ of the free variables $\vec{x}$ possibly appearing in members of Γ and ϕ,

$$\text{if } \mathcal{A} \vDash_{[\vec{x}/\vec{a}]} \gamma \text{ for all } \gamma \in \Gamma, \text{ then } \mathcal{A} \vDash_{[\vec{x}/\vec{a}]} \phi.$$

We write this as $\Gamma \vDash \phi$.

Let T be a set of sentences in the language L. The *first-order theory* of T in the language L is the set of all sentences ϕ in the language L which are logical consequences of T. We shall usually call this the *theory of T* for short or, even shorter, the *theory T*. The sentences in T are called the *axioms* of the theory.

A structure $\mathcal{A}$ which satisfies the axioms of a theory is said to be a *model* of the theory.

As for propositional formulas, we write $\Gamma \nvDash \phi$ when ϕ is not a logical consequence of Γ.

For various technical reasons, it is preferable to use sentences, i.e. formulas with no free variables, in our definition of 'theory', both for the axioms and their consequences.

Exercise 4.59 ──────────────────────────

Suppose T is a set of sentences in a language L regarded as the axioms of a theory and that $\mathcal{A}$ is a model for these axioms. Show that $\mathcal{A}$ satisfies the theory of T, i.e. all logical consequences of the axioms.

Solution

Let ϕ be a sentence of L. By definition of $T \vDash \phi$, if $\mathcal{A}$ satisfies T then $\mathcal{A}$ satisfies ϕ. Thus $\mathcal{A}$ satisfies the theory of T.

Of course, in everyday mathematics we regard the consequences of a set of axioms as the statements that we can prove from the axioms – that is the whole point of the axiomatic approach – so an alternative definition of a theory could be as the set of sentences derivable from the axioms within a formal proof system. We shall provide a suitable proof system in the next chapter and then prove both the soundness and completeness theorems for it, so that this definition of a theory is equivalent to the one we have given in terms of logical consequence. In any case, to demonstrate logical consequences, we have to use some sort of proof at a higher level than the formal system, which in its way is just as stringent as the formal system will be.

A simple example of a logical consequence of the axioms for an equivalence relation is the sentence

$$\forall x \forall y \forall z \forall w ((R(x,y) \wedge (R(y,z) \wedge R(z,w))) \rightarrow R(x,w)).$$

In any equivalence relation $\mathcal{A} = \langle A, R^* \rangle$, for any $a, b, c, d \in A$ with all of $(a,b), (b,c), (c,d) \in R^*$, transitivity on $(b,c), (c,d) \in R^*$ gives $(b,d) \in R^*$, so that another use of transitivity with $(a,b) \in R^*$ gives $(a,d) \in R^*$ as required.

Logical consequence obviously has all the properties in relation to the propositional connectives that we have already discussed in Section 2.6 of Chapter 2 and we shall not repeat all of these here. One important example is Theorem 2.7, namely that

$$\Gamma \cup \{\phi\} \vDash \psi \quad \text{if and only if} \quad \Gamma \vDash (\phi \to \psi).$$

This result helps turn some of the universally valid formulas in the last section into the properties of logical consequence in relation to the quantifiers. For instance, Theorem 4.3 can be rephrased in terms of logical consequence to say that if ϕ is a formula and the term τ is freely substitutable for x in ϕ, then

$$\forall x \phi(x) \vDash \phi(\tau).$$

Other important results are the analogues of Exercises 2.74 and 2.75, which we ask you to show hold for predicate languages below.

Exercise 4.60

Let Γ be a set of sentences and ϕ a sentence in a language L. Show that $\Gamma \vDash \phi$ and $\Gamma \vDash \neg\phi$ if and only if there are no structures satisfying all of the sentences of Γ.

Exercise 4.61

Let Γ be a set of sentences and ϕ a sentence in a language L. Show that if $\Gamma \vDash \phi$ and $\Gamma \vDash \neg\phi$, then $\Gamma \vDash \psi$, for all sentences ψ in the language.

Our experience with propositional formulas should lead us to expect that for a given set of sentences Γ and any sentence ϕ in a language L, it might not be the case that $\Gamma \vDash \phi$ or $\Gamma \vDash \neg\phi$.

The use of the same symbol $\vDash$ both for 'satisfies' and 'logically implies' can give rise to a delicious confusion here! In the context of 'satisfies', for any sentence ϕ and structure $\mathcal{A}$ for its language, exactly one of $\mathcal{A} \vDash \phi$ and $\mathcal{A} \vDash \neg\phi$ must hold. But in the context of logical consequence, only for special (what are called *complete*) sets Γ does exactly one of $\Gamma \vDash \phi$ and $\Gamma \vDash \neg\phi$ hold.

Exercise 4.62

Take Γ to be the set of axioms for an equivalence relation in the language L with equality and a binary relation symbol R. Find a sentence ϕ in this language such that neither $\Gamma \vDash \phi$ nor $\Gamma \vDash \neg\phi$.

Solution

Take for instance ϕ to be the formula $\forall x \exists y (R(x, y) \wedge \neg x = y)$. With the model $\mathcal{A}$ of Γ given by $\langle \mathbb{Z}, =, = \rangle$ (essentially a special case of Example 4.1(a) which interprets both R and $=$ as actual equality on the set of integers), we have

$$\mathcal{A} \nvDash \forall x \exists y (R(x, y) \wedge \neg x = y),$$

so that $\Gamma \nvDash \phi$. With the model $\mathcal{A}$ (essentially in Example 4.1(d)) given by $\langle \mathbb{Z}, R^*, = \rangle$, where $(m, n) \in R^*$ if $m - n = 6k$ for some $k \in \mathbb{Z}$, we have

$$\mathcal{A} \nvDash \neg\forall x \exists y (R(x, y) \wedge \neg x = y),$$

so that $\Gamma \nvDash \neg\phi$.

For Exercise 4.97 later in the book, we will rely on you knowing something about the connection between equivalence relations and partitions, the latter defined as follows.

> ### Definition Partition
>
> A *partition* of a set A is a set P of subsets of A such that each element of A is in exactly one of the subsets. Equivalently, P is a set of subsets of A with the property that for all $a \in A$, there exists $X \in P$ such that
>
> (i) $a \in X$ and
>
> (ii) for all $Y \in P$, if $a \in Y$, then $X = Y$.

To capture that a is in exactly one subset in P, we can say that a is in at least one subset and in at most one.

The connections between equivalence relations and partitions are as follows. Given an equivalence relation $\langle A, R^* \rangle$ and $a \in A$, the *equivalence class* of a is the subset of A given by

$$[\![a]\!] = \{b \in A : (a, b) \in R^*\}.$$

Then it can be shown that the set of all these subsets, $\{[\![a]\!] : a \in A\}$, forms a partition of A. Conversely, given a partition P of the set A, we can define an equivalence relation R^* on A whose equivalence classes are precisely the subsets in P.

There is a point of interest here regardless of whether you go on to attempt Exercise 4.97. Although we have given axioms above for a partition, we shall have difficulty doing this within the framework of predicate calculus in this book. In the axioms we have given above, the quantifiers $\forall$ and $\exists$ govern two different sorts of object: elements a of A; and elements X, Y of P which are really subsets, not elements, of A. This doesn't fit in with our framework where there's only one type of element, namely a member of the domain of a structure for the language. What we seem to need to axiomatize partitions is a language which allows one to quantify both over elements of a set and over its subsets. This is called a *second-order language* and is beyond the scope of this book. There are some very important examples of mathematical theories which are axiomatized in a second-order language, for instance the theory of the real numbers for which we have given axioms in Chapter 1. We shall show in Chapter 6 that this particular theory cannot be axiomatized in a first-order language.

Another related point is that what in this book we are calling the theory of equivalence relations is, using its full name, its first-order theory, i.e. all logical consequences of the axioms expressible within the given first-order language. It doesn't include all the consequences of the axioms deducible in everyday mathematics, as many of these simply aren't expressible within the first-order language. This will be true in general for most of the remaining axiom systems in the book.

Our next examples are various theories of order relations.

Order relations

> ### *Definitions   Weak orders*
>
> Let L be a language with equality with a binary relation symbol R. The theory of *weak partial order* has the following axioms.
>
> 1. $\forall x R(x, x)$ (R is reflexive)
> 2. $\forall x \forall y((R(x, y) \wedge R(y, x)) \to x = y)$ (R is anti-symmetric)
> 3. $\forall x \forall y \forall z((R(x, y) \wedge R(y, z)) \to R(x, z))$ (R is transitive)
>
> A model $\langle A, R^*, = \rangle$ of the axioms is said to be a *weak partial order*.
>
> The addition to the above of the following axiom:
>
> 4 $\forall x \forall y(R(x, y) \vee R(y, x))$ (R is linear)
>
> gives the axioms for *weak linear order* and a model of the axioms is called a *weak linear order*.

Note that we make consistent use of the names for certain properties of 2-place relations: 'reflexive' and 'transitive' mean the same in the weak order axioms as they do for equivalence relations.

As with equivalence relations, in everyday mathematics we express order relations using infix notation, $x R y$ instead of $R(x, y)$.

The same will apply for properties like commutativity and associativity in axioms involving a 2-place function symbol.

Example 4.3

(a) $\langle A, \leq, = \rangle$ for A any of $\mathbb{N}$, $\mathbb{Z}$, $\mathbb{Q}$ and $\mathbb{R}$ is a weak linear order (and thus also a weak partial order). The axioms are clearly satisfied (and are arguably motivated by these examples).

(b) $\langle A, \geq, = \rangle$ for A any of $\mathbb{N}$, $\mathbb{Z}$, $\mathbb{Q}$ and $\mathbb{R}$ is a weak linear order. Note that there is no bias within the axioms that prefers interpreting R by $\leq$ rather than $\geq$.

(c) If A is any open interval of the form (a, b) of $\mathbb{R}$, or any closed interval of the form $[a, b]$, then $\langle A, \leq, = \rangle$ is a weak linear order. (It may be useful for you later to have examples of linear orders like these, some of which contain maximum and minimum elements, and some of which don't!)

It will usually be clear from the context whether (a, b) refers to an open interval or to an ordered pair.

(d) $\langle \mathscr{P}(\mathbb{R}), \subseteq, = \rangle$, where $\mathscr{P}(\mathbb{R})$ is the set of all subsets of $\mathbb{R}$, is a weak partial order. It is not a linear order as, for instance, taking the open intervals $A = (0, 1)$ and $B = (2, 3)$, neither $A \subseteq B$ nor $B \subseteq A$ is true.

(e) $\langle \{0, 1, 2\}, \{(0, 0), (1, 1), (2, 2), (0, 1)\}, = \rangle$ is a weak partial, but not linear, order.   ♦

Exercise 4.63

Let $\langle A, R^*, = \rangle$ be a weak partial order. Can there be $a, b \in A$ with $a \neq b$ and both (a, b) and (b, a) in R^*?

Solution

While not giving you the answer, it may explain the quaint terminology 'anti-symmetric' used to describe the second weak order axiom. The relation R^* is as far as possible not symmetric, in the sense of the second axiom for equivalence relations, given the limited symmetry entailed by the reflexive property.

We hope that the answer is very straightforward, thinking of logical consequences of the axioms.

There is some interest, and often considerable challenge, in deciding whether some axioms of a system are logical consequences of the remaining axioms.

When we have provided our formal proof system, the equivalent problem is whether one axiom is derivable from the rest.

Exercise 4.64

Is the linearity axiom (axiom 4) for a weak linear order a logical consequence of the remaining axioms (i.e. the three axioms for a weak partial order)?

Solution

Example 4.3(d) above provides an example of a structure in which the three partial order axioms are true and in which the linearity axiom is false. So the latter is not a logical consequence of the former.

It happens to be the case that none of the four axioms for linear order is a logical consequence of the other three, so there is no redundancy. (We leave the remaining cases for you as Exercise 4.96.) However, we won't insist that all axiomatizations of interesting theories should have this sort of property. In real mathematics, a spot of redundancy in the axioms often makes for a much easier life when deriving their consequences!

Exercise 4.65

For each of the following sentences, decide whether it is a logical consequence of the axioms for a weak linear order.

(a) $\exists x \forall y R(x, y)$

(b) $\neg \exists x \forall y R(x, y)$

(c) $\forall x \exists y R(x, y)$

(d) $\forall x \exists y (\neg x = y \land R(x, y))$

Solution

(a) If we imagine that $R(x, y)$ is interpreted by 'x is less than or equal to y', then this sentence says 'there is a least element x in the order'. This is not a logical consequence of the axioms, as it is false in the model $\langle \mathbb{Z}, \leq, = \rangle$ of the axioms.

(b) If we imagine that $R(x, y)$ is interpreted by 'x is less than or equal to y', then this sentence says 'there is no least element x in the order'. This is not a logical consequence of the axioms, as it is false in the model $\langle [0, 1], \leq, = \rangle$ of the axioms, taking the closed interval $[0, 1]$ of the real numbers.

(c) Not given.

(d) Not given

Associated with each weak order $\leq$ is a strict order $<$, meaning '$\leq$ and not $=$'. The axioms for strict orders are as follows.

Likewise, with each $\geq$ there is an associated $>$.

Definitions **Strict orders**

Let L be a language with equality with a binary relation symbol S. The theory of *strict partial order* has the following axioms.

1. $\forall x \neg S(x, x)$ (S is irreflexive)

2. $\forall x \forall y \forall z ((S(x, y) \wedge S(y, z)) \to S(x, z))$ (S is transitive)

A model $\langle A, S^*, = \rangle$ of the axioms is said to be a *strict partial order*.

The addition to the above of the following axiom:

3   $\forall x \forall y (S(x, y) \vee x = y \vee S(y, x))$ (S is linear)

gives the axioms for *strict linear order* and a model of the axioms is called a *strict linear order*.

Example 4.4

(a) $\langle A, <, = \rangle$ for A any of $\mathbb{N}$, $\mathbb{Z}$, $\mathbb{Q}$ and $\mathbb{R}$ is a strict linear order (and thus also a strict partial order).

(b) $\langle A, >, = \rangle$ for A any of $\mathbb{N}$, $\mathbb{Z}$, $\mathbb{Q}$ and $\mathbb{R}$ is a strict linear order.

(c) If A is any open interval or closed interval of $\mathbb{R}$, then $\langle A, <, = \rangle$ is a strict linear order.

(d) $\langle \mathbb{Z}, S^*, = \rangle$, where $(m, n) \in S^*$ if m is a proper divisor of n, is a strict partial order. It is not a linear order as, for instance, taking the integers 2 and 3, neither of these is a proper divisor of the other, nor are they equal.

(e) $\langle \{0, 1, 2\}, \{(0, 1)\}, = \rangle$ is a strict partial, but not linear, order.   ♦

Exercise 4.66

Show that the sentence

$$\forall x \forall y (S(x, y) \to \neg S(y, x))$$

is a logical consequence of the axioms for a strict partial order S.

This says that the relation is asymmetric.

Exercise 4.67

Is $\langle \{0, 1, 2\}, \varnothing, = \rangle$ any of the following: a weak partial order; a weak linear order; a strict partial order; a strict linear order?

Exercise 4.68

(a) Let $\langle A, R^*, = \rangle$ be a weak partial order. Define the subset S^* of $A \times A$ by $S^* = \{(a, b) \in R^* : a \neq b\}$. Show that $\langle A, S^*, = \rangle$ is a strict partial order.

(b) Let $\langle A, S^*, = \rangle$ be a strict partial order. Show how to exploit S^* to define a subset R^* of $A \times A$ such that $\langle A, R^*, = \rangle$ is a weak partial order.

(c) Does the result of part (a) extend to cover the linear property when $\langle A, R^*, = \rangle$ is a weak linear order, i.e. is $\langle A, S^*, = \rangle$ then a strict linear order? Does your answer to part (b) extend in a similar way to cover linearity?

Given the results of Exercise 4.68, we may well refer to a partial or linear order without specifying whether the order is weak or strict. In any specific application we will use whichever way of describing the order is most convenient. We will often use the more familiar infix notation $x \leq y$ for weak orders and $x < y$ for strict orders, rather than the prefix notation $R(x, y)$.

There are various more specialised theories of order which are of mathematical interest. We will describe some of these in the following exercise.

Exercise 4.69 ───

For each of the following theories, give a model of the theory and also an example of a linearly ordered set which is not a model of the theory.

(a) The theory of *unbounded* linear order, which has axioms for a strict linear order along with

$$\forall x \exists y \exists z (y < x \wedge x < z) \quad (< \text{ is unbounded}).$$

(b) The theory of linear order with a *maximum* element, which has axioms for a weak linear order along with

$$\exists x \forall y \ y \leq x.$$

(c) The theory of *dense* linear order, which has axioms for a strict linear order along with

$$\forall x \forall y (x < y \rightarrow \exists z (x < z \wedge z < y)) \quad (< \text{ is dense}).$$

(d) The theory of *discrete* linear order, which has axioms for a strict linear order along with the axioms

$$\forall x (\exists y \ x < y \rightarrow \exists z (x < z \wedge \forall w (x < w \rightarrow (z = w \vee z < w)))),$$
$$\forall x (\exists y \ y < x \rightarrow \exists z (z < x \wedge \forall w (w < x \rightarrow (w = z \vee w < z)))).$$

Exercise 4.70 ───

Write down extra axioms, besides those for a strict linear order, whose models are linear orders with a minimum element but with no maximum element.

Exercise 4.71 ───

(a) Let $\mathcal{A}$ be a model of the theory of unbounded linear order in Exercise 4.69(a). Can the domain of $\mathcal{A}$ be finite?

(b) Let $\mathcal{A}$ be a model of the theory of dense linear order in Exercise 4.69(c). Can the domain of $\mathcal{A}$ be finite?

───

Not all interesting theories of order can be axiomatized using a first-order language. One important example of a theory which cannot be axiomatized in this way is the theory of well-ordered sets. A weak linear order $\leq$ on a set A is a *well-order* if every non-empty subset of A contains a least element, i.e. for all $B \subseteq A$ with $B \neq \varnothing$, there is $b_0 \in B$ such that $b_0 \leq b$, for all $b \in B$. This description involves quantifying over subsets of A as well as its elements and is another example of an axiomatization in a second-order language. As such, it cannot directly be translated into a first-order language. This does not mean that there might not be some clever alternative way of axiomatizing the theory in a first-order language. However, we shall show in Chapter 6 that such an axiomatization is in fact impossible.

For instance, $\mathbb{N}$ with the usual $\leq$ is a well-order, while $\mathbb{Z}, \mathbb{Q}, \mathbb{R}$ are not.

Boolean algebras

We shall now look at our first example of an algebraic theory, involving functions on a set, not just relations.

Definitions   *Boolean algebras*

Let L be a language with equality, 2-place function symbols $+$ and $\cdot$, a 1-place function symbol c and constant symbols $\mathbf{0}$ and $\mathbf{1}$. The theory of *Boolean algebras* has the following axioms, using infix notation $x + y$ and $x \cdot y$ for the 2-place function symbols and writing $c(x)$ for the 1-place function symbol.

The notation $\vee, \wedge$ is often used instead of $+, \cdot$.

1. $\forall x \forall y \forall z\, (x + (y + z)) = ((x + y) + z)$ (associativity of $+$)
2. $\forall x \forall y \forall z\, (x \cdot (y \cdot z)) = ((x \cdot y) \cdot z)$ (associativity of $\cdot$)
3. $\forall x \forall y\, (x + y) = (y + x)$ (commutativity of $+$)
4. $\forall x \forall y\, (x \cdot y) = (y \cdot x)$ (commutativity of $\cdot$)
5. $\forall x\, (x + x) = x$ (idempotency of $+$)
6. $\forall x\, (x \cdot x) = x$ (idempotency of $\cdot$)
7. $\forall x \forall y \forall z\, (x + (y \cdot z)) = ((x + y) \cdot (x + z))$ (distributivity)
8. $\forall x \forall y \forall z\, (x \cdot (y + z)) = ((x \cdot y) + (x \cdot z))$ (distributivity)
9. $\forall x \forall y\, (x + (x \cdot y)) = x$ (absorption law)
10. $\forall x \forall y\, (x \cdot (x + y)) = x$ (absorption law)
11. $\forall x \forall y\, c((x + y)) = (c(x) \cdot c(y))$ (de Morgan law)
12. $\forall x \forall y\, c((x \cdot y)) = (c(x) + c(y))$ (de Morgan law)
13. $\forall x\, (x + \mathbf{0}) = x$
14. $\forall x\, (x + \mathbf{1}) = \mathbf{1}$
15. $\forall x\, (x \cdot \mathbf{0}) = \mathbf{0}$
16. $\forall x\, (x \cdot \mathbf{1}) = x$
17. $\neg \mathbf{0} = \mathbf{1}$
18. $\forall x\, (x + c(x)) = \mathbf{1}$
19. $\forall x\, (x \cdot c(x)) = \mathbf{0}$
20. $\forall x\, c(c(x)) = x$

A model $\mathcal{B} = \langle B, +_{\mathcal{B}}, \cdot_{\mathcal{B}}, c^{\mathcal{B}}, 0^{\mathcal{B}}, 1^{\mathcal{B}}, = \rangle$ of the axioms is called a *Boolean algebra*.

Boolean algebras were devised by the English mathematician George Boole (1815–1864), whose publication *An investigation into the Laws of Thought*, published in 1854, launched the modern analysis of this branch of logic. Besides the great importance of Boole's work for logic itself, his work helped mathematicians develop the modern understanding of what is meant by an algebra, by providing a system with algebraic laws different from standard rules for arithmetic.

For a more modern edition, see Boole [2].

Example 4.5

(a) $\langle\{T, F\}, f_\vee, f_\wedge, f_\neg, F, T\rangle$, where T, F are the standard truth values and $f_\vee, f_\wedge, f_\neg$ are the truth functions corresponding to the connectives $\vee, \wedge, \neg$, respectively.

(b) $\langle\mathscr{P}(S), \cup, \cap, C, \varnothing, S\rangle$, for any non-empty set S, where for each $X \in \mathscr{P}(S)$ (i.e. for each subset X of S) $C(X) = S \setminus X$, the complement of X in S. $\blacklozenge$

Exercise 4.72 ______________________________

Show that the following sentences are logical consequences of the axioms for a Boolean algebra.

(a) $c(0) = 1$ and $c(1) = 0$.

(b) $(c(0) + c(1)) = 1$

(c) $\forall x \forall y \forall z\, (x \cdot z) = ((y \cdot (z \cdot x)) + ((x \cdot c(y)) \cdot z))$

Exercise 4.73 ______________________________

Let $\mathcal{B} = \langle B, +_\mathcal{B}, \cdot_\mathcal{B}, c^\mathcal{B}, 0^\mathcal{B}, 1^\mathcal{B}, = \rangle$ be a Boolean algebra.

(a) Show that for any $b, c \in B$,

$$\mathcal{B} \vDash_{x/b, y/c} (x \cdot y) = x \quad \text{if and only if} \quad \mathcal{B} \vDash_{x/b, y/c} (x + y) = y.$$

(b) Define a relation R^* on B by bR^*c if $\mathcal{B} \vDash_{x/b, y/c} (x \cdot y) = x$. Show that R^* is a weak partial order on B with both a maximum and a minimum element.

____

An interesting Boolean algebra underlying propositional formulas is called the Lindenbaum algebra. Consider the set $Form(P)$ of formulas built up from a set P of propositional variables using the connectives $\vee, \wedge, \neg$. Logical equivalence on these formulas is an equivalence relation. Let F be the corresponding set of equivalence classes, so that

$$F = \{\llbracket \phi \rrbracket : \phi \in Form(P)\},$$

where the equivalence class $\llbracket \phi \rrbracket$ of ϕ is defined by

$$\llbracket \phi \rrbracket = \{\psi \in Form(P) : \phi \equiv \psi\}.$$

We shall define the structure $\mathcal{F} = \langle F, +_\mathcal{F}, \cdot_\mathcal{F}, c^\mathcal{F}, 0^\mathcal{F}, 1^\mathcal{F}, = \rangle$, called the *Lindenbaum algebra* of $Form(P)$, as follows. For equivalence classes $\llbracket \phi \rrbracket, \llbracket \psi \rrbracket$, we define

$$\llbracket \phi \rrbracket \cdot_\mathcal{F} \llbracket \psi \rrbracket = \llbracket (\phi \wedge \psi) \rrbracket.$$

Behind this simple-seeming definition is a significant piece of mathematics. For this to define $\cdot_\mathcal{F}$ as a function, it's essential that there is a unique output value for each input pair of elements $(\llbracket \phi \rrbracket, \llbracket \psi \rrbracket)$ in the domain F. What could go wrong is that $\llbracket \phi \rrbracket = \llbracket \phi' \rrbracket$ and $\llbracket \psi \rrbracket = \llbracket \psi' \rrbracket$, where ϕ', ψ' are different from ϕ, ψ, but $\llbracket (\phi \wedge \psi) \rrbracket \neq \llbracket (\phi' \wedge \psi') \rrbracket$. This would mean we had not given $\llbracket \phi \rrbracket \cdot_\mathcal{F} \llbracket \psi \rrbracket$ a unique value. Luckily, by the result of Exercise 2.37, this state of affairs cannot arise. If $\llbracket \phi \rrbracket = \llbracket \phi' \rrbracket$ and $\llbracket \psi \rrbracket = \llbracket \psi' \rrbracket$, we have $\phi \equiv \phi'$ and $\psi \equiv \psi'$. By

Indeed, we would normally expect each formula ϕ to be equivalent to many other formulas ϕ'. For instance, $\phi \equiv \phi'$ where ϕ' is $(\phi \wedge \phi)$.

Exercise 2.37(b), this gives $(\phi \wedge \psi) \equiv (\phi' \wedge \psi')$, so that $[\![(\phi \wedge \psi)]\!] = [\![(\phi' \wedge \psi')]\!]$. Similarly we can define the functions $+_{\mathcal{F}}$ and $c^{\mathcal{F}}$ by

$$[\![\phi]\!] +_{\mathcal{F}} [\![\psi]\!] = [\![(\phi \vee \psi)]\!],$$
$$c^{\mathcal{F}}([\![\phi]\!]) = [\![\neg\phi]\!].$$

We now define the interpretations of the constant symbols by

$$0^{\mathcal{F}} = [\![(p \wedge \neg p)]\!],$$
$$1^{\mathcal{F}} = [\![(p \vee \neg p)]\!],$$

where p is any propositional variable in P.

Exercise 4.74

Explain why $+_{\mathcal{F}}$ and $c^{\mathcal{F}}$ above do truly define functions and why we have unambiguously defined the interpretations of the constant symbols.

Exercise 4.75

Suppose that P consists of finitely many propositional variables $p_1, p_2, \ldots, p_n$. How many elements are in the domain of the Lindenbaum algebra of $Form(P)$? [*Hint*: All truth functions of n variables can be represented by formulas involving $p_1, p_2, \ldots, p_n$ using the connectives $\wedge, \vee, \neg$.]

Infinite sets

Our next example is the theory of infinite sets. Like many, perhaps all, of the examples so far, the interest in this theory lies not so much in the deductive first-order consequences of the axioms, but in its models – infinite sets are of great interest and importance within mathematics. Having said that, the axioms for this theory will generate important results in Chapter 6. We shall leave you to develop axioms for this theory through the following exercise.

Exercise 4.76

Let L be a language with equality.

(a) For each of the following sentences, describe the models of the sentence.

(i) $\exists x_1 \exists x_2 \neg x_1 = x_2$

(ii) $\exists x_1 \exists x_2 \exists x_3 \bigwedge_{1 \leq i < j \leq 3} \neg x_i = x_j$

The large 'and' sign, $\bigwedge$, was introduced earlier as a shorthand for a conjunction of several terms, here all the formulas $\neg x_i = x_j$ where $1 \leq i < j \leq 3$.

(b) Explain how to construct for each $n \in \mathbb{N}$ a sentence (which we will refer to as $\exists_{\geq n}$ in later work) whose models are precisely those sets with at least n elements.

A (correct!) sentence $\exists_{\geq n}$ can be exploited to give a set Σ of axioms for infinite sets, as follows:

$$\Sigma = \{\exists_{\geq n} : n \in \mathbb{N}\}.$$

Any model for Σ has a domain with at least n elements for each $n \in \mathbb{N}$, so is infinite. Likewise any structure with an infinite domain makes each of the sentences in Σ true. Thus Σ axiomatizes the theory of infinite sets.

These axioms can be exploited to give axioms for other theories.

Exercise 4.77 ____

Write down axioms for each of the following theories. (You may use the notation $\exists_{\geq n}$ as a shorthand in your answer!)

(a) The theory of sets with exactly n elements, where n is a given positive integer, in a language with equality.

(b) The theory of infinite strict linear orders, in a language with equality and a binary relation symbol R.

____

Our axiomatization of the theory of infinite sets uses infinitely many axioms. An interesting question arises of whether there is an alternative axiomatization using only finitely many sentences. Another question is whether there are axioms for the theory of finite sets. We shall answer these questions in Chapter 6.

Our next examples of theories will also figure in many applications in Chapter 6. They are all major objects of study in modern algebra: groups, rings and fields.

Groups

____

Definitions   *Groups*

Let L be the language with equality containing the 2-place function symbol $\cdot$ and the constant symbol $\mathbf{e}$. The theory of *groups* has the following axioms. These are written using infix notation $\cdot$, so we write $(x \cdot y)$ instead of $\cdot(x, y)$.

1. $\forall x \forall y \forall z \, (x \cdot (y \cdot z)) = ((x \cdot y) \cdot z)$ ($\cdot$ is associative)

2. $\forall x \, ((x \cdot \mathbf{e}) = x \land (\mathbf{e} \cdot x) = x)$ ($\mathbf{e}$ is an identity element)

3. $\forall x \exists y \, ((x \cdot y) = \mathbf{e} \land (y \cdot x) = \mathbf{e})$ (each element has an inverse)

A model $\mathcal{G} = \langle G, \cdot_{\mathcal{G}}, e^{\mathcal{G}}, = \rangle$ of the axioms is said to be a *group*.

If $\mathcal{G}$ satisfies the additional axiom

4. $\forall x \forall y \, (x \cdot y) = (y \cdot x)$ ($\cdot$ is commutative),

then $\mathcal{G}$ is a said to be a *commutative* group and the theory axiomatized is called the theory of *commutative groups*.

____

For good introductions to the theory of groups, see e.g. Allenby [1], Jordan and Jordan [21], and Ledermann and Weir [23].

A commutative group is also called an *Abelian group* in honour of the Norwegian mathematician Abel.

Example 4.6

(a) $\langle \mathbb{Z}, +, 0, = \rangle$, $\langle \mathbb{Q}, +, 0, = \rangle$, $\langle \mathbb{R}, +, 0, = \rangle$ and $\langle \mathbb{C}, +, 0, = \rangle$ are all commutative groups. (These are often described as the *additive group* of the integers and so on.) For instance, the inverse of 3 is -3 in all of these groups.

(b) $\langle \mathbb{Q} \setminus \{0\}, \times, 1, = \rangle$, $\langle \mathbb{R} \setminus \{0\}, \times, 1, = \rangle$ and $\langle \mathbb{C} \setminus \{0\}, \times, 1, = \rangle$ are all commutative groups. (These are often described as the *multiplicative group* of the non-zero rationals and so on.) The inverse of 3 is $1/3$ in all of these groups.

When writing abstract arguments about groups, the use of the notation $+$, 0 rather than $\cdot$, 1 depends on the context.

(c) For any integer $n \geq 2$, let $\mathbb{Z}_n$ be the set $\{0, 1, \ldots, n-1\}$. The operation $+_n$ of addition modulo n is defined on $\mathbb{Z}_n$ by

$$x +_n y = \text{ the remainder of } x + y \text{ on division by } n,$$

or equivalently

$$x +_n y = \begin{cases} x + y, & \text{if } x + y < n, \\ x + y - n, & \text{if } x + y \geq n. \end{cases}$$

Then $\langle \mathbb{Z}_n, +_n, 0, = \rangle$ is a commutative group with n elements. Note that every element of $\mathbb{Z}_n$ can be obtained from the element 1 by repeated use of $+_n$, e.g. $4 = 1 +_n 1 +_n 1 +_n 1$. The element 1 is said to *generate* the group and the group is said to be *cyclic*. The inverse element of 1, which is $n - 1$, also generates the group, as does any element a for which $\gcd\{a, n\} = 1$.

> This class of examples shows that for positive integer n, there is at least one group with n elements.

(d) For any integer $n \geq 1$, let S_n be the set of all permutations of the n element set $\{1, 2, \ldots, n\}$, i.e. all bijections from this set to itself. S_n has $n!$ elements. Then $\langle S_n, \circ, \mathrm{Id}, = \rangle$ is group, where $\circ$ is the operation of composition of functions and Id is the identity permutation, which maps i to itself for each $i = 1, 2, \ldots, n$. For $n \geq 3$, S_n is a non-commutative group.

> This class of groups illustrates that there are non-commutative groups of arbitrarily large finite orders. For instance, the permutation which swaps 1 and 2 and fixes all other $i \in \{1, 2, \ldots, n\}$ does not commute with the permutation which swaps 2 and 3 and fixes all other i.

(e) The set $\mathrm{GL}(2, \mathbb{R})$ of invertible 2×2 matrices with entries in $\mathbb{R}$ along with matrix multiplication and the identity matrix I gives an example of an infinite non-commutative group. $\blacklozenge$

Exercise 4.78 ______________________________

(a) Show that the sentence

$$(\mathbf{e} \cdot \mathbf{e}) = \mathbf{e}$$

is a logical consequence of the group axioms.

(b) Show that the sentence

$$\forall x \, ((x \cdot x) = x \rightarrow x = \mathbf{e})$$

is a logical consequence of the group axioms.

[*Hints*: If an interpretation g of x in a group $\mathcal{G}$ satisfies $(x \cdot x) = x$, then it also satisfies $((x \cdot x) \cdot y) = (x \cdot y)$, when y is interpreted by the inverse of g in $\mathcal{G}$. Now exploit the group axioms.]

Exercise 4.79 ______________________________

(a) Show that the sentence

$$\forall x \forall y \forall z \, ((x \cdot y) = (x \cdot z) \rightarrow y = z)$$

is a logical consequence of the group axioms.

(b) The property of groups expressed in part (a) is called left cancellation. Write down a sentence that you think should represent right cancellation. Is it a logical consequence of the group axioms?

Exercise 4.80 ______________________________

Show that $\forall x \forall y \exists z \, (x \cdot z) = y$ is a logical consequence of the group axioms.

Exercise 4.81

Show that the commutativity property $\forall x \forall y\,(x \cdot y) = (y \cdot x)$ is not a logical consequence of the group axioms.

We have already given you a relevant piece of information to settle this.

Exercise 4.82

Show that the following sentences are logical consequences of the group axioms.

(a) $\forall y(\forall x\,(x \cdot y) = x \rightarrow y = \mathbf{e})$

(b) $\forall x \forall y \forall z\,((((x \cdot y) = \mathbf{e} \wedge (y \cdot x) = \mathbf{e}) \wedge ((x \cdot z) = \mathbf{e} \wedge (z \cdot x) = \mathbf{e})) \rightarrow y = z)$

The last exercise shows that the identity element and inverses of elements of a group, which exist according to the axioms, are unique. As each element g of a group $\mathcal{G}$ has a unique inverse element, we would be justified in defining a new 1-place function on the group mapping g to its inverse. Indeed there is an axiomatization of group theory in a language including a corresponding 1-place function symbol. The next exercise asks you to give such an axiomatization, as well as turn the remaining axioms into prefix notation.

Exercise 4.83

Let L be the language with equality containing the 2-place function symbol f, the 1-place function symbol g and the constant symbol $\mathbf{e}$. Write down axioms for group theory using prefix notation for the function symbols (so writing $f(x, y)$ rather than $x f y$) such that the interpretation in any model of $g(x)$ will be as the inverse element of x.

Exercise 4.84

The set $\mathbb{Z}_n = \{0, 1, \ldots, n - 1\}$ with the operation $+_n$ of addition modulo n is a group. Give a definition of the interpretation $-_n$ of the 1-place function symbol g in the previous exercise which makes $-_n(j)$ the additive inverse of j for each $j \in \mathbb{Z}_n$.

See Example 4.6(c).

After looking at an axiomatization of the theory of groups in a language with several helpful symbols, it will be instructive for you to show that the theory can in fact be axiomatized using relatively few.

Exercise 4.85

Give axioms for the theory of groups just the language with equality and the single 2-place function symbol $\cdot$.

A useful logical consequence of the associativity axiom is that the combination of x with itself n times results in the same element, however the xs are bracketed. So we can use the shorthand x^n for $\underbrace{(x \cdot (x \cdot (x \cdot \ldots)))}_{n \text{ times}}$. Some of

You were invited to show this in Exercise 2.45 of Section 2.4 of Chapter 2.

the results later in the book will need the idea of the *order* of an element in a group.

Definition *Order of group element*

For an element x of a group G, if there is some positive integer n for which $x^n = e$, the identity element of G, the least such n is called the *order* of x. In such a case, x is said to be of *finite order*. If there are no such n, the element x is said to be of *infinite order*.

The word 'order' is also used (fairly confusingly) to describe the number of elements in a group. We'll avoid this use in this book.

The identity element of any group is always of order 1 and is the only element of order 1 in the group. In the multiplicative group of the non-zero rationals, -1 has order 2 (as $(-1)^2 = 1$, the identity element of the group) and 2 has infinite order (as for no positive n does $2^n = 1$). The group $\mathrm{GL}(2, \mathbb{R})$ contains elements of all finite orders as well as elements of infinite order. In the additive group of the integers, every non-identity element has infinite order.

Results about orders of group elements

Let G be a group with identity element e and let $x \in G$.
1. If $x^k = e$ for some $k \geq 1$, then the order of x divides k (i.e. k is an integer multiple of the order of x).

2. If G is finite with N elements, then each element of G has finite order dividing N.

The converse is also true: $x^k = e$ for any multiple k of the order of x.

This is a corollary of a famous theorem of Lagrange.

In the group $\langle \mathbb{Z}_n, +_n, 0, = \rangle$, for each divisor m of n there is an element of order m (for instance the element n/m). The permutation group S_4 has $4! = 24$ elements with orders 1, 2, 3, 4 and 6, all dividing 24, but no elements of order 8, 12 or 24, so that the second result above cannot be strengthened to say that a group with N elements must contain elements of each order dividing N.

Exercise 4.86 ——————————————————————————

Write down the orders of the elements of each of the groups $\langle \mathbb{Z}_n, +_n, 0, = \rangle$ for $n = 3, 6, 12$.

Exercise 4.87 ——————————————————————————

Show that every non-identity element of $\langle \mathbb{Z}_p, +_p, 0, = \rangle$, where p is prime, has order p.

Solution

The group has p elements, with p prime. By Result 2 above, the order of any element of the group divides p, and can thus only be 1 or p. Only the identity element of a group has order 1, so that the remaining elements each have order p.

The result of Exercise 4.87 will be used in Chapter 6 whenever we require an example of a finite group which has no non-identity elements of orders below a given number N, by taking $\langle \mathbb{Z}_p, +_p, 0, = \rangle$ for a prime $p > N$.

Rings and Fields

Definitions   Rings

Let L be the language with equality containing 2-place function symbols $+$ and $\cdot$, a 1-place function symbol $-$ and constant symbol $\mathbf{0}$. The theory of *rings* has the following axioms, written using infix notation for $+$ and $\cdot$ and using $(-x)$ for $-(x)$.

1. $\forall x \forall y \forall z\, (x + (y + z)) = ((x + y) + z)$ ($+$ is associative)
2. $\forall x((x + \mathbf{0}) = x \wedge (\mathbf{0} + x) = x)$ ($\mathbf{0}$ is an additive identity)
3. $\forall x((x + (-x)) = \mathbf{0} \wedge ((-x) + x) = \mathbf{0})$ (additive inverses exist)
4. $\forall x \forall y\, (x + y) = (y + x)$ ($+$ is commutative)
5. $\forall x \forall y \forall z\, (x \cdot (y \cdot z)) = ((x \cdot y) \cdot z)$ ($\cdot$ is associative)
6. $\forall x \forall y \forall z\, (x \cdot (y + z)) = ((x \cdot y) + (x \cdot z))$ (distributive law)

A model $\mathcal{R} = \langle R, +_{\mathcal{R}}, \cdot_{\mathcal{R}}, -_{\mathcal{R}}, 0^{\mathcal{R}}, = \rangle$ of these axioms is called a *ring*.

With the following additional axioms using an extra constant symbol $\mathbf{1}$

7. $\forall x((x \cdot \mathbf{1}) = x \wedge (\mathbf{1} \cdot x) = x)$ ($\mathbf{1}$ is a multiplicative identity)
8. $\forall x \forall y\, (x \cdot y) = (y \cdot x)$ ($\cdot$ is commutative)

we have axioms for the theory of *commutative rings with a 1*. Models of the above are described as *commutative rings with a 1*.

For a good introduction to the theory of rings and fields, see Allenby [1].

We say that $\cdot$ distributes over $+$.

$+_{\mathcal{R}}$ and $\cdot_{\mathcal{R}}$ are called the *addition* and *multiplication* of the ring. Thanks to axioms 1 to 4, the structure $\langle R, +_{\mathcal{R}}, -_{\mathcal{R}}, 0^{\mathcal{R}} \rangle$ is a commutative group.

The 'commutative' in 'commutative ring' refers to the $\cdot$ function. The $+$ function is always commutative for a ring.

Example 4.7

(a) $\langle \mathbb{Z}, +, \cdot, -, 0, 1, = \rangle$ is a commutative ring with a 1. $\langle \mathbb{Q}, +, \cdot, -, 0, 1, = \rangle$, $\langle \mathbb{R}, +, \cdot, -, 0, 1, = \rangle$ and $\langle \mathbb{C}, +, \cdot, -, 0, 1, = \rangle$ are also commutative rings with a 1.

(b) We have already met the set $\mathbb{Z}_n = \{0, 1, \dots, n-1\}$ with the operation $+_n$ of addition modulo n (in Example 4.6(c)). We can define a function $-_n$ on $\mathbb{Z}_n$ by

$$-_n(j) = n - j \quad \text{for } j \in \{0, 1, \dots, n-1\}.$$

We can also define the operation $\cdot_n$ on $\mathbb{Z}_n$ by

$$x \cdot_n y = \text{ the remainder of } xy \text{ on division by } n,$$

where this *remainder* is defined to be the unique r with $0 \le r < n$ such that for some $q \in \mathbb{Z}$ (called the *quotient* of xy on division by n),

$$xy = qn + r.$$

Then $\langle \mathbb{Z}_n, +_n, \cdot_n, -_n, 0, 1, = \rangle$ is a commutative ring with a 1.

$\mathbb{Q}$, $\mathbb{R}$ and $\mathbb{C}$ all have an important extra property which we shall discuss very soon, so that they are *fields* as well as rings. As $\mathbb{Z}$ lacks this extra property, it is a more classic example of a commutative ring with a 1.

We hope that you obtained something like this in your answer to Exercise 4.84.

For instance, $4 \cdot_9 7 = 1$ (as $4 \times 7 = 28 = 3 \times 9 + 1$).

So there is at least one ring with n elements for each $n \ge 2$.

(c) The set $\mathbb{Q}[t]$ consists of all polynomials in the variable t with coefficients in $\mathbb{Q}$. With the normal rules for adding, subtracting and multiplying polynomials, and using 0 and 1 to represent constant polynomials taking respectively the values 0 and 1, $\langle \mathbb{Q}[t], +, \cdot, -, 0, 1, = \rangle$ is a commutative ring with a 1. Similarly, for any commutative ring R with a 1, the set $R[t]$ of all polynomials in the variable t with coefficients in R can be used as the domain of a commutative ring with a 1.

For instance, for the polynomials $f(t) = 1 + 2t$ and $g(t) = 3 - t^2$,
$$f(t) + g(t) = 4 + 2t - t^2,$$
$$f(t) - g(t) = -2 + 2t + t^2$$
$$f(t) \cdot g(t) = 3 + 6t - t^2 - 2t^3.$$

(d) The set of all 2×2 matrices with integer coefficients with the usual addition, multiplication and subtraction of matrices, and the zero matrix $\begin{pmatrix} 0 & 0 \\ 0 & 0 \end{pmatrix}$ interpreting the constant $\mathbf{0}$ is a non-commutative ring.

For an example of matrices x, y in this ring for which $xy \neq yx$, see the marginal note on page 134.

(e) The ring in the previous example does have a multiplicative identity, namely the identity matrix $\begin{pmatrix} 1 & 0 \\ 0 & 1 \end{pmatrix}$. For an example of a non-commutative ring with no multiplicative identity, take the set of all 2×2 matrices with even integer coefficients, the usual addition, multiplication and subtraction of matrices, and the zero matrix interpreting the constant $\mathbf{0}$.   ♦

Exercise 4.88

Show that the following are logical consequences of the theory of commutative rings with a 1.

(a) $\forall x \, (x \cdot \mathbf{0}) = \mathbf{0}$

(b) $\forall x \, (-(-x)) = x$

(c) $\forall x \, ((-\mathbf{1}) \cdot x) = (-x)$

Exercise 4.89

(a) Show that the sentence $\forall x \forall y \, (x \cdot y) = (y \cdot x)$ is not a logical consequence of the ring axioms.

(b) Show that the sentence $\forall x (\neg\, x = \mathbf{0} \rightarrow \exists y \, (x \cdot y) = \mathbf{1})$ is not a logical consequence of the axioms for a commutative ring with a 1.

Several of our applications in Chapter 6 will involve commutative rings with a 1 which satisfy further axioms, as follows.

Definitions   Fields

Let L be the language with equality containing 2-place function symbols $+$ and $\cdot$, 1-place function symbols $-$ and $^{-1}$ and constant symbols $\mathbf{0}$ and $\mathbf{1}$. The theory of *fields* has the axioms for a commutative ring with a 1 along with the following additional axioms, where we write x^{-1} instead of $^{-1}(x)$.

For any field $\mathcal{F} = \langle F, \ldots \rangle$, the structure $\langle F \setminus \{0^{\mathcal{F}}\}, \cdot_{\mathcal{F}}, ^{-1}_{\mathcal{F}}, 1^{\mathcal{F}} \rangle$ is a commutative group, as well as $\langle F, +_{\mathcal{F}}, -_{\mathcal{F}}, 0^{\mathcal{F}} \rangle$ being a commutative group.

10. $\forall x (\neg\, x = \mathbf{0} \rightarrow ((x \cdot x^{-1}) = \mathbf{1} \wedge (x^{-1} \cdot x) = \mathbf{1})$
                    (multiplicative inverses exist for non-zero elements)

11. $\neg\, \mathbf{0} = \mathbf{1}$

A model $\mathcal{F} = \langle F, +_{\mathcal{F}}, \cdot_{\mathcal{F}}, -_{\mathcal{R}}, ^{-1}_{\mathcal{F}}, 0^{\mathcal{F}}, = \rangle$ of these axioms is called a *field*.

Example 4.8

(a) $\langle \mathbb{Q}, +, \cdot, 0, 1, ^{-1}, = \rangle$, $\langle \mathbb{R}, +, \cdot, 0, 1, ^{-1}, = \rangle$ and $\langle \mathbb{C}, +, \cdot, 0, 1, ^{-1}, = \rangle$ are all fields, where for non-zero elements a we set a^{-1} as the b such that $ab = 1$. Note that these structures have to give 0^{-1} a value, but the axioms only allow for exploitation of a^{-1} when $a \neq 0$, so that the value of 0^{-1} can be chosen arbitrarily, say as 0.

> For $\mathbb{Z}$ with its normal multiplication, there are no multiplicative inverses for elements other than ± 1, so the ring of integers is not a field.

(b) When n is a prime number, the ring $\mathbb{Z}_n$ in Example 4.7 contains multiplicative inverses of non-zero elements, so that $^{-1}$ can be interpreted in a way which makes $\mathbb{Z}_n$ a field. $\quad\blacklozenge$

> For $j \in \mathbb{Z}_n$ with $j \neq 0$, set j^{-1} to be the $i \in \mathbb{Z}_n$ such that $j -_n i = 1$. Also set $0^{-1} = 0$, noting that 0^{-1} is assigned no properties from the field axioms.

Note that the last example shows that there are finite fields of arbitrarily large size, as there are arbitrarily large prime numbers. We shall soon ask you to show that when n is not prime, the ring $\mathbb{Z}_n$ is not a field.

Exercise 4.90

Show that the following sentences are logical consequences of the field axioms.

(a) $\forall x(\neg x = \mathbf{0} \to (x^{-1})^{-1} = x)$

(b) $\forall x \forall y \forall z((x \cdot y) = (x \cdot z) \to y = z)$

(c) $\neg (\mathbf{0}^{-1} \cdot \mathbf{0}) = \mathbf{1}$

Exercise 4.91

(a) Show that the following is a logical consequence of the axioms for a field:
$$\forall x \forall y((x \cdot y) = \mathbf{0} \to (x = \mathbf{0} \lor y = \mathbf{0})).$$

(b) Give an example of an integral domain which is not a field.

> This property is often described by saying that a field has no zero divisors. A commutative ring with a 1 and no zero divisors is called an *integral domain*.

Exercise 4.92

(a) Explain why the ring $\mathbb{Z}_6$ is not a field. [*Hints:* $\mathbb{Z}_6$ is a commutative ring with a 1, so there's only one field axiom where something can go wrong. Or use the result of the last exercise.]

(b) Let n be a composite positive integer, so that $n = ab$ for integers a, b with $a > 1$, $b > 1$. Explain why the ring $\mathbb{Z}_n$ is not a field.

(c) Show that the ring $\mathbb{Z}_p$, where p is a prime, is a field. [*Hint:* One method is as follows. If $a \in \{1, 2, \ldots, p-1\}$, show that the numbers $a \cdot_p k$, for $k \in \{1, 2, \ldots, p-1\}$, are distinct and non-zero. So one of them must equal 1.]

To take us nearer to axioms for the set of real numbers, we can add further axioms to those for fields to give them an order structure which interacts with the arithmetic operations in a suitable way, as follows.

Definitions   *Ordered fields*

Let L be the language with equality containing 2-place function symbols $+$ and $\cdot$, 1-place function symbols $-$ and $^{-1}$ and constant symbols $\mathbf{0}$ and $\mathbf{1}$ used above, with the additional 2-place relation symbol $<$. The theory of *ordered fields* has the axioms for a field with the following additional axioms.

12. $\forall x \, \neg \, x < x$
13. $\forall x \forall y \forall z ((x < y \land y < z) \to x < z)$
14. $\forall x \forall y (x < y \lor (x = y \lor y < x))$
15. $\forall x \forall y \forall z (x < y \to (x + z) < (y + z))$
16. $\forall x \forall y \forall z ((x < y \land \mathbf{0} < z) \to (x \cdot z) < (y \cdot z))$

A model $\mathcal{F} = \langle F, +_{\mathcal{F}}, \cdot_{\mathcal{F}}, -_{\mathcal{R}}, ^{-1}{}_{\mathcal{F}}, \mathbf{0}^{\mathcal{F}}, <_{\mathcal{F}}, = \rangle$ of these axioms is called an *ordered field.*

Example 4.9

$\langle \mathbb{Q}, +, \cdot, 0, 1, ^{-1}, <, = \rangle$ and $\langle \mathbb{R}, +, \cdot, 0, 1, ^{-1}, <, = \rangle$ are classic examples of an ordered field.   ◆

Exercise 4.93 ————————————————————————————

(a) Show that the following are logical consequences of the theory of ordered fields.

  (i) $\mathbf{0} < \mathbf{1}$

  (ii) $\forall x (\mathbf{0} < x \to (-x) < \mathbf{0})$

  (iii) $\forall x \forall y \forall z ((x < y \land z < \mathbf{0}) \to (y \cdot z) < (x \cdot z))$

(b) Show that there is no way of defining an order $<$ on the set $\mathbb{C}$ of complex numbers with its usual arithmetic operations to make $\mathbb{C}$ a complete ordered field. [*Hint*: If there was such a $<$, then exactly one of $i < 0$, $i = 0$ and $0 < i$ would be true. Show that each of these is impossible.]

———————————————————————————————————————

These axioms for an ordered field are equivalent to most of those we gave for the real numbers $\mathbb{R}$ in the Introduction to this book, lacking only the completeness axiom, which says that any non-empty subset A of S which is bounded above has a least upper bound in S. As this statement quantifies over subsets as well as elements of the domain $\mathbb{R}$, it cannot be represented in a first-order language, where only quantification over elements of the domain $\mathbb{R}$ is permitted. The question arises of whether there is an alternative way of representing the completeness axiom in a first-order language; and we shall answer this question in Chapter 6.

Replacing function symbols

We give a final important example, though perhaps not quite of a mathematical theory in the sense of most of those above. Let the language L contain $=$ and a 2-place relation symbol R. What can one say about a model $\mathcal{A} = \langle A, R^*, = \rangle$ of the sentence

$$\forall x(\exists y R(x,y) \wedge \forall z(R(x,z) \rightarrow y = z))?$$

The literal reading is that for each $a \in A$, there is a unique $b \in A$ such that $(a,b) \in R^*$. Within set theory, this is the classic way of representing a function of one variable from A to A as a set, namely as a set of pairs satisfying this sentence. For those, like the author, accustomed to think of functions expressed in terms of domains, codomains and rules, R^* might be described as the *graph* of a function $f^*\colon A \longrightarrow A$, where for each $a \in A$, the value of $f^*(a)$ is the unique b such that $(a,b) \in R^*$. The graph of f^* is the set of pairs $\{(a, f^*(a) : a \in A\}$, which is just R^*.

See e.g. Goldrei [16].

Note that the codomain of f^* is A as $R^* \subseteq A \times A$.

Exercise 4.94

(a) Write down a sentence in the language above in which the interpretation of R in any model is as a function onto the domain A of the model.

(b) Show that all models of the sentences

$$\forall x \exists y (R(x,y) \wedge \forall z(R(x,z) \rightarrow y = z))$$
$$\forall x \forall y \forall z((R(x,y) \wedge R(z,y)) \rightarrow x = z)$$
$$\exists x \forall y \neg R(y,x)$$

are infinite, i.e. have an infinite domain.

Sentences like $\forall x \exists y(R(x,y) \wedge \forall z(R(x,z) \rightarrow y = z))$ provide a way of replacing function symbols within a formal language by relation symbols.

Exercise 4.95

Write down a sentence using an $(n+1)$-place relation symbol R (and $=$ if you wish) whose interpretation in any model $\langle A, \ldots \rangle$ is as the graph of a function from A^n to A.

We have probably given enough axiom systems both to give the impression that the framework of first-order languages does permit the axiomatization of at least some interesting mathematics, and to give some examples to exploit in the remainder of the book. There are some very obvious questions which we should, at least to some extent, address. Given a theory, what are its logical consequences? Is there an algorithmic procedure to decide whether a formula is a logical consequence of a theory? An important part of an answer is to look at the other way, besides logical consequence, discussed in this book of establishing whether one statement is a consequence of others, namely looking for a derivation in a formal proof system. This is what we shall turn to in the next chapter.

Further exercises

Exercise 4.96

Show that each of the three axioms for a weak partial order is not a logical consequence of the remaining axioms.

Exercise 4.97

(a) Let $\langle A, R^* \rangle$ be an equivalence relation. Show that the set of equivalence classes of A, $\{[\![a]\!] : a \in A\}$, forms a partition of A.

(b) Let P be a partition of the set A. A relation R^* on A is defined by

$$(a, b) \in R^* \quad \text{if and only if} \quad a \text{ and } b \text{ belong to the same set in } P.$$

Show that $\langle A, R^* \rangle$ is an equivalence relation.

(c) Construct a formula ϕ in the language L consisting of a binary relation symbol R and equality $=$ with the properties that for all finite sets A,

   (i) if $\langle A, R^* \rangle \models \phi$, then the number of elements of A is a multiple of 5, and

   (ii) if the number of elements of A is a multiple of 5, then it is possible to define a relation R^* on A such that $\langle A, R^* \rangle \models \phi$.

Exercise 4.98

(a) Write down a set of axioms, Σ_n, for a strict linearly ordered set with at least n elements, using a language with equality and a 2-place relation symbol $<$.

(b) Let $\Sigma = \bigcup_{n=1}^{\infty} \Sigma_n$. Show that neither $\Sigma \models \phi$ nor $\Sigma \models \neg\phi$, where ϕ is

$$\forall x (\exists y\, x < y \rightarrow \exists z (x < z \wedge \forall w (x < w \rightarrow (w = z \vee z < w)))).$$

(c) Is there a finite set of sentences Φ such that $\Phi \models \sigma$ for each $\sigma \in \Sigma$? Give such a set or explain why one doesn't exist, as appropriate.

Exercise 4.99

Consider the following sentences (where f is a 2-place function symbol).

$$\forall x \forall y \forall z\, f(x, f(y, z)) = f(f(x, y), z)$$
$$\forall y \exists x\, f(x, x) = y$$
$$\forall y \exists x\, f(x, f(x, f(x, x))) = y$$

For each of these formulas state whether or not it is a logical consequence of the other two. Justify your answers.

Exercise 4.100

Let L be the first-order language with equality which, in addition to $=$, has a single unary relation symbol P and a binary function symbol f. Write down formulas ϕ_i, $i = 1, 2, 3$, in L with the following properties:

(a) ϕ_1 holds precisely when f is interpreted as a one–one and onto function;

(b) ϕ_2 has models with size n^2 for every $n \in \{1, 2, 3, \ldots\}$, but no other finite models;

(c) ϕ_3 has infinite models but no finite ones.

Such a function would be from A^2 to A, where A is the domain of a model.

Exercise 4.101

Let T be a set of sentences in a first-order language L. Show that for all structures $\mathcal{A}$ for L, $\mathcal{A}$ is a model of T if and only if $\mathcal{A}$ satisfies all the sentences in the theory of T.

4.5   Substructures and Isomorphisms

Now that we've seen a few interesting first-order theories, it's worth introducing a couple of important ideas associated with the models of such theories, namely 'substructure' and 'isomorphism'.

This section could possibly be omitted in a first reading of the book.

Let's look first at the idea of a substructure of a structure – a generalization of, for instance, the idea of a subgroup of a group, should you have met this concept before.

Definitions   Substructure

Let L be a first-order language with equality and let $\mathcal{B}$ be a structure for L with domain B. Let A be a non-empty subset of B.

The *closure* of A under the functions and constants of $\mathcal{A}$ is the smallest subset, written as $\overline{A}$, of B containing the elements of A and the constants of $\mathcal{B}$ *closed* under these functions; that is, for each n-place function symbol f in L, if $a_1, a_2, \ldots, a_n \in \overline{A}$ then $f^{\mathcal{B}}(a_1, a_2, \ldots, a_n) \in \overline{A}$.

For brevity, we shall refer to the *functions, relations and constants* of $\mathcal{B}$, meaning the interpretations within $\mathcal{B}$ of respectively the function, relation and constant symbols of L.

The *substructure of $\mathcal{B}$ generated by A* is the structure $\mathcal{A}$ for L with domain $\overline{A}$ and the interpretations of any non-logical symbols of L given as follows:

for any relation symbol R of n arguments,

$$R^{\mathcal{A}} = R^{\mathcal{B}} \cap \overline{A}^{\,n} \text{ (i.e. all } n - \text{tuples in } R^{\mathcal{B}}$$
$$\text{whose elements are in } \overline{A});$$

$R^{\mathcal{A}}$ is called the *restriction* of the relation $R^{\mathcal{B}}$ to $\overline{A}$.

for any function symbol f of m arguments,

$$f^{\mathcal{A}}(a_1, \ldots, a_m) = f^{\mathcal{B}}(a_1, \ldots, a_m) \text{ for all } a_1, \ldots, a_m \in \overline{A};$$

for any constant symbol $\mathbf{c}$, $\mathbf{c}^{\mathcal{A}} = \mathbf{c}^{\mathcal{B}}$.

That is, $f^{\mathcal{A}}$ is the restriction $f^{\mathcal{B}}|_{\overline{A}^m}$.

A *substructure* of $\mathcal{B}$ is any structure that can arise in this way.

For example, let L be the language with equality, a binary relation symbol R, a binary function symbol f and constant symbol $\mathbf{c}$. Let $\mathcal{B}$ be the structure $\langle \mathbb{Z}, <, +, 5, = \rangle$ for L and let A be the subset $\{2\}$ of $\mathbb{Z}$. Then the closure $\overline{A}$ of A and the constant symbols of $\mathcal{B}$, namely of the set $\{2, 5\}$, under the function(s) of $\mathcal{B}$ is the set $\{2, 4, 5, 6, 7, 8, \ldots\}$. The substructure $\mathcal{A}$ generated by A has this set $\overline{A}$ as domain, interprets R by the relation $<$ on this set, f by the function $+$ on this set and interprets $\mathbf{c}$ by the number 5.

Exercise 4.102

Let $\mathcal{B} = \langle B, \ldots \rangle$ be any structure for a language L with equality. For what sort of language L will the domain of the substructure of $\mathcal{B}$ generated by any subset A of B always be equal to A?

Solution

The only sure guarantee is for the language L to contain only relation symbols – no function or constant symbols!

More interestingly (if you have studied some group theory), let L be the language with equality, a binary function symbol f, a unary symbol g and constant symbol $\mathbf{c}$, and let $\mathcal{B}$ be the additive group of integers, expressed as a structure $\langle \mathbb{Z}, +, -, 0, = \rangle$ for L. For any subset A, the substructure generated by A must contain 0 and be closed under both $+$ and $-$, so must be a subgroup of $\mathcal{B}$: it is the subgroup a student of group theory would expect, that generated by A in the normal sense. In general, for any structure $\mathcal{B}$ for L which is a group, interpreting f as the group operation, g by the inverse element function and $\mathbf{c}$ as the identity element of the group, any substructure of it must be a subgroup.

> A *subgroup* of a group G is a subset of G which is a group with the operations of G. Also the subgroup of G generated by a subset A is the smallest subgroup of G containing the elements of A.

Exercise 4.103

Let L be the language with equality, a binary function symbol f and constant symbol $\mathbf{c}$, and let $\mathcal{B}$ be a structure for L which is a group, interpreting f as the group operation and $\mathbf{c}$ as the identity element. Is it true that any substructure of $\mathcal{B}$ is a subgroup of $\mathcal{B}$?

Solution

No, it's not true in general. Take, for instance, $\mathcal{B}$ to be the additive group of integers, expressed as a structure $\langle \mathbb{Z}, +, 0, = \rangle$ for L and let A be the subset $\{2\}$ of $\mathbb{Z}$. Then the domain of the substructure generated by A is the set $\{0, 2, 4, \ldots\}$ which is not a subgroup of $\mathcal{B}$ – it's not closed under inverses, thanks to leaving the unary function symbol for such inverses out of the language L.

> Remember that the domain consists of the closure of A along with the constants of $\mathcal{B}$, so that here 0 must be included.

Exercise 4.104

For each of the following languages L and theories T expressed in the obvious way using the language, decide whether all substructures of models of T are also models of T; and if this is not the case, then could you adjust the language used so that this became true?

(a) $L = \{<, =\}$ and T is the theory of linear orders.

(b) $L = \{<, =\}$ and T is the theory of linear orders with a least element.

(c) $L = \{+, \cdot, -, 0, =\}$ and T is the theory of rings.

(d) $L = \{+, \cdot, -, 0, 1, =\}$ and T is the theory of fields.

(e) $L = \{+, \cdot, -, ^{-1}, 0, 1, =\}$ and T is the theory of fields.

> Using the terminology of this exercise, the results above could be expressed as follows: if $L = \{+, -, 0, =\}$ and T is the theory of groups, then every substructure of a model of T is also a model of T; but if $L = \{+, 0, =\}$ and T is the same theory, then substructures of models of T need not be models of T.

The magic ingredient to ensure that all substructures of a model of T are also models of T is given in the following theorem.

Theorem 4.7

Let L be a first-order language with equality. Let $\mathcal{A}, \mathcal{B}$ be structures for L with domains A, B respectively such that $\mathcal{A}$ is a substructure of $\mathcal{B}$.

(a) For any quantifier-free formula $\phi(x_1, x_2, \ldots, x_n)$ and $a_1, a_2, \ldots, a_n \in A$,

$$\mathcal{A} \vDash_{x_1/a_1, x_2/a_2, \ldots, x_n/a_n} \phi(x_1, x_2, \ldots, x_n) \text{ if and}$$
$$\text{only if } \mathcal{B} \vDash_{x_1/a_1, x_2/a_2, \ldots, x_n/a_n} \phi(x_1, x_2, \ldots, x_n).$$

> Recall the convention that writing a formula as $\phi(x_1, x_2, \ldots, x_n)$ signifies that the free variables in ϕ are some or all of the $x_1, x_2, \ldots, x_n$.

(b) For any formula $\forall y_1 \ldots \forall y_k \phi(y_1, \ldots, y_k, x_1, x_2, \ldots, x_n)$, where $\phi(y_1, \ldots, y_k, x_1, x_2, \ldots, x_n)$ is quantifier-free, and any $a_1, a_2, \ldots, a_n \in A$,

$$\text{if } \mathcal{B} \vDash_{x_1/a_1, x_2/a_2, \ldots, x_n/a_n} \forall y_1 \ldots \forall y_k \phi(y_1, \ldots, y_k, x_1, x_2, \ldots, x_n)$$
$$\text{then } \mathcal{A} \vDash_{x_1/a_1, x_2/a_2, \ldots, x_n/a_n} \forall y_1 \ldots \forall y_k \phi(y_1, \ldots, y_k, x_1, x_2, \ldots, x_n).$$

> Recall that this is called a universal formula.

(c) If a theory T can be axiomatized by *universal sentences*, that is, sentences of the form $\forall y_1 \ldots \forall y_k \phi$ where ϕ is quantifier-free and its free variables are included in the list $y_1, \ldots, y_k$, then for any model of T, all of its substructures are also models of T.

We shall leave the proof of the theorem for you as an exercise. The result of part (c) (which we hope you can see follows immediately from part (b)) means that for many of the theories axiomatized in Section 4.4, all substructures of models of the theory are also models of the theory.

Exercise 4.105

Prove Theorem 4.7.

Exercise 4.106

For which of the theories axiomatized in Section 4.4 are all substructures of models of the theory also models of the theory?

Solution

Several of the theories are axiomatized using universal sentences, for example, the theories of: equivalence relations, partial orders and linear orders (weak and strict), Boolean algebras, rings, fields, ordered fields. All these theories must then have the required property for substructures of models. We hope that your solution to Exercise 4.83 gave an axiomatization for group theory consisting of universal sentences.

Exercise 4.107

Let $\mathcal{B}$ be a structure for a language L and $\mathcal{A}$ a substructure of $\mathcal{B}$. Let θ be an *existential sentence* of L, that is a sentence of the form $\exists y_1 \ldots \exists y_k \phi$, where ϕ is quantifier-free and its free variables are included in the list $y_1, \ldots, y_k$. Show that if $\mathcal{A} \vDash \theta$, then $\mathcal{B} \vDash \theta$. [*Hint*: Exploit Theorem 4.5(b).]

We shall exploit some of these facts later in the book. Perhaps surprisingly there is a converse result to this consequence of Theorem 4.7(c), namely:

> If T is a theory such that for all models of T, every substructure is also a model of T, then T can be axiomatized using universal sentences.

We shall prove this later in the book.

In Theorem 6.14 in Chapter 6.

Now let's look at the idea of 'isomorphism' of two structures for the same language L. We shall extend the familiar ideas from branches of maths like group theory. In the latter, a group is essentially defined as a set G with a binary operation $*$ obeying various axioms. All the information about the group follows from the table, called the *Cayley table*, giving the values of $g * g'$ for all $g, g' \in G$. Two groups, say $G, *$ and $H, \circ$, are isomorphic if there is a bijection θ from G to H via which the Cayley tables of G and H match; that is, for all $g, g' \in G$, $\theta(g * g') = \theta(g) \circ \theta(g')$ – the group H is essentially the group G with different names for its elements. The bijection θ is called an isomorphism. Regarding these groups as structures for a language with equality and a binary function symbol f, interpreted by the function $*$ on G and the function $\circ$ on H, the interpretations of the symbol f match via the bijection θ. An isomorphism between two structures for a more general language L is likewise a bijection between their domains which makes their interpretations of all the function, relation and constant symbols match, as given more formally in the following definition.

Definitions   Isomorphism

Let L be a first-order language with equality and let $\mathcal{A}, \mathcal{B}$ be structures for L with domains A, B respectively. The structures $\mathcal{A}$ and $\mathcal{B}$ are *isomorphic* if there is a function $\theta \colon A \longrightarrow B$, said to be an *isomorphism* from $\mathcal{A}$ to $\mathcal{B}$, with the following properties:

(i) θ is a bijection between A and B;

(ii) for any relation symbol R of n arguments and all $a_1, a_2, \ldots, a_n \in A$,
$(a_1, a_2, \ldots, a_n) \in R^{\mathcal{A}}$ if and only if $(\theta(a_1), \theta(a_2), \ldots, \theta(a_n)) \in R^{\mathcal{B}}$;

(iii) for any function symbol f of m arguments and all $a_1, a_2, \ldots, a_m \in A$,

$$\theta(f^{\mathcal{A}}(a_1, a_2, \ldots, a_m)) = f^{\mathcal{B}}(\theta(a_1), \theta(a_2), \ldots, \theta(a_m));$$

(iv) for any constant symbol $\mathbf{c}$, $\theta(\mathbf{c}^{\mathcal{A}}) = \mathbf{c}^{\mathcal{B}}$.

We hope that you will have encountered the idea of isomorphisms before and appreciate the importance of their use in classifying the different models of a theory – for the purpose of classification, isomorphic structures are 'essentially the same'.

Classification seems to be a key enterprise in the study of just about any subject!

Example 4.10

(a) Let L consist only of the equality symbol. Then $\langle A, = \rangle$ and $\langle B, = \rangle$ are isomorphic when there is a bijection between A and B, i.e. they have the same number of elements.

(b) Let L be a language with equality, a 2-place function symbol f and a constant symbol $\mathbf{c}$. Let A be the set $\mathbb{Z}_4 = \{0, 1, 2, 3\}$ and recall that with the operation $+_4$ of addition modulo 4 this forms a group. Let B be the set $\mathrm{Rot}(\square) = \{r_0, r_{\pi/2}, r_\pi, r_{3\pi/2}\}$, where r_θ is the rotation of the plane $\mathbb{R}^2$ about the origin through an angle θ anti-clockwise: with the operation $\circ$ of composition of functions, this also forms a group.

> $\mathrm{Rot}(\square)$ is the set of rotational symmetries of a square centred on the origin.

(i) The structures $\mathcal{A} = \langle \mathbb{Z}_4, +_4, 0, = \rangle$ and $\mathcal{B} = \langle \mathrm{Rot}(\square), \circ, r_0, = \rangle$ are isomorphic. One isomorphism θ from $\mathcal{A}$ to $\mathcal{B}$ has rule

$$\theta(k) = r_{k\pi/2}, \quad k = 0, 1, 2, 3.$$

These structures are groups and isomorphism as structures coincides with the usual notion of group isomorphism.

(ii) The structures $\mathcal{A} = \langle \mathbb{Z}_4, +_4, 1, = \rangle$ and $\mathcal{B} = \langle \mathrm{Rot}(\square), \circ, r_{3\pi/2}, = \rangle$ are also isomorphic. An isomorphism is the function θ' defined by

$$\theta' : \mathbb{Z}_4 \longrightarrow \mathrm{Rot}(\square)$$
$$0 \longmapsto r_0$$
$$1 \longmapsto r_{3\pi/2}$$
$$2 \longmapsto r_\pi$$
$$3 \longmapsto r_{\pi/2}.$$

> θ and θ' are the two standard group-theoretic isomorphisms between the group $\langle \mathbb{Z}_4, +_4, 0, = \rangle$ and the group of rotations of the square.

In contrast to the previous example, we would probably not describe the structures as groups, because of the perverse interpretations of the constant symbols as non-identity elements; but the definition of isomorphism for structures for the language still applies.

(iii) The structures $\mathcal{A} = \langle \mathbb{Z}_4, +_4, 2, = \rangle$ and $\mathcal{B} = \langle \mathrm{Rot}(\square), \circ, r_{3\pi/2}, = \rangle$ are not isomorphic. The only bijections from $\mathbb{Z}_4$ to $\mathrm{Rot}(\square)$ which preserve the interpretations of the function symbol f are the θ and θ' in the preceding examples, neither of which preserve the interpretation of the constant symbol $\mathbf{c}$. $\qquad\qquad\blacklozenge$

We mentioned earlier the idea of a cyclic group, that is, one for which there is an element $g \in G$ such that every element of G is g^n for some $n \in \mathbb{Z}$. (The element g is then said to generate the cyclic group.) Both the groups in Example 4.10(b) are cyclic. The element 1 (and also the element 3) generates $\langle \mathbb{Z}_4, +_4, 0, = \rangle$ and $r_{\pi/2}$ (and also $r_{3\pi/2}$) generates $\mathcal{B} = \langle \mathrm{Rot}(\square), \circ, r_0, = \rangle$. It is no surprise that these groups are isomorphic as it can be shown that, in general, if two cyclic groups have the same number of elements, they are isomorphic. Thus one can classify cyclic groups by giving, for each possible number of elements, one example of a cyclic group with that number of elements.

> A complete classification of cyclic groups is given by the groups $\langle \mathbb{Z}_n, +_n, 0, = \rangle$ for each integer $n \geq 1$ and the additive group of integers $\langle \mathbb{Z}, +, 0, = \rangle$.

Exercise 4.108 ___________

Let L be a language with equality, a 2-place relation symbol R and a constant symbol **c**. All the structures for L in this exercise interpret R as a linear order.

(a) Show that the structures $\langle \mathbb{N}, <, 0, = \rangle$ and $\langle \{2 - \frac{1}{n+1} : n \in \mathbb{N}\}, <, 1, = \rangle$ are isomorphic.

(b) Which pairs, if any, of the following structures are isomorphic?

$$\langle \mathbb{N}, <, 0, = \rangle$$
$$\langle \mathbb{N}, >, 0, = \rangle$$
$$\langle \{2\} \cup \{2 - \tfrac{1}{n+1} : n \in \mathbb{N}\}, <, 2, = \rangle$$
$$\langle \{-1\} \cup \{2 - \tfrac{1}{n+1} : n \in \mathbb{N}\}, <, -1, = \rangle$$
$$\langle \{2\} \cup \{2 + \tfrac{1}{n+1} : n \in \mathbb{N}\}, >, 2, = \rangle$$

Exercise 4.109 ___________

(a) Suppose that $\theta \colon A \longrightarrow B$ is an isomorphism from $\mathcal{A}$ to $\mathcal{B}$. Show that the inverse function θ^{-1} is an isomorphism from $\mathcal{B}$ to $\mathcal{A}$.

(b) Show that '$\mathcal{A}$ is isomorphic to $\mathcal{B}$' is an equivalence relation on the class of all structures for L.

As you'd expect, a major result about isomorphism is that isomorphic structures satisfy the same formulas, in the following sense.

Theorem 4.8

Suppose that $\theta \colon A \longrightarrow B$ is an isomorphism between structures $\mathcal{A} = \langle A, \ldots \rangle$ and $\mathcal{B} = \langle B, \ldots \rangle$ for a language L. Then for any formula $\phi(x_1, x_2, \ldots, x_n)$ with free variables amongst $x_1, x_2, \ldots, x_n$,

$$\mathcal{A} \vDash_{x_1/a_1 \; x_2/a_2 \; \ldots x_n/a_n} \phi(x_1, x_2, \ldots, x_n)$$

if and only if

$$\mathcal{B} \vDash_{x_1/\theta(a_1) \; x_2/\theta(a_2) \; \ldots x_n/\theta(a_n)} \phi(x_1, x_2, \ldots, x_n),$$

for all $a_1, a_2, \ldots, a_n \in A$.

In particular $\mathcal{A}$ and $\mathcal{B}$ satisfy the same set of sentences.

Remember that a sentence is a formula with no free variables.

Exercise 4.110 ____

Prove Theorem 4.8 for a language L with equality, a 3-place relation symbol R, a 2-place function symbol f and a constant symbol $\mathbf{c}$.

[*Hints*: First deal with terms. Show that for any term $\tau(x_1, x_2, \ldots, x_k)$ involving variables out of the list $x_1, x_2, \ldots, x_k$ and $a_1, a_2, \ldots, a_k \in A$, we have $\theta(\tau^A(a_1, a_2, \ldots, a_k)) = \tau^B(\theta(a_1), \theta(a_2), \ldots, \theta(a_k))$: this requires an induction on the number of function symbols in the term τ. This will contribute to showing that the required result holds for all atomic formulas ϕ, remembering that these include all formulas of the form $\tau_1 = \tau_2$, where τ_1, τ_2 are terms, as well as formulas of the form $R(\tau_1, \tau_2, \tau_3)$, where τ_1, τ_2, τ_3 are terms. The full result can then be proved by mathematical induction on the number of logical connectives in ϕ.]

We hope that proving this for a limited number of symbols will give you enough of an idea of how to prove it for a more general language.

As already mentioned, isomorphism is a natural idea for classifying the models of a theory. Pretty well all the theories in Section 4.4 have lots of non-isomorphic models. For instance, in Chapter 6 we shall see that once such a theory has an infinite model, it must have models of all large enough infinite cardinalities; and once two models have different cardinalities, there can be no bijection between them, hence no isomorphism. This means that two of the most important mathematical theories, those of the natural numbers and the real numbers, cannot be axiomatized in a first-order language, as both theories have axiomatizations in a richer language with exactly one infinite model up to isomorphism.

Two sets have the same cardinality (i.e. size) when there is a bijection from one to the other.

However, there are some first-order theories which have just one model up to isomorphism of a particular infinite cardinality. We would like to end this section by looking at one such theory, the theory of *unbounded dense linear order*. Following the terminology introduced in parts (a) and (c) of Exercise 4.69 of Section 4.4, this theory, in the language with equality and the 2-place relation symbol $<$, has axioms for a strict linear order along with

$$\forall x \exists y \exists z (y < x \wedge x < z) \quad (< \text{ is unbounded})$$

and

$$\forall x \forall y (x < y \rightarrow \exists z (x < z \wedge z < y)) \quad (< \text{ is dense}).$$

Examples of models of this theory include each of the reals $\mathbb{R}$, the open interval $(0, 1)$ of $\mathbb{R}$, the rationals $\mathbb{Q}$, the open interval $\mathbb{Q} \cap (0, 1)$ of the rationals and the subset $\mathbb{Q} \cap ((0, 1) \cup (2, 3))$ of $\mathbb{Q}$, all with their usual order, of course. All models of the theory are infinite, thanks to the unboundedness axiom. What is surprising is that all countable models, like the last three of the examples we've just given, are isomorphic, which we'll state as a theorem.

We ask you to construct isomorphisms to show that these three countable examples are isomorphic in Exercise 4.111.

Theorem 4.9

Let $\langle X, <_X, = \rangle$ and $\langle Y, <_Y, = \rangle$ be two unbounded dense linear orders with X and Y both countably infinite. Then these structures are isomorphic.

The proof of this theorem is left as an exercise for the interested reader in Exercise 4.112 below. The result has an interesting impact on the logical consequences of this theory, as we shall discuss at the end of Chapter 6.

Further exercises

Exercise 4.111 ____

In this exercise, we ask you to show that the countable unbounded dense linear orders $\langle \mathbb{Q}, <, = \rangle$, $\langle \mathbb{Q} \cap (-1, 1), <, = \rangle$ and $\langle \mathbb{Q} \cap ((0, 1) \cup (2, 3)), <, = \rangle$ are isomorphic by constructing suitable isomorphisms. (Although this result follows from Theorem 4.9, the proof of the theorem, which we shall guide you through in Exercise 4.112, is not entirely helpful in constructing suitable isomorphisms in a nice form!)

(a) Show that the function

$$f \colon \mathbb{R} \longrightarrow (-1, 1)$$
$$x \longmapsto \frac{x}{1 + |x|}$$

is an isomorphism between $\langle \mathbb{R}, <, = \rangle$ and $\langle (-1, 1), <, = \rangle$, where both sets are ordered by the usual $<$.

(b) Exploit this function f to find an isomorphism between $\langle \mathbb{Q}, <, = \rangle$ and $\langle \mathbb{Q} \cap (-1, 1), <, = \rangle$.

(c) Find an isomorphism between $\langle \mathbb{Q}, <, = \rangle$ and $\langle \mathbb{Q} \cap (0, 1), <, = \rangle$.

(d) Find an isomorphism between $\langle \mathbb{Q}, <, = \rangle$ and $\langle \mathbb{Q} \cap ((0, 1) \cup (2, 3)), <, = \rangle$. [This part requires some sneakiness compared to the previous parts!]

Exercise 4.112 ____

In this exercise we ask you to prove Theorem 4.9. The theorem was first proved by Cantor, the founder of modern set theory (and some more!). His method is quite nicely described as a 'back and forth' argument. We shall take you through it, asking you to fill in some of the gaps.

We are given two countable unbounded dense linear orders $\langle X, <_X, = \rangle$ and $\langle Y, <_Y, = \rangle$. As the sets X and Y are countable, their elements can be listed as

$$x_0, x_1, x_2, \ldots, x_n, \ldots \quad \text{and} \quad y_0, y_1, y_2, \ldots, y_m, \ldots,$$

where every element of X appears exactly once in the list as an x_n for some $n \in \mathbb{N}$ and each element of Y appears exactly once in the list as a y_m for some $m \in \mathbb{N}$.

The method of proof is to construct both an isomorphism $f \colon X \longrightarrow Y$ and its inverse function $g \colon Y \longrightarrow X$ by recursion. That is to say, for the function f, we shall define the initial value of $f(x_0)$ and then, for each $n \in \mathbb{N}$, define $f(x_{n+1})$ in terms of the values of $f(x_0)$, $f(x_1)$, ..., $f(x_n)$. As every element of X appears exactly once as an x_n, this will define f as a function on X. We shall define g as a function on Y in a similar way. In each step of the process, we must ensure that f is order-preserving on $\{x_0, x_1, \ldots, x_n\}$, i.e. if $x_i <_X x_j$ then $f(x_i) <_Y f(x_j)$ for $i, j \in \{0, 1, \ldots, n\}$, and likewise that g is

We'll use the notation g for the inverse function rather than f^{-1} as we think it might avoid a bit of confusion.

order-preserving on $\{y_0, y_1, \ldots, y_n\}$. At the same time we must ensure that g is constructed to be the inverse function of f, so that if $f(x_i)$ is defined to be y_j, then $g(y_j)$ is defined to be x_i; and likewise if $g(y_j)$ is defined to be x_i, then $f(x_i)$ is defined to be y_j.

To get started, define $f(x_0)$ to be y_0 and $g(y_0)$ to be x_0.

Now suppose that $f(x_0), f(x_1), \ldots, f(x_n)$ and $g(y_0), g(y_1), \ldots, g(y_n)$ have both been defined so that the following are all satisfied:

(i)   the order structure of the elements $f(x_0), f(x_1), \ldots, f(x_n)$ in Y matches that of the corresponding elements $x_0, x_1, \ldots, x_n$ in X, i.e. if $x_i <_X x_j$ then $f(x_i) <_Y f(x_j)$ for $i, j \in \{0, 1, \ldots, n\}$;

(ii)   the order structure of the elements $g(y_0), g(y_1), \ldots, g(y_n)$ of X matches that of the corresponding elements $y_0, y_1, \ldots, y_n$ in Y, i.e. if $y_i <_Y y_j$ then $g(y_i) <_X g(y_j)$ for $i, j \in \{0, 1, \ldots, n\}$;

(iii) if for an $i \in \{0, 1, \ldots, n\}$ it happens to be the case that $f(x_i)$ is defined to be y_j for some $j \in \{0, 1, \ldots, n\}$, then $g(y_j)$ is defined to be x_i;

(iv) if for a $j \in \{0, 1, \ldots, n\}$ it happens to be the case that $g(y_j)$ is defined to be x_i for some $i \in \{0, 1, \ldots, n\}$, then $f(x_i)$ is defined to be y_j.

Now for your task! Show how to define first $f(x_{n+1})$ so that (i) and (iii) above are true with $n + 1$ in place of n; and then define $g(y_{n+1})$ so that (ii) and (iv) are true with $n + 1$ in place of n. Plainly it is at this stage that you will need to exploit the unbounded dense linear orders $<_X$ and $<_Y$. Then tie up the ends of the argument to explain why $\langle X, <_X, = \rangle$ and $\langle Y, <_Y, = \rangle$ are isomorphic.

Defining $g(y_0)$ as anything other than x_0 would, of course, destroy any prospect of g being f^{-1}.

Note that we are not saying that for each $i \in \{0, 1, \ldots, n\}$ the value of $f(x_i)$ has to be y_j for some $j \in \{0, 1, \ldots, n\}$. For general listings $x_0, x_1, x_2, \ldots, x_n, \ldots$ and $y_0, y_1, y_2, \ldots, y_m, \ldots$ this would be impossible to achieve. So if $f(x_i) = y_j$ where $i \leq n$ and $j > n$, then at the stage of the construction defining $g(y_j)$ we have to remember to define it as x_i to keep our aim that g should be f^{-1} on track.

Defining $f(x_{n+1})$ is the 'forth' and $g(y_{n+1})$ is the 'back' of the 'back and forth' argument.

5 FORMAL PREDICATE CALCULUS

5.1 Introduction

We shall now look at a formal proof system which matches our idea of logical consequence in the previous chapter. This particular system is the main focus of our discussion of formal systems. We're going for a single system which will match logical consequence as dealt with in Chapter 4 and provide soundness and completeness theorems similar to those for the propositional calculus, so that for all sets of sentences Γ and sentences ϕ,

$$\Gamma \vDash \phi \quad \text{if and only if} \quad \Gamma \vdash \phi.$$

Although the notions of logical consequence and formal proof are connected by the soundness and completeness theorems, they are very different concepts. To test whether $\Gamma \vDash \phi$, we have to investigate each model of Γ to see if it also satisfies ϕ. This could involve checking infinitely many models and plainly wouldn't give a practical algorithmic procedure for deciding whether ϕ is a logical consequence of Γ. In everyday maths we do know some mathematical consequences of theories described by axioms, so our knowledge must come from other means. The other means – the whole point of the approach described in the book – is some form of proof involving small steps using inferences regarded as acceptable by the mathematical community (or at least enough of it!). That there is some relatively easy to describe formal proof system of this sort which matches logical consequence for first-order languages is a major result.

Let's look at an example of a fairly informal mathematical proof from axioms. We shall take the first three of our axioms for rings on page 202, in the language L with equality including the 2-place function symbol $+$, the 1-place function symbol $-$ and constant symbol $\mathbf{0}$, adapting the axioms so that they only use the connectives $\neg, \rightarrow$ and the quantifier $\forall$, giving the following five axioms.

1. $\forall x \forall y \forall z\, (x + (y + z)) = ((x + y) + z)$
2. $\forall x\, (x + \mathbf{0}) = x$
3. $\forall x\, (\mathbf{0} + x) = x$
4. $\forall x\, (x + (-x)) = \mathbf{0}$
5. $\forall x\, ((-x) + x) = \mathbf{0}$

A consequence of these axioms is $\forall x((x + x) = x \rightarrow x = \mathbf{0})$ and we might demonstrate this informally as follows.

Suppose that x is any element satisfying

$$(x + x) = x.$$

Then 'adding' $-x$ to the right on each side of the equation gives

$$((x + x) + (-x)) = (x + (-x)).$$

Historically, formal systems have been a major object of study in their own right, as a framework for the formal treatment of mathematics and now for the automation of proofs. Establishing completeness was a non-trivial task and indeed the definition of satisfiability used in the text came even later in the development of the subject.

For instance, the theory of groups has infinitely many non-isomorphic models.

These sentences can be used as axioms for the theory of groups.

Using Axiom 1 the lefthand side of this equation equals $(x + (x + (-x)))$, so that the equation becomes

$$(x + (x + (-x))) = (x + (-x)).$$

Using Axiom 4, this becomes

$$(x + \mathbf{0}) = \mathbf{0},$$

which using Axiom 2 gives

$$x = \mathbf{0}.$$

As we have shown $x = \mathbf{0}$ from the assumption that $(x + x) = x$, this means we have shown that

$$((x + x) = x \rightarrow x = \mathbf{0}).$$

As x is a typical element, this must hold for all x, so that

$$\forall x((x + x) = x \rightarrow x = \mathbf{0}).$$

To help see some of the features a formal proof system might require to handle such a proof, we shall rewrite this argument a bit more formally, numbering lines and giving reasons for each line, just as our formal system for propositional calculus would lead us to expect.

(1)	$(x + x) = x$	Assumption
(2)	$((x + x) + (-x)) = (x + (-x))$	Adding $(-x)$ to the right of each side of the equality in 1
(3)	$(x + (x + (-x))) = ((x + x) + (-x))$	Special case of Axiom 1
(4)	$(x + (x + (-x))) = (x + (-x))$	Substituting one side of 2 for itself in 3
(5)	$(x + (-x)) = \mathbf{0}$	Special case of Axiom 4
(6)	$(x + \mathbf{0}) = \mathbf{0}$	Substituting one side of 5 for itself in 4
(7)	$(x + \mathbf{0}) = x$	Special case of Axiom 2
(8)	$x = \mathbf{0}$	Substituting one side of 7 for itself in 6

We conclude that $((x + x) = x \rightarrow x = \mathbf{0})$ and as x was a typical element we have $\forall x((x + x) = x \rightarrow x = \mathbf{0})$.

This informal proof gives us some ideas of what we should expect our formal system to handle. First of all, to prove something holds 'for all x', we normally show it holds for a typical x and then conclude it holds for all x. So our system will probably need a rule or axiom which allows us to make this step generalizing a result to cover all x. Next, guided by the presence of $\rightarrow$ in the statement of what we wanted to show for each typical x, we assumed $(x + x) = x$ and tried to prove from this assumption that $x = \mathbf{0}$, aiming to conclude that $((x + x) = x \rightarrow x = \mathbf{0})$ held using no assumptions. So we want the deduction theorem to hold for our system. Next, given that our axioms all have the character of 'for all x something holds', we wanted to conclude that this something held for a specific x – another feature on the wish list for our system! Then we want to be able to make all sorts of substitutions of equal quantities, one for the other, in statements we have already derived.

Here, from $\forall x\, (x + \mathbf{0}) = x$, we happened to want to infer only that $(x + \mathbf{0}) = x$. But we might have wanted to replace x by a more complicated term, say $(y + z)$ and infer that $((y + z) + \mathbf{0}) = (y + z)$.

Lastly, as on line 2, we want to be able to do the same thing to each side of an equality to get a new equality.

These considerations give some idea of what our shopping list will be when designing our formal system, which will be an extension of our system S for propositional calculus in Chapter 3. Many of the issues discussed in relation to propositional calculus will be helpful here.

We will clearly need extra axioms and/or rules to cope with interesting properties of quantifiers. We will go initially for a system just using $\forall$ along with propositional connectives $\neg, \rightarrow$. Given what we know about logically equivalent formulas, we can use $\exists x$ as a shorthand for $\neg\forall\neg x$. Similarly we can use $(\phi \wedge \psi)$, $(\phi \vee \psi)$ and so on for well-known equivalents expressed using the adequate set of connectives $\neg$ and $\rightarrow$. With the soundness theorem in mind, we should obviously go for axioms/rules that are valid. But however we do it, there are awkward pitfalls to be avoided. For instance, we saw in our simple proof above that to handle the universal quantifier $\forall$, we need ways of moving both from the general to the particular (from $\forall x\phi$ infer that ϕ holds for a particular x), and from the particular to the general (from ϕ holding for a typical x, infer that $\forall x\phi$ holds). Plainly, given that $\forall$ is the principal extra logical symbol added to those used for propositional calculus, rules for handling $\forall$ will be of the greatest importance. A first stab at formal rules for handling the universal quantifier $\forall$ might be the following:

(i)   from $\Gamma \vdash \forall x\phi(x)$ infer $\Gamma \vdash \phi(\tau)$ for any term τ;

(ii) from $\Gamma \vdash \phi$ infer $\Gamma \vdash \forall x\phi$.

Both of these rules go awry unless there's some sort of restriction.

As with propositional calculus, the formal system will have a mechanical aspect. Derivations will be generated by, and checked against, syntactic rules. So we shall have a framework in which questions about decidability are natural, for instance whether there is an algorithmic procedure for deciding whether a given formula is derivable.

Rule (i) tells us how to eliminate $\forall x$ from a formula and Rule (ii) tells us how to introduce it into one.

Exercise 5.1

By asking whether Rule (i) above is valid, we really mean the following: is it the case that

$$\text{if } \Gamma \vDash \forall x\phi(x) \text{ then } \Gamma \vDash \phi(\tau),$$

for any term τ? Show that this is not the case and suggest how the rule might be adjusted to give a valid rule.

Solution

You might well suspect from the discussion in Section 4.3 of Chapter 4, that we could run into trouble if the term τ isn't freely substitutable for x in ϕ. This is indeed the case. For instance, if $\phi(x)$ is the formula $\exists y \neg x = y$ and Γ is the set $\{\forall x\phi(x)\}$, then trivially

$$\Gamma \vDash \forall x\phi(x).$$

But it is not the case that

$$\Gamma \vDash \phi(y),$$

where the term τ has be taken to be the variable y – you will recall that the variable y cannot be freely substituted for x in the formula $\exists y \neg x = y$. Any (normal) structure with domain consisting of two or more elements satisfies Γ, but does not satisfy $\phi(y)$.

Given our discussion in Section 4.3, adding the requirement that the term τ can be freely substituted for x in ϕ is likely to produce a valid rule.

Rule (ii) corresponds to saying that

> if $\Gamma \vDash \phi$ then $\Gamma \vDash \forall x \phi$.

This rule corresponds to the normal way of arguing in mathematics that starts by proving a statement about a typical or general element x of some set A and concludes that the statement holds for all x in A. This process is called *generalization* and is a vital requirement for any proof system handling predicates.

Alas, as it stands this rule isn't valid. For instance, suppose that the language includes a constant symbol $\mathbf{c}$. Then trivially

$$x = \mathbf{c} \vDash x = \mathbf{c},$$

but

$$x = \mathbf{c} \nvDash \forall x\, x = \mathbf{c}.$$

In any structure $\mathcal{A}$ with domain consisting of two or more elements, the formula $x = \mathbf{c}$ is satisfied by interpreting x as $\mathbf{c}^{\mathcal{A}}$, but the formula $\forall x\, x = \mathbf{c}$ is not satisfied.

The real problem with the rule is that if $\Gamma \vDash \phi$ and the variable x occurs free in both ϕ and some of the assumptions in Γ, this is tantamount to assuming some special properties of x and inferring a further special property $\phi(x)$ of that x. The x isn't a general element and we cannot then infer $\forall x \phi(x)$. To rescue the rule, we shall require the variable x not to occur free in any of the assumptions in Γ.

An easy way to avoid problems with generalization is to make all the assumptions in Γ *sentences*, i.e. formulas with no free variables. This will rarely impoverish the theory being developed and we shall often do it.

We have given a hint of both some of the complications which we are going to have to resolve with our formal proof system and some of the properties that we want it to have. It's now time to plunge into the icy water of the system itself! In Section 5.2 we shall describe the formal proof system and do some derivations within it. In Section 5.3 we shall look at the soundness theorem for the system. In Section 5.4 we shall grasp the nettle of structures which do not interpret the $=$ symbol by actual equality on their domains (so are not *normal* structures). Finally in Section 5.5 we shall prove the completeness theorem for the system.

5.2   A formal system for predicate calculus

Our formal proof system, building on the system S of Chapter 3, is described in the following definitions. First we give the definition for a first-order language which doesn't involve equality and then we extend this to cover the use of equality.

Definitions   Formal system for predicate calculus

Let Γ be a (possibly empty) set of formulas and let ϕ be a formula, all in a first-order language L. A *derivation* of $\Gamma \vdash \phi$ is a finite sequence of formulas

$$\phi_1, \phi_2, \phi_3, \ldots, \phi_n,$$

where ϕ_n is the formula ϕ and the inclusion of each formula ϕ_i can be explained in one of the following ways:

(i) $\phi_i \in \Gamma$;

(ii) ϕ_i is a formula (an axiom) of one of the following forms:

  (Ax 1) $(\phi \to (\psi \to \phi))$,

  (Ax 2) $((\phi \to (\psi \to \theta)) \to ((\phi \to \psi) \to (\phi \to \theta)))$,

  (Ax 3) $((\neg\phi \to \neg\psi) \to (\psi \to \phi))$,

  (Ax 4) $(\forall x\phi(x) \to \phi(\tau))$, where τ is a term freely substitutable for x in ϕ,

  (Ax 5) $(\forall x(\phi \to \psi) \to (\phi \to \forall x\psi))$, where x is not free in ϕ,

  where ϕ, ψ, θ are any formulas of L;

(iii) there are two previous formulas in the sequence, ϕ_j and ϕ_k with $j, k < i$, where ϕ_k is the formula $(\phi_j \to \phi_i)$;

(iv) ϕ_i is of the form $\forall x\phi_j$, where ϕ_j is a previous formula in the sequence, i.e. where $j < i$, and the quantified variable x does not appear free in any formula in the set Γ.

As with propositional calculus, we shall use $\Gamma \vdash \phi$ as a shorthand for 'there is a derivation of $\Gamma \vdash \phi$'.

In addition, if we have a derivation $\Gamma_0 \vdash \phi$ obeying the conditions (i) to (iv) above for some subset Γ_0 of Γ, we shall also declare that $\Gamma \vdash \phi$. Thanks to this, if we have a derivation of $\Gamma \vdash \phi$, this means that there is a derivation following conditions (i) to (iv) of $\Gamma_0 \vdash \phi$ for some subset Γ_0 of Γ.

This is just the Rule of Assumptions, Ass, as before. Unlike propositional calculus, the set Γ could be something interesting, like the axioms for a mathematical theory!

The interesting case of Ax 4 is when x is actually free in ϕ.

This is just the rule of Modus Ponens, MP, as before.

This is called the rule of *generalization*, abbreviated as Gen.

This is the *thinning rule* for the system. For propositional calculus this was a metatheorem about the system, but for predicate calculus it is built into the system. We shall discuss the reason for this in a moment.

As you can see, axiom Ax 4 and the rule Gen seem to address the issues of respectively how to eliminate a universal quantifier from, and introduce one to, a formula; and axiom Ax 5 tells us how $\forall$ interacts with $\to$.

The extra axioms and rule for handling $\forall$ are much more natural than the propositional calculus axioms!

What we have given above deals with how to use $\neg$, $\to$ and $\forall$ in formal proofs, but doesn't tell us anything about how to deal with the special relation symbol $=$ for equality. If the language L includes $=$ and we want the symbol to correspond as much as possible to equality, our formal system has the following additional features.

Definition   *Axioms for equality*

If the language is one with equality, then the formal system has the following additional axioms:

(Ax 6) $\forall x\, x = x$,

(Ax 7) $(x = y \to (\phi(x, x) \to \phi(x, y)))$, for all atomic formulas ϕ, where $\phi(x, y)$ is obtained by substituting the variable y for some (not necessarily all) of the occurrences of the variable x in $\phi(x, x)$.

Axiom Ax 7 attempts to capture the principle that if two objects are equal, then any property of one is a property of the other. Examples of instances of axiom Ax 7 in a language with equality including a 3-place relation symbol R and a 2-place function symbol f are

$$(x = y \to (R(t, x, f(x, y)) \to R(t, y, f(y, y)))),$$

where all occurrences of x are substituted by y, and

$$(x = z \to (f(x, x) = x \to f(z, x) = x)),$$

where just one occurrence of x is substituted by z.

In Section 5.4 we shall discuss the limitations of these extra axioms for equality. Basically it is possible to interpret the symbol $=$ within a structure by a relation on the domain which isn't true equality, but which satisfies these axioms. That's why we introduced the concept of a normal interpretation which does always interpret $=$ by equality; and for almost all purposes in this book we consider only normal interpretations. A key part of the soundness theorem for our formal system, which we shall deal with in the next section, is of course that the equality axioms Ax 6 and Ax 7 are true in any normal interpretation.

While we hope that you are expecting both a soundness and completeness theorem for this system, we also hope that you find it remarkable that there is any system at all, let alone one with such a short description, for which the completeness theorem holds. These rules and axioms really are sufficient to give a formal derivation $\Gamma \vdash \phi$ for any logical consequence $\Gamma \vDash \phi$, for all sets of sentences Γ and sentences ϕ. To take a very simple example of how surprising this is, the axiom scheme Ax 7 allows equal quantities to be substituted for each other only in atomic formulas. This axiom doesn't say that this can be done for more complicated formulas. Yet we shall see later that it can be done for all of these, with some sensible restrictions, achieving the goal that if two objects are equal, then any property of one expressible in the formal language is a property of the other.

The formula ϕ is allowed to involve variables besides x.

For a penetrating and witty analysis of the caution needed in the application of this principle, that of substituting equals for equals, see Frege [15].

For certain technical reasons, we shall confine the completeness theorem to apply to sentences, i.e. formulas with no free variables.

As an example, we shall give a derivation of

$$\forall x\phi(x) \vdash \forall y\phi(y),$$

for any formula $\phi(x)$ in which x appears as a free variable and for any variable y which doesn't appear in $\phi(x)$,

$$
\begin{array}{lll}
(1) & \forall x\phi(x) & \text{Ass} \\
(2) & (\forall x\phi(x) \to \phi(y)) & \text{Ax 4} \\
(3) & \phi(y) & \text{MP}, 1, 2 \\
(4) & \forall y\phi(y) & \text{Gen}, 3
\end{array}
$$

Note that line 2 is a legitimate instance of axiom $\text{Ax}\,4$ as the variable y is freely substitutable for x in ϕ (because y doesn't appear in $\phi(x)$, so cannot become bound by a hidden $\forall y$). Also as the variable y is not free in the one assumption $\forall x\phi(x)$ (again because y doesn't appear in $\phi(x)$ and thus not in $\forall x\phi(x)$), the use of Gen on line 4 is correct.

Let's use this derivation as an example to explain why we have built the thinning rule into the definition of $\Gamma \vdash \phi$. If x appears as a free variable in $\phi(x)$ and y doesn't appear in $\phi(x)$, then we have

$$\forall x\phi(x) \vDash \forall y\phi(y).$$

This will follow from the soundness theorem for the system, but it is also easily verified directly from the definition of logical consequence. It follows straightforwardly from the properties of logical consequence that

$$\Delta, \forall x\phi(x) \vDash \forall y\phi(y)$$

for any set of formulas Δ – we might say that the thinning rule applies to logical consequence – and it would then be desirable, given that we want a completeness theorem for the system, that

$$\Delta, \forall x\phi(x) \vdash \forall y\phi(y).$$

Without the thinning rule for the formal system, we would run into trouble if the set Δ contained any formulas involving y as a free variable, for instance just the formula $\phi(y)$. Our derivation of $\forall x\phi(x) \vdash \forall y\phi(y)$ wouldn't give a derivation of $\phi(y), \forall x\phi(x) \vdash \forall y\phi(y)$ because the use of Gen on line 4 would then quantify a variable appearing free in one of the assumptions, albeit an assumption not exploited anywhere in the derivation! But thanks to the thinning rule, we can say that

$$\phi(y), \forall x\phi(x) \vdash \forall y\phi(y),$$

because there is a derivation of $\forall y\phi(y)$ using a subset of the set of assumptions.

We shall plainly have to take a bit of care with the rule Gen, which is such a vital feature of the system. Because of the potential complication of free variables appearing in assumptions when trying to use Gen, we shall often confine our attention to sets of assumptions consisting entirely of sentences, i.e. formulas with no free variables.

Let's now get back to the formal proof system and what can be done with it. Rather than leave you to flounder with proofs from first principles, we shall get straight into some of the metatheorems which help show what is derivable and which give recipes for constructing an actual derivation within

Any interpretation satisfying Δ and $\forall x\phi(x)$ automatically satisfies $\forall x\phi(x)$ and must thus satisfy $\forall y\phi(y)$, as $\forall x\phi(x) \vDash \forall y\phi(y)$.

As we can always give axioms for the mathematical theories in which we are interested using sentences, this is no great restriction.

Metatheorems like proof by contradiction and the deduction theorem would be handy for our new system!

the original system. Given our discussion about possible complications with
the rule Gen and how in practice we shall avoid them by using sentences as
assumptions, our first metatheorem will explain just how this works.

Theorem 5.1

Suppose that Γ is a set of sentences and that $\Gamma \vdash \phi$. Then for any
variable x, we have $\Gamma \vdash \forall x \phi$.

Proof

Suppose that we have a derivation of $\Gamma \vdash \phi$ using k lines with ϕ appearing as
the kth line. Then we adjust the derivation by adding an extra line as follows.

$$
\begin{array}{lll}
\vdots & \vdots & \vdots \\
(k) & \phi & \cdots \\
(k+1) & \forall x \phi & \text{Gen}
\end{array}
$$

As there are no free variables in any assumptions from Γ appearing in the
derivation, the use of Gen on line $k + 1$ is correct, so that we have a derivation
of $\Gamma \vdash \forall x \phi$. $\blacksquare$

One often uses the following special case of Theorem 5.1, the proof of which
we leave as an exercise for you.

Exercise 5.2

(a) Suppose that $\vdash \phi$, where ϕ may include free variables. Show that $\vdash \forall x \phi$,
for any variable x.

(b) Suppose that $\vdash \phi(x_1, x_2, \ldots, x_n)$, where ϕ has free variables in the list
$x_1, x_2, \ldots, x_n$. Show that

$$\vdash \forall x_1 \forall x_2 \ldots \forall x_n \phi(x_1, x_2, \ldots, x_n).$$

Just as for propositional calculus, we have an analogue of Theorem 3.1(i) of
Chapter 3:

Theorem 5.2

Suppose that a derivation of $\Gamma \vdash \phi$ involves uses of the Rule of Assumptions only with the formulas $\theta_1, \theta_2, \ldots, \theta_k$ from Γ. Then it is also a
derivation of

$$\{\theta_1, \theta_2, \ldots, \theta_k\} \vdash \phi.$$

Proof

If $\Gamma \vdash \phi$, then there is a derivation following conditions (i) to (iv) of $\Gamma_0 \vdash \phi$ for
some subset Γ_0 of Γ. The sequence of formulas in this derivation involves only
finitely many formulas, and in particular involves only finitely formulas using
the rule Ass, say $\theta_1, \theta_2, \ldots, \theta_k$. The same sequence then gives a derivation of
$\{\theta_1, \theta_2, \ldots, \theta_k\} \vdash \phi.$ $\blacksquare$

Next we turn to formulas which are instances of tautologies. All tautologies can be derived within our system S for propositional calculus, according to the completeness theorem in Chapter 3. What happens with our system for predicate calculus?

Exercise 5.3

Show that for any formula ψ, $\quad \vdash (\forall x\psi \to \forall x\psi)$.

Solution

The formula is a substitution instance of the tautology $(p \to p)$ obtained by substituting $\forall x\phi$ for p. We know from Chapter 3 that there is a derivation of $(p \to p)$ within the system S consisting of Ax 1, Ax 2, Ax 3 and Modus Ponens, all of which are part of our proof system for predicate calculus. By replacing all occurrences of p in this derivation by $\forall x\psi$, we thereby obtain a predicate calculus derivation of $\vdash (\forall x\psi \to \forall x\psi)$.

See Exercise 3.7 of Chapter 3 for a derivation.

The result of this exercise is a special case of the following more general result.

Theorem 5.3

Let ϕ be a substitution instance of a tautology built up using $\to, \neg$. Then $\vdash \phi$.

We shall ultimately allow the use of other standard propositional connectives, to be treated as abbreviations for the usual complicated logical equivalents built up using $\to, \neg$. As $\{\to, \neg\}$ is an adequate set of connectives, Theorem 5.3 tells us that we can derive all instances of tautologies involving connectives like $\land, \lor, \leftrightarrow$ as well as $\to, \neg$.

We shall ask you to prove this theorem in a moment – not much more is involved than in our solution to Exercise 5.3 exploiting results about the system S for propositional calculus in Chapter 3, but a certain amount of caution is needed, as we shall explain in the context of other metatheorems for our system for predicate calculus.

To what extent do other results about the formal system for propositional calculus in Chapter 3 hold for the system for predicate calculus? Obviously, it would be highly desirable that many of them do hold. Well, there's good news. For instance, the deduction theorem will hold for this system: if $\Gamma, \phi \vdash \psi$ then $\Gamma \vdash (\phi \to \psi)$ with all its benefits in terms of finding derivations. But there's slightly bad news ...!

Exercise 5.4

Does the deduction theorem hold for our system of predicate calculus simply because it holds for the fragment of propositional calculus that lives inside it?

Solution

Alas, no! We have to look at how the deduction theorem for propositional calculus was proved, as Theorem 3.3 of Chapter 3. The method was to look at each line in the derivation of $\Gamma, \phi \vdash \psi$, say the ith line $\Gamma, \phi \vdash \psi_i$, and show how to derive a corresponding line $\Gamma \vdash (\phi \to \psi_i)$ in a derivation of $\Gamma \vdash (\phi \to \psi)$. This meant accounting for whatever justification was used to obtain the line $\Gamma, \phi \vdash \psi_i$ in the original proof. Our system for predicate calculus includes extra ways for justifying lines within a derivation, namely additional axioms and the extra rule Gen. So the deduction theorem needs a new proof.

Exercise 5.5 ____

Prove the deduction theorem for our system of predicate calculus: for all formulas ϕ, ψ and sets of formulas Γ, if $\Gamma, \phi \vdash \psi$ then $\Gamma \vdash (\phi \to \psi)$. [*Hints*: It's not at all bad! It's really only the new rule of Gen that needs work. The proof of Theorem 3.3 does almost all that's required and the extra axioms turn out not to require extra work – do it for yourself and you'll see!]

Exercise 5.6 ____

Prove Theorem 5.3. [*Hints*: Adapt our solution to Exercise 5.3. Why can the completeness theorem for propositional calculus (Theorem 3.9 of Chapter 3) be exploited without worries about Gen?]

So the general message is that if a result about our formal system for the propositional calculus required an induction on the length of a derivation, accounting for all the ways in which a line in a derivation can be justified, then a new proof will be needed for our predicate calculus system.

Exercise 5.7 ____

Does the derived rule of proof by contradiction hold for our system of predicate calculus? Recall the statement of the rule for our system of propositional calculus in Theorem 3.2 of Chapter 3: if $\Gamma, \neg\phi \vdash \psi$ and $\Gamma, \neg\phi \vdash \neg\psi$, then $\Gamma \vdash \phi$.

Solution

The answer is yes! But why?

Similarly, the rule of proof by contradiction holds in the form: if $\Gamma, \phi \vdash \psi$ and $\Gamma, \phi \vdash \neg\psi$, then $\Gamma \vdash \neg\phi$.

A whole host of results about the formal system S for propositional calculus must also hold for the predicate calculus. If all the formulas involved are sentences, the statements of these results will be essentially identical. Take for instance all the results of Exercise 3.8 of Chapter 3, which we state as a theorem.

Theorem 5.4

Suppose that Γ, Δ are sets of sentences, and ϕ, ψ sentences, in a language L.

(a) If $\Gamma \vdash (\phi \to \psi)$, then $\Gamma, \phi \vdash \psi$.

(b) If $\Gamma \vdash \phi$ and $\Delta, \phi \vdash \psi$, then $\Gamma, \Delta \vdash \psi$.

(c) If, for some ψ, $\Gamma \vdash \psi$ and $\Gamma \vdash \neg\psi$, then $\Gamma \vdash \phi$, for any formula ϕ.

Just as for propositional calculus, if both $\Gamma \vdash \psi$ and $\Gamma \vdash \neg\psi$ for some ψ, we shall call the set Γ *inconsistent*.

Proof

The proofs of these results essentially require suitable adjustment and joining together of given derivations. As all the formulas given as assumptions, or being turned into assumptions, are sentences, no problem can arise with any use of the rule Gen when these derivations are adjusted. ∎

There can be complications with the proofs of each of the above if we drop the constraint that all the formulas involved are sentences. We shall ask you to investigate this in the next exercise, but reiterate that for most of our uses of the predicate calculus we shall stick to sentences to avoid these pitfalls.

Exercise 5.8 ____

For each part of Theorem 5.4, give an example why a proof of the corresponding result in Exercise 3.8 of Chapter 3 could become more complicated if the requirement that all formulas are sentences is dropped.

Solution

We shall deal with part (a) and leave the rest for you.

Suppose that the formulas in Γ, ϕ, ψ involve free variables. It is conceivable that a derivation of $\Gamma \vdash (\phi \rightarrow \psi)$ involves a use of Gen on a variable free in ϕ but not in any formula in Γ. Then the sort of straightforward adjustment of the derivation to give a derivation of $\Gamma, \phi \vdash \psi$ given in our solution to Exercise 3.8 of Chapter 3 would involve a use of Gen where the variable is now free in one of the assumptions (namely ϕ), so wouldn't give a derivation. For instance, take a language including the one-place relation symbol P, the empty set for Γ, and the following derivation of $\vdash ((P(x) \rightarrow P(x)) \rightarrow \forall x(P(x) \rightarrow P(x)))$.

$$(1) \quad (P(x) \rightarrow P(x)) \qquad\qquad\qquad\qquad\qquad \text{instance of tautology}$$

$$(2) \quad \forall x(P(x) \rightarrow P(x)) \qquad\qquad\qquad\qquad\qquad\qquad \text{Gen}$$

$$(3) \quad (\forall x(P(x) \rightarrow P(x)) \rightarrow ((P(x) \rightarrow P(x)) \rightarrow \forall x(P(x) \rightarrow P(x)))) \quad \text{Ax}\,1$$

$$(4) \quad ((P(x) \rightarrow P(x)) \rightarrow \forall x(P(x) \rightarrow P(x))) \qquad\qquad\qquad \text{MP}, 2, 3$$

Following the method given in our solution to Exercise 3.8(a) entails simply adding the extra lines

$$(5) \quad (P(x) \rightarrow P(x)) \qquad \text{Ass}$$

$$(6) \quad \forall x(P(x) \rightarrow P(x)) \quad \text{MP}, 4, 5$$

While this might at first sight look like a derivation, it now involves an incorrect use of the rule Gen – what was correct in the derivation given by lines 1 to 4 is no longer correct within the longer derivation with its extra assumption – funny old world, eh!

In this case, lines 1 and 2 give a correct derivation of $\vdash \forall x(P(x) \rightarrow P(x))$, so by the thinning rule for our system there *is* a derivation of $(P(x) \rightarrow P(x)) \vdash \forall x(P(x) \rightarrow P(x))$, just not the derivation that the solution to Exercise 3.8(a) would lead one to expect.

____

We shall not harp on these problems, but will usually avoid them in some way, for instance (as we said earlier) by having assumptions which are sentences or sometimes by not using assumptions at all – another sure way of not worrying about the use of the rule Gen! Speedily moving on from all this awkwardness, let's give another example of what is derivable, using some of these metatheorems and the calculus with equality. We shall show that

$$\vdash \exists x\, x = x,$$

recalling that $\exists x$ is an abbreviation for $\neg \forall x \neg$. First we produce the following derivation.

$$
\begin{array}{lll}
(1) & \forall x \,\neg\, x = x & \text{Ass} \\
(2) & (\forall x \,\neg\, x = x \to \neg\, x = x) & \text{Ax}\,4 \\
(3) & \neg\, x = x & \text{MP}, 1, 2 \\
(4) & \forall x \, x = x & \text{Ax}\,6 \\
(5) & (\forall x \, x = x \to x = x) & \text{Ax}\,4 \\
(6) & x = x & \text{MP}, 4, 5
\end{array}
$$

On both lines 2 and 5, the term x is guaranteed to be freely substitutable for x, so we have correct instances of Ax 4.

This derivation shows that both $\forall x \,\neg\, x = x \vdash \neg\, x = x$ (from lines 1 to 3) and $\forall x \,\neg\, x = x \vdash x = x$, so that by the rule of contradiction there is a derivation of

$$\vdash \neg\forall x \,\neg\, x = x,$$

i.e. of $\ \vdash \exists x \, x = x$, as required. This example suggests that to derive something of the form $\Gamma \vdash \exists x \phi$, a useful strategy is to try to derive a contradiction from the assumptions $\Gamma, \forall x \neg \phi$ and then use the rule of contradiction to obtain that $\Gamma \vdash \neg\forall x \neg \phi$ is derivable.

Feel free to use the metatheorems we have already justified in subsequent exercises!

Exercise 5.9

Suppose that the variable x is not free in any of the formulas $\gamma_1, \gamma_2, \ldots, \gamma_n$ and that

$$\vdash (\gamma_1 \to (\gamma_2 \to (\ldots \to (\gamma_{n-1} \to (\gamma_n \to \phi))\ldots))).$$

Show that

$$\vdash (\gamma_1 \to (\gamma_2 \to (\ldots \to (\gamma_{n-1} \to (\gamma_n \to \forall x\phi))\ldots))).$$

Exercise 5.10

Show that there are derivations of each of the following.

(a) $\vdash (\phi(\tau) \to \exists x \phi(x))$, where the term τ is freely substitutable for x in ϕ.

(b) $\vdash (\forall x \phi \to \exists x \phi)$, for any ϕ.

(c) $(\exists x \phi \to \psi) \vdash \forall x (\phi \to \psi)$, where x is not free in ψ.

(d) $\forall x (\phi \to \psi) \vdash (\exists x \phi \to \psi)$, where x is not free in ψ.

(e) $\vdash \exists x (\phi \to \forall x \phi)$, for any formula ϕ.

As ever in this chapter, $\exists x$ is an abbreviation for $\neg\forall x\neg$.

Solution

(a) We wish to derive $\vdash (\phi(\tau) \to \neg\forall x\neg\phi(x))$ and would like to relate the formula to something we already know about. When trying to derive something of the form $(\theta \to \neg\psi)$, it sometimes pays to exploit the tautology $((\psi \to \neg\theta) \to (\theta \to \neg\psi))$, although it often leaves one no better off than when one started! Here, however, it does get one onto a useful track. As the term τ is freely substitutable for x in ϕ, we can derive

$$\vdash \forall x\neg\phi(x) \to \neg\phi(\tau)$$

in just one line, as the formula is an instance of axiom Ax 4. As we can derive the following substitution instance of the tautology above,

$$((\forall x \neg \phi(x) \to \neg \phi(\tau)) \to (\phi(\tau) \to \neg \forall x \neg \phi(x))),$$

one use of Modus Ponens gives us

$$\vdash (\phi(\tau) \to \neg \forall x \neg \phi(x)).$$

(b) Not given. Our derivation exploits proof by contradiction, as with several of the other parts of this exercise.

(c) First we shall show that

$$(\neg \forall x \neg \phi \to \psi), \neg \psi \vdash \neg \phi.$$

This is given by the following derivation.

$$
\begin{array}{lll}
(1) & (\neg \forall x \neg \phi \to \psi) & \text{Ass} \\
(2) & ((\neg \forall x \neg \phi \to \psi) \to (\neg \psi \to \forall x \neg \phi)) & \text{instance of tautology} \\
(3) & (\neg \psi \to \forall x \neg \phi) & \text{MP}, 1, 2 \\
(4) & \neg \psi & \text{Ass} \\
(5) & \forall x \neg \phi & \text{MP}, 3, 4 \\
(6) & (\forall x \neg \phi \to \neg \phi) & \text{Ax}\,4 \\
(7) & \neg \phi & \text{MP}, 5, 6
\end{array}
$$

Then by the deduction theorem there is a derivation of

$$(\neg \forall x \neg \phi \to \psi) \vdash (\neg \psi \to \neg \phi),$$

which by using the instance of a tautology $((\neg \psi \to \neg \phi) \to (\phi \to \psi))$ can be turned into a derivation of

$$(\neg \forall x \neg \phi \to \psi) \vdash (\phi \to \psi). \tag{$*$}$$

As the variable x doesn't appear free in ψ (given) and of course doesn't appear free in $\neg \forall x \neg \phi$ (thanks to the presence of the $\forall x$), x doesn't appear free in the assumption $(\neg \forall x \neg \phi \to \psi)$, so that the rule Gen can be used to turn the derivation at $(*)$ into one of

$$(\neg \forall x \neg \phi \to \psi) \vdash \forall x(\phi \to \psi),$$

i.e. of $(\exists x \phi \to \psi) \vdash \forall x(\phi \to \psi)$, as required.

(d) Not given.

(e) Not given.

We shall need the following result when we prove the completeness theorem.

> **Theorem 5.5**
>
> Let Γ be a set of formulas and ϕ, ψ formulas in a language L. Suppose that the variable x does not appear free in any formula in Γ or in ψ, but appears free in ϕ. Then
>
> $$\text{if } \Gamma, \phi(x) \vdash \psi \text{ then } \Gamma, \exists x \phi(x) \vdash \psi.$$

Proof

We shall suppose that $\Gamma, \phi(x) \vdash \psi$ and then show that $\Gamma, \neg\forall x\neg\phi(x) \vdash \psi$. What we shall actually aim to show first is that

$$\Gamma, \neg\psi \vdash \forall x\neg\phi(x),$$

as then the deduction theorem gives

$$\Gamma \vdash (\neg\psi \to \forall x\neg\phi(x)),$$

so that using the substitution instance

$$((\neg\psi \to \forall x\neg\phi(x)) \to (\neg\forall x\neg\phi(x) \to \psi))$$

of the tautology $((\neg\theta \to \chi) \to (\neg\chi \to \theta))$ and Modus Ponens gives

$$\Gamma \vdash (\neg\forall x\neg\phi(x) \to \psi),$$

from which we obtain

$$\Gamma, \neg\forall x\neg\phi(x) \vdash \psi,$$

as required.

To get from $\Gamma, \phi(x) \vdash \psi$ to $\Gamma, \neg\psi \vdash \forall x\neg\phi(x)$, we use the deduction theorem to obtain

$$\Gamma \vdash (\phi(x) \to \psi)$$

and then use the substitution instance

$$((\phi(x) \to \psi) \to (\neg\psi \to \neg\phi(x)))$$

of the tautology $((\theta \to \chi) \to (\neg\chi \to \neg\theta))$ and Modus Ponens to obtain

$$\Gamma \vdash (\neg\psi \to \neg\phi(x)).$$

As the variable x doesn't appear free in any formula in Γ or in ψ, we can use the rule Gen to obtain

$$\Gamma \vdash \forall x(\neg\psi \to \neg\phi(x)).$$

A (somewhat rare!) exploitation of axiom Ax 5, using the fact that x is not free in ψ, gives

$$\Gamma \vdash (\forall x(\neg\psi \to \neg\phi(x)) \to (\neg\psi \to \forall x\neg\phi(x))),$$

and Modus Ponens gives

$$\Gamma \vdash (\neg\psi \to \forall x\neg\phi(x)),$$

from which we obtain

$$\Gamma, \neg\psi \vdash \forall x\neg\phi(x).$$

Putting this all together, we obtain $\Gamma, \exists x\phi(x) \vdash \psi$,   ∎

So far we have looked mostly at derivations which don't make special use of the axioms for the equality symbol. Given that almost all the axiom systems for mathematical theories in Section 4.4 of Chapter 4 involved $=$, we should now turn to derivations involving the symbol.

While axioms Ax 6 and Ax 7 are plausibly valid, they seem (to the author!) very weak. For instance, Ax 7 only caters for atomic formulas. This axiom represents the informal notion that if x equals y and ϕ holds for x, then ϕ holds for y too. Surely we want this to hold for all formulas ϕ, however complex, subject to what we hope by now seems sensible caution about what is freely substitutable for x in ϕ. What's more, surely we are interested in the case where x equals some complicated term τ, not simply a variable y. It turns out that all these desires are fulfilled within our formal system, as the next few exercises and theorems will establish.

We might express this use of $=$ as the substitution of equal quantities, one for another, preserving the truth of a statement.

Theorem 5.6

The following are derivable within the formal system with equality.

(i) $\vdash \tau = \tau$, for all terms τ.

(ii) $\vdash (x = y \rightarrow y = x)$.

(iii) $\vdash (x = y \rightarrow (y = z \rightarrow x = z))$

Proof

For (i), the following simple derivation based on eliminating the $\forall x$ in the axiom $\forall x\, x = x$ establishes the result.

$$
\begin{array}{lll}
(1) & \forall x\, x = x & \text{Ax}\,6 \\
(2) & (\forall x\, x = x \rightarrow \tau = \tau) & \text{Ax}\,4 \\
(3) & \tau = \tau & \text{MP}, 1, 2
\end{array}
$$

We hope that by now you are becoming accustomed to the use of Ax 4 and MP to eliminate the $\forall x$ from a formula $\forall x \phi$ previously obtained in a derivation.

For (ii), given that $\rightarrow$ is the principal connective of what we want to derive, the chances are that we should first try to derive $x = y \vdash y = x$ and then use the deduction theorem to conclude that $\vdash (x = y \rightarrow y = x)$; and this is what we shall do. We are going to exploit Ax 7 and will have to use a little bit of cunning in our choice of the atomic formula ϕ in it.

$$
\begin{array}{lll}
(1) & x = y & \text{Ass} \\
(2) & (x = y \rightarrow (x = x \rightarrow y = x)) & \text{Ax}\,7 \\
(3) & (x = x \rightarrow y = x) & \text{MP}, 1, 2 \\
(4) & \forall x\, x = x & \text{Ax}\,6 \\
(5) & (\forall x\, x = x \rightarrow x = x) & \text{Ax}\,4 \\
(6) & x = x & \text{MP}, 4, 5 \\
(7) & y = x & \text{MP}, 3, 6
\end{array}
$$

As you can see, the formula ϕ in Ax 7 was chosen to be $x = x$, which we know is easily derivable from Ax 6. As we are assuming $x = y$, the formula on line 2, is a correct instance of axiom Ax 7 likely to lead by use of Modus Ponens to the desired $y = x$.

Note that using the formula $y = y$ (also easily derivable from Ax 6) would not have worked. $(x = y \rightarrow (y = y \rightarrow y = x))$ is not a correct instance of Ax 7. The order of the variables in the initial $x = y$ matters!

Thus we have derived $x = y \vdash y = x$ and by the deduction theorem there is a derivation of $\vdash (x = y \rightarrow y = x)$.

(iii) is left as an exercise for you. ∎

Exercise 5.11 ________________________________

Show that there are derivations of each of the following.

(a) $\vdash (x = y \rightarrow (y = z \rightarrow x = z))$

(b) $\vdash (x = y \rightarrow f(x, y, x) = f(y, x, y))$, where f is a 3-place function symbol.

(c) $\vdash \exists x\, x = \tau$, for any term τ not involving x.

Solution

(a) The appearance of $\rightarrow$ as principal connective (and its further occurrence) suggests trying to derive $x = y$, $y = z \vdash x = z$ and two uses of the deduction theorem. Here is such a derivation. Note the choice of the formula $x = y$ for ϕ in the instance of Ax 7 on line 3.

$$
\begin{array}{llll}
(1) & x = y & & \text{Ass} \\
(2) & y = z & & \text{Ass} \\
(3) & (y = z \rightarrow (x = y \rightarrow x = z)) & & \text{Ax 7} \\
(4) & (x = y \rightarrow x = z) & & \text{MP}, 2, 3 \\
(5) & x = z & & \text{MP}, 1, 4
\end{array}
$$

The formula $y = z$ would have been no good as ϕ as the formula $(x = y \rightarrow (y = z \rightarrow x = z))$ isn't an instance of Ax 7.

(b) Not given. A spot of cunning is needed for the choice of ϕ in Ax 7, just as in the proof of Theorem 5.6(ii).

(c) Not given.

____

Now we shall extend Ax 7 to cover all formulas ϕ, not just the atomic formulas covered by the axiom.

Theorem 5.7

Suppose that L is a language with equality. Then

$$\vdash (x = y \rightarrow (\phi(x, x) \rightarrow \phi(x, y))),$$

for all formulas ϕ, where $\phi(x, y)$ is obtained by substituting the variable y for some (not necessarily all) of the free occurrences of the variable x in $\phi(x, x)$, provided that y is freely substitutable for these occurrences of x.

Proof

We shall use mathematical induction on the length of ϕ, i.e. the number of connectives and quantifiers in ϕ.

The base case is when ϕ is an atomic formula, i.e. of length 0, in which case the derivation consists of just the one line, namely the relevant instance of Ax 7.

For the inductive step we suppose that the result holds for all ϕ of length $\leq n$ and look at each possibility for ϕ of length $n + 1$. There are three cases to consider, but before we do this we need the following cunning observation. The inductive hypothesis is that for all ϕ of length $\leq n$ with some free occurrences of x for which y is freely substitutable, we have

$$\vdash (x = y \rightarrow (\phi(x, x) \rightarrow \phi(x, y))).$$

The x and y here stand for any variables; and for $\phi(x, x)$ of length $\leq n$, the formula $\phi(x, y)$ has the same length. This means that the inductive hypothesis also includes the supposition that

$$\vdash (y = x \rightarrow (\phi(x, y) \rightarrow \phi(x, x))),$$

as the xs are freely substitutable for the free ys in such a $\phi(x, y)$. This follows from the condition on the substitutability of y for the free x in the original $\phi(x, x)$. This means that for any ϕ of length $\leq n$ with some free occurrences of x for which y is freely substitutable, not only that

$$\vdash (x = y \rightarrow (\phi(x, x) \rightarrow \phi(x, y))),$$

but, as we have a derivation of $\vdash (x = y \rightarrow y = x)$, that by combining derivations appropriately we can obtain

$$\vdash (x = y \rightarrow (\phi(x, y) \rightarrow \phi(x, x))),$$

for such a formula ϕ. This will be of great help, as you will now see.

Case when ϕ is of form $\neg\psi$

In this case, as ψ has length n and the $\neg$ doesn't alter whether any occurrences of variables in $\neg\psi$ are free or bound, the inductive hypothesis gives us that

$$\vdash (x = y \rightarrow (\psi(x, x) \rightarrow \psi(x, y))),$$

and our cunning observation tells us that

$$\vdash (x = y \rightarrow (\psi(x, y) \rightarrow \psi(x, x))).$$

Routine use of this and the instance of a propositional tautology

$$((\psi(x, y) \rightarrow \psi(x, x)) \rightarrow (\neg\psi(x, x) \rightarrow \neg\psi(x, y)))$$

then tells us that we can derive

$$\vdash (x = y \rightarrow (\neg\psi(x, x) \rightarrow \neg\psi(x, y))),$$

as required.

Case when ϕ is of form $(\theta \rightarrow \psi)$

This is left as an exercise.

Case when ϕ is of form $\forall z\psi$

How we deal with this case depends on which variable z is.

If the variable z is x, then $\phi(x, x)$ contains no free occurrences of x and so $\phi(x, y)$ is then just the same formula as $\phi(x, x)$, namely $\forall x\psi$. As $(\forall x\psi \rightarrow \forall x\psi)$ is a tautology it's very easy to show that $\vdash (x = y \rightarrow (\forall x\psi \rightarrow \forall x\psi))$.

If the variable z is y, then there are no occurrences of x in $\forall y\psi$ for which y can be freely substituted for x, so that again the formula $\phi(x, y)$ is just the same formula as $\phi(x, x)$ and it's easy to show that $\vdash (x = y \rightarrow (\phi(x, x) \rightarrow \phi(x, y)))$.

Finally there's the case when z is neither of x and y. We shall write $\phi(x, x)$ as $\forall z\psi(x, x)$. As z is neither x nor y, so doesn't affect which free occurrences of x may be freely substituted by y, the result of replacing some of the free occurrences of x in $\phi(x, x)$ by y, where y is freely substitutable for these occurrences of x, is $\forall z\psi(x, y)$. We now take the assumptions $x = y$ and

The combination exploits the standard propositional calculus result that if $\vdash (\theta \rightarrow \psi)$ and $\vdash (\psi \rightarrow \chi)$, then $\vdash (\theta \rightarrow \chi)$.

$\forall z\psi(x,x)$ and using the instance $(\forall z\psi(x,x) \rightarrow \psi(x,x))$ of Ax 4 (that is, we remove the $\forall z$ and leave the zs in $\psi(x,x)$ as z), a straightforward use of MP gives a derivation of

$$x = y, \ \forall z\psi(x,x) \vdash \psi(x,x).$$

By the induction hypothesis, there is a derivation of

$$\vdash (x = y \rightarrow (\psi(x,x) \rightarrow \psi(x,y))).$$

Gluing these together appropriately and using MP twice gives a derivation of

$$x = y, \ \forall z\psi(x,x) \vdash \psi(x,y).$$

As there are no zs free in the assumptions, we can use Gen to obtain a derivation of

$$x = y, \ \forall z\psi(x,x) \vdash \forall z\psi(x,y),$$

and two uses of the deduction theorem give

$$\vdash (x = y \rightarrow (\forall z\psi(x,x) \rightarrow \forall z\psi(x,y))),$$

as required. ∎

We can extend the result of Theorem 5.7 to cover substitution in formulas by a general term τ, not just a variable y.

Theorem 5.8

For all formulas ϕ and terms τ

$$\vdash (x = \tau \rightarrow (\phi(x,x) \rightarrow \phi(x,\tau))),$$

where $\phi(x,\tau)$ is obtained by substituting the term τ for some (not necessarily all) of the free occurrences of the variable x in $\phi(x,x)$, provided that τ is freely substitutable for these occurrences of x.

Proof

Let y be a variable which may be freely substituted for x in ϕ. Then Theorem 5.7 tells us that there is a derivation of

$$\vdash (x = y \rightarrow (\phi(x,x) \rightarrow \phi(x,y)))$$

involving no assumptions. We shall add a few extra lines to this derivation as follows.

$$
\begin{array}{lll}
 & \vdots \quad \vdots & \qquad\qquad \vdots \\
(k) & (x = y \rightarrow (\phi(x,x) \rightarrow \phi(x,y))) & \\
(k+1) & \forall y(x = y \rightarrow (\phi(x,x) \rightarrow \phi(x,y))) & \text{Gen}, k \\
(k+2) & (\forall y(x = y \rightarrow (\phi(x,x) \rightarrow \phi(x,y))) & \\
 & \quad \rightarrow (x = \tau \rightarrow (\phi(x,x) \rightarrow \phi(x,\tau)))) & \text{Ax 4} \\
(k+3) & (x = \tau \rightarrow (\phi(x,x) \rightarrow \phi(x,\tau))) & \text{MP}, k+1, k+2
\end{array}
$$

> Taking y to be a variable not occurring in ϕ is always a safe choice here!

> As there are no assumptions, Gen can be used with no problems.

The condition on τ guarantees that the use of Ax 4 on line $k+2$ is correct. ∎

Exercise 5.12

Fill in the detail of the inductive step in the proof of Theorem 5.7 for the case when ϕ is of form $(\theta \to \psi)$.

For the sort of theorem that we discussed in the introduction to this chapter, we really want to be able to exploit results about what follows when two terms τ_1 and τ_2 are equal, not just when two variables (or a variable x and a term τ as in Theorem 5.8) are equal. We state these results as theorems and leave their proof to you as exercises.

Theorem 5.9

For any terms τ_1, τ_2, τ_3 in a language L with equality,

$$\vdash (\tau_1 = \tau_2 \to (\tau_2 = \tau_3 \to \tau_1 = \tau_3)).$$

Exercise 5.13

Prove Theorem 5.9. [*Hint*: Look for a use of Gen similar to that in the proof of Theorem 5.8 applied to the result of Exercise 5.11(a).]

Theorem 5.10

Let τ_1, τ_2 be terms and ϕ a formula with x amongst its free variables in a language L with equality. Suppose that the formula $\phi(x, \tau_1)$ is obtained by substituting the term τ_1 for some (not necessarily all) of the free occurrences of x in $\phi(x, x)$ and that the formula $\phi(x, \tau_2)$ is obtained by substituting the term τ_2 for the same occurrences of x. If both τ_1 and τ_2 are freely substitutable for these occurrences of x in ϕ, then

$$\vdash (\tau_1 = \tau_2 \to (\phi(x, \tau_1) \to \phi(x, \tau_2))).$$

Just in case the statement of this theorem seems a bit mysterious, we shall give an example of its use, from which we hope it will become clearer just how valuable it is. In everyday maths we might want to exploit an identity like

$$x^2 - 1 = (x - 1)(x + 1)$$

to show that

$$x^2(x^2 - 1) = x^2(x - 1)(x + 1).$$

In the formal system it is Theorem 5.10 which permits us to do this. Taking T to be the axioms for the theory of commutative rings with 1 on page 202, and using the formal language for rings $(+, \cdot, -, \mathbf{0}, \mathbf{1}, =)$ we would be able to derive

$$T \vdash ((x \cdot x) + (-\mathbf{1})) = ((x + (-\mathbf{1})) \cdot (x + \mathbf{1})).$$

To make the step to

$$T \vdash ((x \cdot x) \cdot ((x \cdot x) + (-\mathbf{1}))) = ((x \cdot x) \cdot ((x + (-\mathbf{1})) \cdot (x + \mathbf{1}))),$$

When the set of assumptions is the set of axioms of a theory T and $T \vdash \phi$, where ϕ is a sentence, we shall often call ϕ a *theorem* of T. The soundness and completeness theorems will guarantee that the theorems of T coincide with its logical consequences.

we proceed as follows.

First we use Theorem 5.6(a) to derive

$$T \vdash ((x \cdot x) \cdot ((x \cdot x) + (-\mathbf{1}))) = ((x \cdot x) \cdot ((x \cdot x) + (-\mathbf{1}))).$$

You'll see why we chose this formula in a moment!

Now for the use of Theorem 5.10! Take $\phi(x, x)$ to be the magic formula

$$((x \cdot x) \cdot ((x \cdot x) + (-\mathbf{1}))) = ((x \cdot x) \cdot \underline{x})$$

where it's the underlined x which we are going to substitute separately by the terms τ_1 and τ_2, where τ_1 is $((x \cdot x) + (-\mathbf{1}))$ and τ_2 is $((x + (-\mathbf{1})) \cdot (x + \mathbf{1}))$. Theorem 5.10 then says that there is a derivation of

$$\vdash (\tau_1 = \tau_2 \rightarrow (((x \cdot x) \cdot ((x \cdot x) + (-\mathbf{1}))) = ((x \cdot x) \cdot \tau_1)$$
$$\rightarrow ((x \cdot x) \cdot ((x \cdot x) + (-\mathbf{1}))) = ((x \cdot x) \cdot \tau_2)))$$

that is,

$$\vdash (((x \cdot x) + (-\mathbf{1})) = ((x + (-\mathbf{1})) \cdot (x + \mathbf{1}))$$
$$\rightarrow (((x \cdot x) \cdot ((x \cdot x) + (-\mathbf{1}))) = ((x \cdot x) \cdot ((x \cdot x) + (-\mathbf{1})))$$
$$\rightarrow ((x \cdot x) \cdot ((x \cdot x) + (-\mathbf{1}))) = ((x \cdot x) \cdot (x + (-\mathbf{1})) \cdot (x + \mathbf{1})))).$$

Combining these derivations with a couple of uses of Modus Ponens gives the required result, that

$$T \vdash ((x \cdot x) \cdot ((x \cdot x) + (-\mathbf{1}))) = ((x \cdot x) \cdot (((x + (-\mathbf{1})) \cdot (x + \mathbf{1})))).$$

See whether you can see how to exploit Theorem 5.10 in the following exercise and then see if you can prove the theorem.

Exercise 5.14

For the theory T above, we have

$$T \vdash (-(x + 1)) = ((-x) + (-1)).$$

Explain how to use Theorem 5.10 to show that

$$T \vdash ((-(x + 1)) \cdot y) = (((-x) + (-1)) \cdot y).$$

Exercise 5.15

Prove Theorem 5.10. [*Hints*: Replace by a new variable y those occurrences of x which are to be substituted, apply Theorem 5.8 to this new formula with the term τ_2 and then use the trick in the proof of Theorem 5.8.]

Solution

Given the formula $\phi(x, x)$, replace those occurrences of x for which the terms τ_1, τ_2 are to be substituted by a variable, y say, not occurring in any of $\phi(x, x), \tau_1, \tau_2$, and thus obtain the formula $\phi(x, y)$. Then by Theorem 5.8 we have a derivation of

$$\vdash (y = \tau_2 \rightarrow (\phi(x, y) \rightarrow \phi(x, \tau_2)))$$

involving no assumptions. To this derivation, we add the lines

$$
\begin{array}{lll}
(k) & (y = \tau_2 \rightarrow (\phi(x, y) \rightarrow \phi(x, \tau_2))) & \\
(k+1) & \forall y (y = \tau_2 \rightarrow (\phi(x, y) \rightarrow \phi(x, \tau_2))) & \text{Gen}, k \\
(k+2) & (\forall y (y = \tau_2 \rightarrow (\phi(x, y) \rightarrow \phi(x, \tau_2))) & \\
& \quad \rightarrow (\tau_1 = \tau_2 \rightarrow (\phi(x, \tau_1) \rightarrow \phi(x, \tau_2)))) & \text{Ax}\,4 \\
(k+3) & (\tau_1 = \tau_2 \rightarrow (\phi(x, \tau_1) \rightarrow \phi(x, \tau_2))) & \text{MP}, k+1, k+2
\end{array}
$$

As there are no assumptions, Gen can be used with no problems.

obtaining the required result.

We are now in a position to return to the example used in the introduction to this chapter and show how the sentence $\forall x ((x + x) = x \rightarrow x = \mathbf{0})$ is derivable within our system from various axioms. The axioms in question were as follows, in the language L with equality including the 2-place function symbol $+$, the 1-place function symbol $-$ and constant symbol $\mathbf{0}$.

1. $\forall x \forall y \forall z\, (x + (y + z)) = ((x + y) + z)$
2. $\forall x\, (x + \mathbf{0}) = x$
3. $\forall x\, (\mathbf{0} + x) = x$
4. $\forall x\, (x + (-x)) = \mathbf{0}$
5. $\forall x\, ((-x) + x) = \mathbf{0}$

Let's call the set of these axioms GP, as they do axiomatize the theory of groups. We wish to show that $\text{GP} \vdash \forall x ((x + x) = x \rightarrow x = \mathbf{0})$. The axioms of the theory can of course be used as assumptions in a derivation. When we use them in this way, we shall justify the line by saying that the formula is an axiom of GP; but really it is just an assumption out of the set GP. First we derive GP, $(x + x) = x \vdash x = \mathbf{0}$ as follows.

(1)	$(x + x) = x$	Ass
(2)	$((x + x) = x \rightarrow (((x + x) + (-x)) = ((x + x) + (-x))$	
	$\rightarrow ((x + x) + (-x)) = (x + (-x))))$	Theorem 5.10
(3)	$(((x + x) + (-x)) = ((x + x) + (-x)) \rightarrow ((x + x) + (-x)) = (x + (-x)))$	MP, 1, 2
(4)	$((x + x) + (-x)) = ((x + x) + (-x))$	Theorem 5.6(a)
(5)	$((x + x) + (-x)) = (x + (-x))$	MP, 3, 4
(6)	$\forall x \forall y \forall z\, (x + (y + z)) = ((x + y) + z)$	GP axiom 1
(7)	$(\forall x \forall y \forall z\, (x + (y + z)) = ((x + y) + z) \rightarrow \forall y \forall z\, (x + (y + z)) = ((x + y) + z))$	Ax 4
(8)	$\forall y \forall z\, (x + (y + z)) = ((x + y) + z)$	MP, 6, 7
(9)	$(\forall y \forall z\, (x + (y + z)) = ((x + y) + z) \rightarrow \forall z\, (x + (x + z)) = ((x + x) + z))$	Ax 4
(10)	$\forall z\, (x + (x + z)) = ((x + x) + z)$	MP, 8, 9
(11)	$(\forall z\, (x + (x + z)) = ((x + x) + z) \rightarrow (x + (x + (-x))) = ((x + x) + (-x)))$	Ax 4
(12)	$(x + (x + (-x))) = ((x + x) + (-x))$	MP 10, 11
(13)	$((x + x) = x \rightarrow ((x + (x + (-x))) = ((x + x) + (-x))$	
	$\rightarrow (x + (x + (-x))) = (x + (-x))))$	Theorem 5.10
(14)	$((x + (x + (-x))) = ((x + x) + (-x)) \rightarrow (x + (x + (-x))) = (x + (-x)))$	MP, 1, 13
(15)	$(x + (x + (-x))) = (x + (-x))$	MP, 12, 14
(16)	$\forall x\, (x + (-x)) = \mathbf{0}$	GP axiom 4
(17)	$(\forall x\, (x + (-x)) = \mathbf{0} \rightarrow (x + (-x)) = \mathbf{0})$	Ax 4
(18)	$(x + (-x)) = \mathbf{0}$	MP, 16, 17
(19)	$((x + (-x)) = \mathbf{0} \rightarrow ((x + (x + (-x))) = (x + (-x)) \rightarrow (x + \mathbf{0}) = \mathbf{0}))$	Theorem 5.10
(20)	$((x + (x + (-x))) = (x + (-x)) \rightarrow (x + \mathbf{0}) = \mathbf{0})$	MP, 18, 19
(21)	$(x + \mathbf{0}) = \mathbf{0}$	MP, 15, 20
(22)	$\forall x\, (x + \mathbf{0}) = x$	GP axiom 2
(23)	$(\forall x\, (x + \mathbf{0}) = x \rightarrow (x + \mathbf{0}) = x)$	Ax 4
(24)	$(x + \mathbf{0}) = x$	MP, 22, 23
(25)	$((x + \mathbf{0}) = x \rightarrow ((x + \mathbf{0}) = \mathbf{0} \rightarrow x = \mathbf{0}))$	Theorem 5.10
(26)	$((x + \mathbf{0}) = \mathbf{0} \rightarrow x = \mathbf{0})$	MP, 24, 25
(27)	$x = \mathbf{0}$	MP, 21, 26

Now by the deduction theorem, there is a derivation of

$$\mathrm{GP} \vdash ((x + x) = x \rightarrow x = \mathbf{0}),$$

and as all the axioms of GP are sentences, the rule Gen can be used to give
a derivation of

$$\mathrm{GP} \vdash \forall x((x + x) = x \rightarrow x = \mathbf{0}).$$

You might like to try one or two formal derivations from these axioms GP for
groups yourself.

Exercise 5.16 ______________

Show that there are derivations of each of the following.

(a) $\mathrm{GP} \vdash (\mathbf{0} + \mathbf{0}) = \mathbf{0}$

(b) $\mathrm{GP} \vdash \forall y \forall z((\mathbf{0} + y) = (\mathbf{0} + z) \rightarrow y = z)$

(c) $\mathrm{GP} \vdash \forall x \forall y \forall z((x + y) = (x + z) \rightarrow y = z)$

Please continue for the rest of this
section to use results about the
formal proof system obtained so
far, rather than the completeness
theorem, which we haven't yet
proved (and which would save you
many agonies for this exercise!).

Let us look at more examples of formal proofs within other mathematical
theories axiomatized in Section 4.4 of Chapter 4.

Exercise 5.17 ──

Let SPO be the following set of sentences in a language with equality and binary relation symbol S:

$$\{\forall x \neg S(x,x), \quad \forall x \forall y \forall z (S(x,y) \rightarrow (S(y,z) \rightarrow S(x,z)))\}.$$

Then SPO gives axioms for the theory of strict partial order as introduced on page 193. Show that there are derivations of each of the following.

(a) $\text{SPO} \vdash \forall x \forall y (S(x,y) \rightarrow \neg x = y)$

(b) $\text{SPO} \vdash \forall x \forall y \forall z (S(x,y) \rightarrow (S(y,z) \rightarrow \neg x = z))$

(c) $\text{SPO}, \forall x \exists y S(x,y) \vdash \neg \forall x \forall y\, x = y$

SPO expresses the transitive property without using the connective $\wedge$.

Solution

(a) We shall assume $S(x,y)$ and $x = y$ and derive a contradiction, so that we can conclude that $\text{SPO} \vdash (S(x,y) \rightarrow \neg x = y)$. Two uses of the rule Gen will then give the required result.

First we have the derivation

$$
\begin{array}{lll}
(1) & S(x,y) & \text{Ass} \\
(2) & x = y & \text{Ass} \\
(3) & (x = y \rightarrow (S(x,y) \rightarrow S(y,y))) & \text{Ax}\,7 \\
(4) & (S(x,y) \rightarrow S(y,y)) & \text{MP}, 2, 3 \\
(5) & S(y,y) & \text{MP}, 1, 4
\end{array}
$$

which shows that

$$S(x,y),\ x = y \vdash S(y,y),$$

so that by the thinning rule

$$\text{SPO},\ S(x,y),\ x = y \vdash S(y,y).$$

We also have the derivation

$$
\begin{array}{lll}
(1) & \forall x \neg S(x,x) & \text{axiom of SPO} \\
(2) & (\forall x \neg S(x,x) \rightarrow \neg S(y,y)) & \text{Ax}\,4 \\
(3) & \neg S(y,y) & \text{MP}, 2, 3
\end{array}
$$

which shows, again using the thinning rule, that

$$\text{SPO},\ S(x,y),\ x = y \vdash \neg S(y,y).$$

Proof by contradiction thus tells us that there is a derivation of

$$\text{SPO},\ S(x,y) \vdash \neg x = y,$$

and by the deduction theorem we then have a derivation of

$$\text{SPO} \vdash (S(x,y) \rightarrow \neg x = y).$$

As the assumptions in SPO are all sentences, two applications of Theorem 5.1, first with the variable y and then with x, give

$$\text{SPO} \vdash \forall x \forall y (S(x,y) \rightarrow \neg x = y),$$

as required.

(b) Not given.

(c) Not given.

Exercise 5.18 ____

Let Equiv be the following set of sentences in the language L with equality
and a binary relation symbol R:

$$\{\forall x R(x,x),\ \forall x \forall y (R(x,y) \rightarrow R(y,x)),\ \forall x \forall y \forall z (R(x,y) \rightarrow (R(y,z) \rightarrow R(x,z)))\}.$$

Then Equiv gives axioms for the theory of equivalence relations as introduced
on page 186. Use results about the formal proof system (i.e. not the completeness theorem, which we are pretending we don't know about) to show
that there are derivations of each of the following.

(a) $\vdash (\forall x \neg \forall y R(x,y) \rightarrow \neg \forall x \forall y\, x = y)$

(b) $\vdash (\neg \forall x \neg \forall y R(x,y) \rightarrow \forall z \forall y R(z,y))$

Solution

We shall give a solution to (a) and leave (b) to you.

With a use of the deduction theorem in mind, we shall assume $\forall x \neg \forall y R(x,y)$
and try to derive $\neg \forall x \forall y\, x = y$. As what we are trying to derive starts with
a $\neg$, we shall also assume $\forall x \forall y\, x = y$ and try to derive a contradiction. Here
goes!

(1)	$\forall x \neg \forall y R(x,y)$	Ass
(2)	$\forall x \forall y\, x = y$	Ass
(3)	$\forall x R(x,x)$	Axiom of Equiv
(4)	$(\forall x \neg \forall y R(x,y) \rightarrow \neg \forall y R(x,y))$	Ax 4
(5)	$\neg \forall y R(x,y)$	MP, 1, 4
(6)	$(\forall x \forall y\, x = y \rightarrow \forall y\, x = y)$	Ax 4
(7)	$\forall y\, x = y$	MP, 2, 6
(8)	$(\forall y\, x = y \rightarrow x = y)$	Ax 4
(9)	$x = y$	MP, 7, 8
(10)	$(\forall x R(x,x) \rightarrow R(x,x))$	Ax 4
(11)	$R(x,x)$	MP, 3, 10
(12)	$(x = y \rightarrow (R(x,x) \rightarrow R(x,y)))$	Ax 7
(13)	$(R(x,x) \rightarrow R(x,y))$	MP, 9, 12
(14)	$R(x,y)$	MP, 11, 13
(15)	$\forall y R(x,y)$	Gen

Note that as all assumptions are sentences, we can't run into trouble with
Gen! From this derivation, with judicious use of the thinning rule, we can
conclude that there are derivations of both

$$\text{Equiv}, \forall x \neg \forall y R(x,y), \forall x \forall y\, x = y \vdash \neg \forall y R(x,y)$$

and

$$\text{Equiv}, \forall x \neg \forall y R(x,y), \forall x \forall y\, x = y \vdash \forall y R(x,y).$$

Proof by contradiction then gives us a derivation of

$$\text{Equiv}, \forall x \neg \forall y R(x,y) \vdash \neg \forall x \forall y\, x = y$$

and the deduction theorem then gives that there is a derivation of

$$\text{Equiv} \vdash (\forall x \neg \forall y R(x,y) \rightarrow \neg \forall x \forall y\, x = y).$$

As for SPO, we have written the
transitive axiom avoiding the use of
$\wedge$.

These formulas are logically
equivalent to
$(\forall x \exists y \neg R(x,y) \rightarrow \exists x \exists y \neg x = y)$
and $(\exists x \forall y R(x,y) \rightarrow \forall z \forall y R(z,y))$.

As we have already hinted, just as with the propositional calculus we shall step outside our formal system and see what is derivable within it by looking at the completely different notion of logical consequence and connecting this to formal derivation by soundness and completeness theorems. But it is valuable to have a brief look at a few more metatheorems that provide shortcuts to establish what can be derived within the system and which give us a recipe for actually constructing a derivation using just the basic features of the system.

From our experience of derivations so far, we could nominate a couple of derived rules of inference as contained in the following exercise.

Exercise 5.19 ─────────────────────────────

In both parts of this exercise, you should explain how to obtain the desired derivation from any given derivations. (So no cheating by saying 'it follows from the soundness and completeness theorems', which in any case we're pretending you don't know about yet!)

(a) Let Γ be a set of formulas, $\phi(x)$ a formula and τ a term which can be freely substituted for x in $\phi(x)$. Show that if $\Gamma \vdash \forall x \phi(x)$, then there is a derivation of $\Gamma \vdash \phi(\tau)$.

This is often called the *universal elimination* rule.

(b) Let Γ be a set of formulas and ϕ a formula in a language with equality, and let τ_1, τ_2 be terms. Suppose that the formula $\phi(x, \tau_1)$ is obtained by substituting the term τ_1 for some (not necessarily all) of the free occurrences of x in $\phi(x, x)$ and that the formula $\phi(x, \tau_2)$ is obtained by substituting the term τ_2 for the same occurrences of x. Then suppose that both τ_1 and τ_2 are freely substitutable for these occurrences of x in ϕ. Show that if $\Gamma \vdash \tau_1 = \tau_2$ and $\Gamma \vdash \phi(x, \tau_1)$, then there is a derivation of $\Gamma \vdash \phi(x, \tau_2)$.

This shortcut based on Theorem 5.10 is often called the *substitution* rule.

───

As well as sensible shortcuts such as these, the system could be expanded to cope more directly with connectives besides $\neg, \rightarrow$ and with the quantifier $\exists$. You can find rules for dealing with $\vee$ and $\wedge$ in Exercises 3.46 and 3.47 at the end of Chapter 3, as well as some further rules, besides various forms of proof by contradiction, for dealing with $\neg$. As for rules for handling $\exists$, one rule, stems from the result of Exercise 5.10(a) and can be stated as follows.

> If the term τ is freely substitutable for x in ϕ, and $\Gamma \vdash \phi(\tau)$,
> then $\Gamma \vdash \exists x \phi(x)$.

This is often called the *existential introduction* rule.

A matching rule for $\exists$ is given by Theorem 5.5.

> Let Γ be a set of formulas and ϕ, ψ formulas in a language L. Suppose that the variable x does not appear free in any formula in Γ or in ψ, but appears free in ϕ. If $\Gamma, \phi(x) \vdash \psi$ then $\Gamma, \exists x \phi(x) \vdash \psi$.

This is often called the *existential hypothesis* rule.

We are not going to pursue the use of these rules any further in this book, despite the fact that most of the axiom systems we discussed in Section 4.4 of Chapter 4 used the extra symbols within the axioms. Plainly, if you were going to use the formal proof system for real mathematical derivations, you would probably want to exploit these rules and more.

Further exercises

Exercise 5.20

Under the standard interpretations of the usual connectives and quantifiers, some of the following attempts at rules of inference are valid and some are invalid. Sort out which are which, giving reasons for your answers. In the invalid cases try to suggest reasonable corrections.

(a) If $\Gamma, \phi \vDash \psi$, then $\Gamma, \neg\phi \vDash \neg\psi$.

(b) If $\Gamma \vDash (\phi \vee \psi)$, then $\Gamma \vDash \phi$ and $\Gamma \vDash \psi$.

(c) If $\Gamma, \phi \vDash \psi$ and $\Gamma, \psi \vDash \theta$, then $\Gamma \vDash (\phi \to \theta)$.

(d) If $\Gamma \vDash \forall x \phi(x)$ and $\Gamma, \phi(\tau) \vDash \psi$, where τ is a term, then $\Gamma \vDash \psi$.

(e) If $\Gamma, \forall x \phi(x) \vDash \forall x \psi(x)$, then $\Gamma \vDash \forall x (\phi(x) \to \psi(x))$.

(f) If $\Gamma \vDash \exists x \phi(x)$, then $\Gamma \vDash (\exists x \exists y \neg x = y \ \vee \ \forall x \phi(x))$.

Exercise 5.21

Suppose that ϕ is a substitution instance of a tautology. Show that $\vdash \forall x \phi$ and $\vdash \exists x \phi$.

Exercise 5.22

Show that there are derivations of each of the following.

(a) $\vdash (\forall x \forall y \phi \to \forall y \forall x \phi)$

(b) $\vdash \forall x (\phi \to \psi) \vdash (\forall x \neg \psi \to \forall x \neg \phi)$

(c) $\exists x \neg \psi \vdash (\forall x (\neg \phi \to \psi) \to \exists x \phi)$

(d) $\vdash (\forall z (P(z, z) \to Q(z)) \to \forall x (\forall y P(x, y) \to Q(x)))$, where P is a 2-place relation symbol and Q is a 1-place relation symbol.

(e) $\vdash ((\phi \to \forall x \psi) \to \forall x (\phi \to \psi))$, where x is not free in ϕ.

5.3   The soundness theorem

In this section we shall prove the soundness theorem for our formal proof system for predicate calculus. We shall also investigate structures for a language with equality which satisfy the equality axioms but are not normal, that is, they don't interpret the $=$ symbol by actual equality on their domains.

The statement of the soundness theorem is the same as for the propositional calculus, namely that if $\Gamma \vdash \phi$ then $\Gamma \vDash \phi$, for any formula ϕ and set of formulas Γ. The structure of its proof is just the same as for the propositional calculus in Section 3.3 of Chapter 3. We take a structure $\mathcal{A}$ for the relevant language and an interpretation $\vec{a}$ of all the variables $\vec{x}$ and show by mathematical induction on the length of formal derivations that if $\mathcal{A}$ satisfies Γ with this interpretation, then it satisfies ϕ. The details of the proof are very similar to the propositional calculus case, but the argument must deal with the extra ways in which lines can be derived using the axioms and rule (generalization) which do things with the universal quantifier $\forall$.

Theorem 5.11 Soundness theorem

For all formulas ϕ and sets Γ of formulas in a first-order language L, if $\Gamma \vdash \phi$ then $\Gamma \vDash \phi$.

Proof

We shall give an outline of the proof and leave you to fill in the details as an exercise.

If $\Gamma \vdash \phi$, then there is a subset Γ_0 of Γ for which there is a derivation of $\Gamma_0 \vdash \phi$ following conditions (i) to (iv) in the definition of $\Gamma \vdash \phi$. We shall prove that $\Gamma_0 \vDash \phi$, from which it follows, using the definition of logical consequence, that $\Gamma \vDash \phi$.

The point about taking the subset Γ_0 is that there is no risk that the rule Gen has been used with a variable free in some formula in Γ_0.

Just as with the result for propositional calculus, we take a structure $\mathcal{A}$ for L and an interpretation $\vec{a}$ of the free variables occurring in Γ_0 and ϕ such that $\mathcal{A} \vDash_{\vec{x}/\vec{a}} \gamma$, for all $\gamma \in \Gamma$. We take a derivation

$$\phi_1, \phi_2, \ldots, \phi_n = \phi$$

of $\Gamma_0 \vdash \phi$ and use induction to show that $\mathcal{A} \vDash_{\vec{x}/\vec{a}} \phi_i$ for all $i = 1, 2, \ldots, n$.

The cases when a line is an assumption in Γ_0 or an instance of one of Ax 1, Ax 2, Ax 3 or follows from an application of MP are dealt with just as in the proof of the soundness theorem for propositional calculus in Chapter 3.

Theorem 4.3 of Section 4.3 of Chapter 4 deals with the case when a line is an instance of axiom Ax 4. Exercise 4.47(c) of the same section effectively deals with the case when a line is an instance of axiom Ax 5 – make sure that you see why!

If the language L includes equality, our definition of universally valid and logical consequence is in terms of normal structures. For a normal structure, it is immediate that

$$\mathcal{A} \vDash_{\vec{x}/\vec{a}} x = x,$$

which deals with any use of axiom Ax 6 in the derivation. We shall leave dealing with instances of Ax 7 as an exercise for you.

Probably the most interesting case to deal with is a use of the rule of generalization. That means that the line has the form

$$\Gamma_0 \vdash \forall x \phi,$$

where there is an earlier line of the form $\Gamma_0 \vdash \phi$ with the variable x not free in any formula in Γ_0. By the inductive hypothesis for this earlier line, we have

$$\Gamma_0 \vDash \phi$$

and from this we have to show $\Gamma_0 \vDash \forall x \phi$. If x is not free in ϕ, this is particularly easy to show using the application of Theorem 4.1 of Section 4.2 of Chapter 4 discussed immediately following Exercise 4.19 on page 160. We shall leave you to think about the more interesting case when x is free in ϕ. ∎

Exercise 5.23

(a) Let L be a language with equality, $\mathcal{A} = \langle A, \ldots, = \rangle$ a normal structure for L and $\vec{a}$ any interpretation of the variables $\vec{x}$ by elements of A. Show that $\mathcal{A}$ satisfies any instance of axiom Ax 7 with this interpretation, that is,

$$\mathcal{A} \vDash_{\vec{x}/\vec{a}} (x = y \to (\phi(x, x) \to \phi(x, y))),$$

for any atomic formula ϕ, where $\phi(x, y)$ is obtained by substituting the variable y for some (not necessarily all) of the occurrences of the variable x in $\phi(x, x)$. [*Hints*: As L might include some function symbols, you have to allow for involvement of more complicated terms within ϕ than simply variables or constant symbols. The result of Exercise 4.23 of Section 4.2 of Chapter 4 will be very helpful.]

(b) Write out a proper proof of Theorem 5.11. [*Hint*: We have filled one major gap in part (a) of this exercise. The most significant remaining gap requires showing that if $\Gamma_0 \vDash \phi$, where x is not free in any formula in Γ_0, then $\Gamma_0 \vDash \forall x \phi$.]

As with propositional calculus, the soundness theorem gives a very valuable way of showing when $\Gamma \nvdash \phi$. When Γ axiomatizes a specific mathematical theory and one knows a few models of the theory, this often gives us a quick way of dismissing some statements as theorems. For instance, let SLO be the theory of strict linear orders with axioms given on page 193 using a binary relation symbol S. We might ask whether

$$\forall x (\exists t S(x, t) \to \exists y (S(x, y) \land \forall z (S(x, z) \to (y = z \lor S(y, z)))))$$

is a theorem of SLO, that is, whether

$$\text{SLO} \vdash \forall x (\exists t S(x, t) \to \exists y (S(x, y) \land \forall z (S(x, z) \to (y = z \lor S(y, z))))).$$

This sentence essentially says that for each x that is not a maximal element, there is a least y greater than x. Recalling that $\langle \mathbb{Q}, <, = \rangle$ is a model of the theory and noting that this property fails for $\mathbb{Q}$, we have

$$\text{SLO} \nvDash \forall x (\exists t S(x, t) \to \exists y (S(x, y) \land \forall z (S(x, z) \to (y = z \lor S(y, z))))),$$

so that by the soundness theorem

$$\text{SLO} \nvdash \forall x (\exists t S(x, t) \to \exists y (S(x, y) \land \forall z (S(x, z) \to (y = z \lor S(y, z))))).$$

If x, y are rationals with $x < y$, there are infinitely many rationals z such that $x < z < y$.

Exercise 5.24

Let SLO be the theory of strict linear order as above. Show that none of the following statements are theorems of SLO.

(a) $\forall x \exists y S(x, y)$

(b) $\exists x \forall y (\neg x = y \to S(x, y))$

(c) $\forall x \forall y (S(x, y) \to \exists z (S(x, z) \land S(z, y)))$

Exercise 5.25 ───────────────────────────────

Let GPs be the theory of groups as axiomatized on page 198. Show that none of the following statements are theorems of GPs.

(a) $\forall x \forall y\, (x \cdot y) = (y \cdot x)$

(b) $\exists x(\neg x = \mathbf{e} \wedge (x \cdot x) = \mathbf{e})$

───────────────────────────────

As for the propositional calculus, there is another way of phrasing the soundness theorem in terms of consistency.

> **Definitions   Consistency**
>
> A set Γ of first-order formulas is *inconsistent* if there is a formula θ for which both
>
> $$\Gamma \vdash \theta \quad \text{and} \quad \Gamma \vdash \neg\theta.$$
>
> In the case that Γ is the empty set, we say that *the system S is inconsistent*.
>
> We say that the set Γ is *consistent* if it is not inconsistent.

Exercise 5.26 ───────────────────────────────

Suppose that Γ is a set of formulas and ϕ a formula such that $\Gamma \vdash \forall x \phi$ and $\Gamma \vdash \forall x \neg \phi$. Show that Γ is inconsistent.

───────────────────────────────

There are several useful results about consistency similar to those we obtained for the propositional calculus in Section 3.3 of Chapter 3. These are left for you as the following exercises.

Exercise 5.27 ───────────────────────────────

Suppose that Γ is a set of sentences and ϕ is a sentence. Show the following.

(a) $\Gamma \cup \{\neg\phi\}$ is consistent if and only if $\Gamma \nvdash \phi$.

(b) $\Gamma \cup \{\phi\}$ is consistent if and only if $\Gamma \nvdash \neg\phi$.

(c) If Γ is consistent and $\Gamma \vdash \phi$, then $\Gamma \cup \{\phi\}$ is consistent.

These results could have been stated in terms of formulas rather than sentences. However, we shall only need the results for sentences and their proofs could well be simpler without the worry of free variables!

───────────────────────────────

Given that the set of sentences Γ might now represent something quite interesting mathematically (in comparison to propositional formulas), these results are correspondingly more interesting. For instance, we showed earlier that

$$\text{SLO} \nvdash \forall x(\exists t S(x,t) \rightarrow \exists y(S(x,y) \wedge \forall z(S(x,z) \rightarrow (y = z \vee S(y,z))))),$$

so that the set of sentences

$$\text{SLO} \cup \{\neg \forall x(\exists t S(x,t) \rightarrow \exists y(S(x,y) \wedge \forall z(S(x,z) \rightarrow (y = z \vee S(y,z)))))\}$$

is consistent.

We can now rephrase the soundness theorem in terms of consistency, just as we did for the propositional calculus.

Theorem 5.12

The following general statements about formulas in a first-order language L are equivalent.

(A) For all formulas ϕ and all sets of formulas Γ, if $\Gamma \vdash \phi$ then $\Gamma \vDash \phi$.

(B) For all sets of formulas Δ, if Δ is satisfiable then Δ is consistent.

This is the soundness theorem.

Recall that Δ is satisfiable if there is a structure for L and an interpretation of any variables appearing free in Δ satisfying all $\delta \in \Delta$.

Exercise 5.28

Prove Theorem 5.12.

As was the case with our formal proof system for propositional calculus, we can exploit the soundness theorem in the form (B) of Theorem 5.12 to infer that our formal proof system for predicate calculus is consistent.

Theorem 5.13

The formal proof system for predicate calculus is consistent, i.e. there is no formula ϕ for which there are derivations $\vdash \phi$ and $\vdash \neg\phi$.

Proof

Take the set Δ in (B) of Theorem 5.12 to be empty. Then Δ is satisfied by any structure (in the sense that no structure makes any of the sentences in Δ false). Thus Δ is consistent, using form (B) of the soundness theorem, i.e. the formal system is consistent. ∎

This chapter will conclude with a proof of the converse to the soundness theorem, namely the completeness theorem for our system, showing that every consistent set of sentences Δ in a language L has a model. When L is a language with equality, our construction will first show that Δ has a model which satisfies the equality axioms Ax 6 and Ax 7, but which isn't a normal structure, that is, it doesn't interpret the $=$ symbol by actual equality. However, from this non-normal structure we can create a normal model satisfying the same sentences, ia particular those in Δ. In the next section, we shall look at non-normal structures which satisfy the equality axioms and at how to construct from them these corresponding normal structures.

5.4 The equality axioms and non-normal structures

Throughout the book so far we have chosen to interpret languages with equality by normal structures. These are structures of the form $\langle A, \ldots \rangle$ where the symbol $=$ is interpreted by actual equality on the set A, that is the relation given by the set of pairs $\{(a, a) : a \in A\}$. As part of the proof of the soundness theorem, we have shown that all instances of the equality axioms Ax 6 and Ax 7 of our formal proof system for such languages are satisfied by all normal structures $\mathcal{A}$ for all interpretations $\vec{a}$ of any free variables $\vec{x}$ in the formulas. However, there are also structures which aren't normal, i.e. in which the equality symbol is not interpreted by actual equality on the domain, which also satisfy these axioms, and this is what we shall discuss in this section.

For the purpose of formalising everyday mathematics, it's entirely natural to constrain the interpretation of the equality symbol in this way.

Rather than the mouthful '$\mathcal{A}$ satisfies θ for all interpretations $\vec{a}$ of any free variables $\vec{x}$ in the formula θ', we shall say that a structure $\mathcal{A}$ *satisfies* a formula θ if for all interpretations $\vec{a}$ of the free variables in L,

$$\mathcal{A} \vDash_{\vec{x}/\vec{a}} \theta.$$

Note the subtle difference between '$\mathcal{A}$ satisfies θ' here and 'θ is satisfiable', meaning that there is some structure $\mathcal{A}$ such that

$$\mathcal{A} \vDash_{\vec{x}/\vec{a}} \theta$$

for *some* sequence $\vec{a}$.

We shall mainly use the terminology for formulas θ which are instances of axioms, especially the two equality axioms. With this terminology, it is straightforward to show that if the free variables in θ are included in the list $x_1, x_2, \ldots, x_n$, then

$$\mathcal{A} \text{ satisfies } \theta \text{ if and only if } \mathcal{A} \vDash \forall x_1 \forall x_2 \ldots \forall x_n \theta.$$

Then we can rephrase the definition of a formula θ being universally valid by saying that all structures $\mathcal{A}$ for the relevant language satisfy θ.

Let us give an example of a non-normal structure which satisfies the equality axioms. Suppose that the language L contains only the relation symbol $=$. For any set A with more than one element, let $\mathcal{A}$ be the structure $\langle A, \ A \times A \rangle$, so that the interpretation $=^{\mathcal{A}}$ of the relation symbol $=$ is the set of all pairs (a, b) with $a, b \in A$. Axiom Ax 6 is satisfied by $\mathcal{A}$ as the set $=^{\mathcal{A}}$ includes the pair (a, a) for each $a \in A$. As the only atomic formulas are of the form $x = y$ for some variables x, y, it is easy to see that all instances of axiom Ax 7 are also satisfied by $\mathcal{A}$.

If A has just one element and $=^{\mathcal{A}}$ is $A \times A$, i.e. $\{(a, a)\}$, then $=^{\mathcal{A}}$ is actual equality on A, so that $\mathcal{A}$ fails to be an example of a non-normal structure!

Let's investigate non-normal structures for more complicated languages. Suppose that $\mathcal{A} = \langle A, \ldots \rangle$ is a non-normal structure for a language L with equality satisfying the axioms for equality, Ax 6 and Ax 7. We shall investigate the properties of this structure and show that from it one can create a normal structure satisfying the same sentences. This has the consequence that if a theory has a model, perhaps one that isn't normal, then it also has a normal model.

$\mathcal{A}$ non-normal means that there are at least two distinct elements $a, b \in A$ with (a, b) in the set of pairs $=^{\mathcal{A}}$.

Our first observation is that although it is not normal, as $\mathcal{A}$ satisfies Ax 6 and Ax 7 it is still the case that if $\Gamma \vdash \phi$ and $\mathcal{A}$ satisfies Γ, then $\mathcal{A}$ satisfies ϕ. The details of the proof of the soundness theorem still work for such a structure. In the proof of the soundness theorem, all structures satisfy axioms Ax 1 to Ax 5 – these axioms don't rely on any specific properties of the relation symbol $=$ or on how it is interpreted in a structure. Likewise the validity of the rules MP and Gen doesn't rely on how a structure interprets $=$. Also we are assuming that this structure $\mathcal{A}$ does satisfy axioms Ax 6 and Ax 7 (which of course we

A non-normal structure need not satisfy axioms Ax 6 and Ax 7. For instance, if $=^{\mathcal{A}}$ doesn't include (a, a) for some $a \in A$, then Ax 6 is not satisfied. However, in this book we are only interested in non-normal structures which do satisfy the equality axioms.

had to make efforts to justify for a normal structure). Thus we can exploit the soundness theorem for this $\mathcal{A}$.

Let's write $=^{\mathcal{A}}$ for the interpretation of $=$ in the structure $\mathcal{A}$, and write $a =^{\mathcal{A}} b$ for $(a, b) \in =^{\mathcal{A}}$. First observe that $=^{\mathcal{A}}$ is an equivalence relation on A, as follows. We are assuming that $\mathcal{A}$ satisfies Ax 6, which is the reflexive property for $=^{\mathcal{A}}$. As we can exploit the soundness theorem and we are assuming that $\mathcal{A}$ satisfies Ax 7 as well as Ax 6, $\mathcal{A}$ satisfies any formula derived within the formal system. So, by Theorem 5.6(ii) and (iii), $=^{\mathcal{A}}$ also has the symmetric and transitive properties and is thus an equivalence relation. Equivalent elements of A have a very strong relationship with each other as given by the following theorem.

Theorem 5.14

Let $\mathcal{A} = \langle A, \ldots, =^{\mathcal{A}} \rangle$ be a non-normal structure satisfying axioms Ax 6 and Ax 7. Let $\phi(x_1, x_2, \ldots, x_n)$ be a formula of L with free variables included in the list $x_1, x_2, \ldots, x_n$. Let $a_1, a_2, \ldots a_n$ and $b_1, b_2, \ldots, b_n$ be elements of A such that $a_i =^{\mathcal{A}} b_i$ for each i. Then

$$\mathcal{A} \vDash_{x_1/a_1, x_2/a_2, \ldots, x_n/a_n} \phi(x_1, x_2, \ldots, x_n)$$

if and only if

$$\mathcal{A} \vDash_{x_1/b_1, x_2/b_2, \ldots, x_n/b_n} \phi(x_1, x_2, \ldots, x_n).$$

Proof

We shall give an outline of the proof.

Suppose that

$$\mathcal{A} \vDash_{x_1/a_1, x_2/a_2, \ldots, x_n/a_n} \phi(x_1, x_2, \ldots, x_n).$$

We shall show that

$$\mathcal{A} \vDash_{x_1/b_1, x_2/b_2, \ldots, x_n/b_n} \phi(x_1, x_2, \ldots, x_n)$$

by making repeated use of Theorem 5.7. Let $y_1, y_2, \ldots, y_n$ be variables appearing nowhere in the formula $\phi(x_1, x_2, \ldots, x_n)$, so that each y_i can be freely substituted for x_i in ϕ. By Theorem 5.7 we can derive

$$\vdash (x_1 = y_1 \rightarrow (\phi(x_1, x_2, \ldots, x_n) \rightarrow \phi(y_1, x_2, \ldots, x_n))),$$

so that by the soundness theorem

$$\mathcal{A} \vDash_{x_1/a_1, x_2/a_2, \ldots, x_n/a_n, y_1/b_1} (x_1 = y_1 \rightarrow (\phi(x_1, x_2, \ldots, x_n)$$
$$\rightarrow \phi(y_1, x_2, \ldots, x_n))).$$

As $a_1 =^{\mathcal{A}} b_1$, we have

$$\mathcal{A} \vDash_{x_1/a_1, x_2/a_2, \ldots, x_n/a_n, y_1/b_1} x_1 = y_1,$$

so that as (by our initial supposition)

$$\mathcal{A} \vDash_{x_1/a_1, x_2/a_2, \ldots, x_n/a_n, y_1/b_1} \phi(x_1, x_2, \ldots, x_n)$$

it follows that

$$\mathcal{A} \vDash_{x_1/a_1, x_2/a_2, \ldots, x_n/a_n, y_1/b_1} \phi(y_1, x_2, \ldots, x_n).$$

As there are no free occurrences of x_1 in $\phi(y_1, x_2, \ldots, x_n)$, we have

$$\mathcal{A} \vDash_{y_1/b_1, x_2/a_2, \ldots, x_n/a_n} \phi(y_1, x_2, \ldots, x_n).$$

We now replace the y_1s in $\phi(y_1, x_2, \ldots, x_n)$ by the original x_1s and conclude that

$$\mathcal{A} \vDash_{x_1/b_1, x_2/a_2, \ldots, x_n/a_n} \phi(x_1, x_2, \ldots, x_n).$$

Next, by using the result

$$\vdash (x_2 = y_2 \rightarrow (\phi(x_1, x_2, \ldots, x_n) \rightarrow \phi(x_1, y_2, \ldots, x_n)))$$

given by Theorem 5.7 and the fact that $a_2 =^{\mathcal{A}} b_2$ we can show that

$$\mathcal{A} \vDash_{x_1/b_1, x_2/b_2, x_3/a_3 \ldots, x_n/a_n} \phi(x_1, x_2, \ldots, x_n).$$

By continuing in this way we show that all the a_is can be replaced by b_is so that

$$\mathcal{A} \vDash_{x_1/b_1, x_2/b_2, \ldots, x_n/b_n} \phi(x_1, x_2, \ldots, x_n).$$

Plainly the argument is symmetrical in the a_is and b_is, so that the required if and only if result is thus proved.　∎

We shall use this result to show that given any non-normal structure $\mathcal{A}$ satisfying the equality axioms, we can construct a normal structure with properties strongly associated to those of $\mathcal{A}$ as follows.

Definition　Normal contraction

Let $\mathcal{A} = \langle A, \ldots, =^{\mathcal{A}} \rangle$ be a non-normal structure for a language L with equality which satisfies the equality axioms, so that $=^{\mathcal{A}}$ is an equivalence relation on A. The *normal contraction* of $\mathcal{A}$ is the structure $[\mathcal{A}]$ for L defined as follows. Its domain, which we shall write as $[A]$, is the set of all equivalence classes of $=^{\mathcal{A}}$, that is

$$[A] = \{[a] : a \in A\}.$$

We shall interpret the $=$ symbol in $[\mathcal{A}]$ by actual equality on the set $[A]$. For each remaining n-place relation symbol R, we shall specify its interpretation by

$$([a_1], [a_2], \ldots [a_n]) \in R^{[\mathcal{A}]} \text{ if and only if } (a_1, a_2, \ldots, a_n) \in R^{\mathcal{A}},$$

for all $a_1, a_2, \ldots, a_n \in A$. For each m-place function symbol f, we shall specify its interpretation as the function $f^{[\mathcal{A}]} : [A]^m \longrightarrow [A]$ given by

$$f^{[\mathcal{A}]}([a_1], [a_2], \ldots, [a_m]) = [f^{\mathcal{A}}(a_1, a_2, \ldots, a_m)],$$

for all $a_1, a_2, \ldots, a_m \in A$. For each constant symbol $\mathbf{c}$ let $\mathbf{c}^{[\mathcal{A}]}$ be the element $[c^{\mathcal{A}}]$.

The equivalence class $[a]$ of $a \in A$ is the set $\{b \in A : (a, b) \in =^{\mathcal{A}}\}$. Equivalence classes were mentioned in Section 4.4 on page 190.

There's a massive sleight of hand above as these definitions of the interpretations of the symbols only make sense thanks to Theorem 5.14. For instance, suppose that R is a 2-place relation symbol and a_1, a_2, b_1, b_2 are elements of A such that $a_1 =^A b_1$ and $a_2 =^A b_2$. That means that the ordered pair $([a_1], [a_2])$ of elements of $[A]$ is the same as the ordered pair $([b_1], [b_2])$. For our definition of $R^{[A]}$ to make sense, we need it to be the case that

$$(a_1, a_2) \in R^A \text{ if and only if } (b_1, b_2) \in R^A.$$

Otherwise, supposing that one has $(a_1, a_2) \in R^A$ and $(b_1, b_2) \notin R^A$, the first gives that

$$([a_1], [a_2]) \in R^{[A]},$$

while the second gives

$$([b_1], [b_2]) \notin R^{[A]},$$

which is impossible given that $([a_1], [a_2]) = ([b_1], [b_2])$. But by Theorem 5.14 applied to the atomic formula $R(x_1, x_2)$, this situation cannot arise and we have genuinely defined interpretations in $[A]$ of all the symbols.

We then have the following theorem connecting the structures A and $[A]$.

Theorem 5.15

Let $A = \langle A, \ldots \rangle$ be a non-normal structure for a language L with equality satisfying the equality axioms and let $[A]$ be its normal contraction. Then for all formulas $\phi(x_1, x_2, \ldots, x_n)$ with free variables included in the list $x_1, x_2, \ldots, x_n$ and all $a_1, a_2, \ldots a_n \in A$,

$$[A] \vDash_{x_1/[a_1], x_2/[a_2], \ldots, x_n/[a_n]} \phi(x_1, x_2, \ldots, x_n)$$

if and only if

$$A \vDash_{x_1/a_1, x_2/a_2, \ldots, x_n/a_n} \phi(x_1, x_2, \ldots, x_n).$$

Exercise 5.29 ⎯⎯⎯⎯⎯⎯⎯⎯⎯⎯⎯⎯⎯⎯⎯⎯⎯⎯⎯⎯⎯⎯⎯⎯

Prove Theorem 5.15. [*Hints*: For the sake of economy assume that all formulas have been built up using a small adequate set of connectives, say $\neg$ and $\wedge$, and $\forall$. It's then a straightforward mathematical induction on the length of all formulas ϕ, with some use of Theorem 5.14.]

An important corollary of Theorem 5.15 is that A and its normal contraction satisfy the same sentences. This proves the following theorem.

Theorem 5.16

Let T be a theory in a language L with equality. If T has a non-normal model satisfying the equality axioms, then T has a normal model.

Recall that a theory is a set of sentences, namely all the logical consequences of some given set of (non-logical) axioms.

The construction of a normal contraction and Theorem 5.15 also give an idea of how to construct non-normal models of a theory T. We start with a normal model B of the theory, add extra elements to its domain and make these mimic the behaviour of some of the original elements in such a way that the new structure $\mathcal{A}$ is non-normal and that its normal contraction $[\![\mathcal{A}]\!]$ is isomorphic to the original $\mathcal{B}$. Then by Theorem 5.15 $\mathcal{A}$ and $\mathcal{B}$ satisfy the same sentences, so that $\mathcal{A}$ is a non-normal model of T.

We shall illustrate the process with the theory of strict partial order in the language with binary relation symbol S and equality (see page 193) and the normal model $\mathcal{B} = \langle \{0,1\}, <, = \rangle$. This just has the two element set $\{0,1\}$ as domain with the usual $<$ on this set, so that $S^{\mathcal{B}}$ consists of just the one pair $(0,1)$, and of course the usual $=$.

We shall create a non-normal structure $\mathcal{A}$ as follows. Let its domain be the set $A = \{0, 0', 0'', 1, 1'\}$. Let $S^{\mathcal{A}}$ be the set

$$\{(0,1), (0',1), (0'',1), (0,1'), (0',1'), (0'',1')\}$$

and $=^{\mathcal{A}}$ the set

$$\{(0,0), (0',0'), (0'',0''), (1,1), (1',1'),$$
$$(0,0'), (0',0), (0',0''), (0'',0'), (0'',0), (0,0''), (1,1'), (1',1)\}.$$

Given that the only essentially different atomic formulas in this language are $x = x$, $x = y$, $S(x,x)$ and $S(x,y)$ it is easy to check that $\mathcal{A}$ satisfies the equality axioms Ax 6 and Ax 7 – we have rigged it so that each of the 'equal' elements $0, 0', 0''$ has the same relationship under $S^{\mathcal{A}}$ to the 'equal' elements 1 and $1'$. So $=^{\mathcal{A}}$ is an equivalence relation and the normal contraction $[\![\mathcal{A}]\!]$ of $\mathcal{A}$ is the set $\{[\![0]\!], [\![1]\!]\}$ with the one pair, namely $([\![0]\!], [\![1]\!])$, in $S^{[\![\mathcal{A}]\!]}$, so that $[\![\mathcal{A}]\!]$ is isomorphic to the original $\mathcal{B}$.

Exercise 5.30 ____

Create a non-normal model for the axioms of group theory in the language L with equality containing the 2-place function symbol $\cdot$ and the constant symbol **e** (see page 198) which has a domain with 5 elements and with a normal contraction isomorphic to the cyclic group $\mathbb{Z}_2$ with two elements.

____

The discussion illustrates the limitations of the axioms for equality. They are sufficient to ensure that elements of a structure $\mathcal{A} = \langle A, \ldots, = \rangle$ satisfying Ax 6 and Ax 7 which have different properties expressible within the language are not equal. That is, if $a, b \in A$ and there is some formula $\phi(x)$ such that

$$\mathcal{A} \vDash_{x/a} \phi(x) \text{ and } \mathcal{A} \nvDash_{x/b} \phi(x),$$

then

$$\mathcal{A} \vDash_{x/a, y/b} \neg x = y,$$

In this case, not only are a and b plainly unequal, but this shows up within the interpretation $=^{\mathcal{A}}$ of $=$ within $\mathcal{A}$.

regardless of whether $\mathcal{A}$ is normal. But without the insistence that the structure $\mathcal{A}$ is normal, the axioms can't truly distinguish between unequal elements a, b of A which have exactly the same properties expressible within the language. That is, if $\mathcal{A}$ is not normal, one can have $a, b \in A$ with

$$\mathcal{A} \vDash_{x/a, y/b} x = y$$

such that for all formulas $\phi(x, x_1, \ldots, x_n)$ and all $c_1, \ldots, c_n \in A$,

$$\mathcal{A} \vDash_{x/a, x_1/c_1, \ldots, x_n/c_n} \phi(x, x_1, \ldots, x_n)$$

if and only if

$$\mathcal{A} \vDash_{x/b, x_1/c_1, \ldots, x_n/c_n} \phi(x, x_1, \ldots, x_n).$$

5.5 The completeness theorem

In this section we shall prove the completeness theorem for the formal system, which we state as follows.

Theorem 5.17 Completeness theorem

For all sentences ϕ and sets Γ of sentences in a first-order language L, if $\Gamma \vDash \phi$ then $\Gamma \vdash \phi$.

Also called the *adequacy* theorem.

We have stated the result for sentences (formulas with no free variables) to avoid complications that might arise in derivations using the rule of generalization if free variables occur in assumptions. We shall lose some information as a consequence about what is derivable from assumptions involving free variables, but our applications of the result will always be to sentences, so no matter. As with the propositional calculus, we have a result connecting two ways of viewing the completeness theorem, given in the next exercise.

Exercise 5.31 ____

Prove that the following statements are equivalent.

(C) If $\Gamma \vDash \phi$ then $\Gamma \vdash \phi$, for all sentences ϕ and sets Γ of sentences.

(D) If Δ is consistent then Δ has a model, for all sets Δ of sentences.

[*Hint*: The result of Exercise 5.27(a) in Section 5.3 will be of use for part of the argument.]

____

Version (C) above of the completeness theorem can be paraphrased as saying that the formal proof system has got enough axioms and rules of proof to be able to prove everything that's valid. This is pretty remarkable, given the relative simplicity of the formal proof system. It also serves as a reminder that, just as with the proof of the completeness theorem for the propositional calculus, there will be key points of the proof where we rely on the formal proof system having enough power to produce certain derivations.

Proof of the completeness theorem

Let's now look at a proof of the completeness theorem. The result was proved by the Austrian mathematician and logician Kurt Gödel (1906–1978) in 1930, the first of his results which have had a major impact on modern mathematics and logic. We shall base our proof on one produced subsequently by the American logician Leon Henkin in 1949. The method extends what we used earlier to prove the completeness theorem for propositional calculus. Importantly, we shall prove the theorem in detail only for sets of sentences in a *countable* language. As for the propositional calculus, we shall prove the theorem in the version (D) given in Exercise 5.31 above, showing that a consistent set of sentences Δ has a model.

In this book, we are not assuming knowledge of the theory of infinite sets beyond routine manipulation of countable sets. However, we do discuss uncountable sets and the completeness theorem for uncountable languages in Section 6.4 of Chapter 6.

It is worth recapping the method we used to prove the theorem for propositional calculus. Given a consistent set Δ of propositional formulas in a language L, we first extended Δ to a complete set of formulas Σ in the same language. We then used this set Σ to define a truth assignment v by

$$v(p) = \begin{cases} T, & \text{if } p \in \Sigma, \\ F, & \text{if } \neg p \in \Sigma, \end{cases}$$

and then showed that for all formulas ϕ,

$$v(\phi) = T \quad \text{if and only if} \quad \phi \in \Sigma.$$

As the original set Δ is a subset of Σ, this truth assignment v satisfies Δ, proving the theorem.

We shall adapt this method for the predicate calculus, extending a consistent set Δ of sentences in a language L to a complete set of sentences Σ from which we define a structure – the equivalent of a truth assignment – which satisfies all sentences in Σ, including all those in Δ. Unsurprisingly there's a complication for the predicate calculus in that we shall have extra requirements on Σ besides being a complete set, usually requiring that Σ uses a larger language than the original language L, as you shall soon see.

As our method will involve several enlargements of the original language and several extensions of the original set of sentences, it will make sense for the rest of this discussion to call the original language and set of sentences in it L_0 and Δ_0 respectively, leaving the symbols L and Δ available for any of the languages or sets along the way. We shall need the same concepts of a complete and maximal consistent set as for the propositional calculus, but we shall use them for *sentences* in each language L, rather than for formulas, which in the current context could involve inconvenient free variables.

<table>
<tr><td>

Definitions   *Complete, maximal consistent*

The set Σ of sentences in L is *complete* for L if it is consistent and for each sentence ϕ in L, exactly one of ϕ and $\neg\phi$ belongs to Σ. Also Σ is *maximal consistent* for L if Σ is consistent and for any consistent set of sentences Σ' in L with $\Sigma \subseteq \Sigma'$, we have $\Sigma = \Sigma'$.

</td><td>

Or, equivalently, Σ is consistent and if ϕ is a sentence in L with $\phi \notin \Sigma$, then $\Sigma \cup \{\phi\}$ is inconsistent.

</td></tr>
</table>

As for propositional calculus, these definitions are equivalent, as you can now show.

Exercise 5.32 ———————————————————

Let Σ be a set of sentences in a language L. Show that Σ is complete if and only if it is maximal consistent (for the same language). [*Hints*: Think about whether you can exploit the proof of the equivalent result for propositional calculus, Theorem 3.12 of Section 3.3 of Chapter 3. See also Exercise 5.27 in this section. Using sentences rather than formulas of the predicate language L is helpful in that we thereby avoid complications which could otherwise occur in the use of the rule Gen.]

Exercise 5.33 ____

Let Σ be a complete set of sentences and ϕ a sentence in a language L such that $\Sigma \vdash \phi$. Show that $\phi \in \Sigma$.

____

Given our original consistent set of sentences Δ_0 in a countable language L_0, we shall show how to add extra constant symbols to L_0 to give a countable language L^* and construct a set of sentences Δ^* in L^* with the following properties:

(1) $\Delta_0 \subseteq \Delta^*$ and Δ^* is complete (and thus also maximal consistent) for L^*;

(2) If $\phi(x)$ is a formula in L^* with one free variable x such that the sentence $\exists x \phi(x)$ is in Δ^*, then there is a corresponding constant symbol $\mathbf{c}_\phi$ in L^* such that the sentence $\phi(\mathbf{c}_\phi)$ is also in Δ^*.

The ϕ appearing as a subscript in the constant symbol $\mathbf{c}_\phi$ is short for $\phi(x)$ for whichever variable x stands for. We shall regard e.g. $\exists x_1 \phi(x_1)$ and $\exists x_2 \phi(x_2)$, where x_1 is free in $\phi(x_1)$ and x_2 is free in $\phi(x_2)$, as giving rise to distinct constants $\mathbf{c}_{\phi(x_1)}$ and $\mathbf{c}_{\phi(x_2)}$, for those with very sharp eyesight. In general, each different formula gives rise to a distinct constant symbol.

We shall call the second property for Δ^* the *witness property* on the grounds that the constant symbol $\mathbf{c}_\phi$ provides a specific witness to the existence of some x such that $\phi(x)$ is in the set Δ^*.

We shall show how to find the language L^* and sentences Δ^* later, but for the moment let's concentrate on how to create from them a special structure $\mathcal{A}$ which satisfies Δ^* and thus also satisfies the original set Δ_0. In the case that the original language L_0 includes the $=$ symbol, our structure $\mathcal{A}$ will not necessarily be a normal model. But we will ensure that it satisfies the equality axioms $\mathrm{Ax}\,6$ and $\mathrm{Ax}\,7$, so that by Theorem 5.16 in Section 5.4, the normal contraction of $\mathcal{A}$ is a normal model of Δ^*.

To start, we shall get a feeling for how many new constant symbols will have been introduced to the language L^* for Δ^* to have the witness property. We shall assume that the original language L_0 includes at least one relation symbol R, as if it has no relation symbols, there will be no formulas to disturb our peace! Take $\phi(x)$ to be the formula

$$(R(x, x, \ldots, x) \rightarrow R(x, x, \ldots, x)).$$

Using x for all the arguments of R.

As this is a substitution instance of a tautology, we have $\vdash \exists x \phi(x)$, so that $\Delta^* \vdash \exists x \phi(x)$. As Δ^* is complete, the result of Exercise 5.33 gives that $\exists x \phi(x) \in \Delta^*$. Then by the witness property the language L^* includes a corresponding constant symbol $\mathbf{c}_\phi$. By taking other formulas $\psi(x)$ involving R and a free variable x which are substitution instances of tautologies, for which we will also have $\exists x \psi(x) \in \Delta^*$, the witness property ensures that the language L^* includes infinitely many corresponding constant symbols $\mathbf{c}_\psi$.

For instance, using the same $\phi(x)$ as earlier, we can take
$$(\phi(x) \rightarrow \phi(x)),$$
$$((\phi(x) \rightarrow \phi(x)) \rightarrow (\phi(x) \rightarrow \phi(x))),$$
and so on.

We shall now specify the domain A of our special structure $\mathcal{A}$ corresponding to Δ^* by

$$A = \{\tau : \tau \text{ is a term of } L^* \text{ containing no variables}\}.$$

A term containing no variables is often called a *closed term*. As L^* contains infinitely many constant symbols, each of which is a closed term, the domain A is an infinite set. There is a risk that one might become confused between closed terms τ regarded as members of the set A and their use within formulas of L^*; but we hope that the context will make it clear in which sense they are being used.

In fact, with the original language L_0 being countable, our construction will make L^* a countably infinite set.

Observe that if f is an n-place function symbol of the language L^* and $\tau_1, \tau_2, \ldots, \tau_n$ are closed terms, then $f(\tau_1, \tau_2, \ldots, \tau_n)$ is also a closed term. This enables us to define the interpretation of the function symbol f in the structure $\mathcal{A}$ by

$$f^{\mathcal{A}}(\tau_1, \tau_2, \ldots, \tau_n) =_{\text{def}} f(\tau_1, \tau_2, \ldots, \tau_n),$$

for all closed terms $\tau_1, \tau_2, \ldots, \tau_n \in A$. As the righthand side is in A whenever the terms τ_i are in A, this does indeed define a function $f^{\mathcal{A}}$ on A^n.

This means that f is a function symbol of the original language L_0, as L^* differs from L_0 only by having extra constant symbols.

We shall now define the interpretation in $\mathcal{A}$ of relation symbols of L^*. For an n-place relation symbol R, we define $R^{\mathcal{A}}$ by

$$(\tau_1, \tau_2, \ldots, \tau_n) \in R^{\mathcal{A}} \text{ if and only if } R(\tau_1, \tau_2, \ldots, \tau_n) \in \Delta^*,$$

for all closed terms $\tau_1, \tau_2, \ldots, \tau_n \in A$. Note that as the τ_i are closed terms, the formula $R(\tau_1, \tau_2, \ldots, \tau_n)$ is a sentence, so that it makes sense to talk in terms of whether it belongs to Δ^*, which is a set of sentences. For the moment we are treating the symbol $=$, should L_0 be a language with equality, simply as a 2-place relation symbol, so that the interpretation $=^{\mathcal{A}}$ might not be true equality on the set A.

As for function symbols, the relation symbols of L^* are those of the original language L_0.

Lastly we must define the interpretation of constant symbols in the language L^*. We adopt the obvious interpretation of each constant symbol by itself (which is of course a closed term), that is,

$$\mathbf{c}^{\mathcal{A}} =_{\text{def}} \mathbf{c},$$

There are inevitably extra constant symbols in L^* compared to the original language L_0.

for each such constant symbol – the righthand side is indeed an element of A as constant symbols are closed terms.

The structure

$$\mathcal{A} = \langle A, R^{\mathcal{A}} \ldots, f^{\mathcal{A}} \ldots, \mathbf{c}^{\mathcal{A}} \ldots \rangle$$

defined in this way is often called the *canonical structure* corresponding to the set Δ^* of sentences.

As we have indicated, the point of the structure $\mathcal{A}$ is given in the following theorem.

Theorem 5.18

For all formulas $\phi(x_1, x_2, \ldots, x_n)$ of L^* with free variables in the list $x_1, x_2, \ldots, x_n$, and elements $\tau_1, \tau_2, \ldots, \tau_n$ of A,

$$\mathcal{A} \vDash_{x_1/\tau_1, x_2/\tau_2, \ldots, x_n/\tau_n} \phi(x_1, x_2, \ldots, x_n)$$

if and only if

$$\phi(\tau_1, \tau_2, \ldots, \tau_n) \in \Delta^*.$$

Note that as the τ_i are closed terms, $\phi(\tau_1, \tau_2, \ldots, \tau_n)$ is a sentence, which might thus be a candidate for membership of the set Δ^* of sentences.

Proof

The result is analogous to Theorem 3.13 for propositional calculus in Chapter 3 and the proof follows similar lines, using mathematical induction on the length of formulas. Every formula is logically equivalent to one built up using only the connectives $\neg, \rightarrow$ and the quantifier $\exists$, and it will be sufficient to prove the result for formulas of this more limited type. Our induction hypothesis for $k \geq 0$ is that the result holds for all formulas with $\leq k$ connectives and quantifiers and all $\tau_1, \tau_2, \ldots, \tau_n \in A$ for *all* $n \geq 0$. Let's give an example to explain why the hypothesis allows for all $\tau_1, \tau_2, \ldots, \tau_n \in A$ for all possible n. Suppose that the language includes a relation symbol R with 3 arguments and that $\phi(x_1)$ is the formula $\exists x_2 \exists x_3 R(x_1, x_2, x_3)$ with just one free variable. Then to test whether

$$\mathcal{A} \vDash_{x_1/\tau_1} \phi(x_1),$$

we shall need to know whether for some $\tau_2 \in A$, $\mathcal{A}$ satisfies a formula with fewer connectives, namely $\exists x_3 R(x_1, x_2, x_3)$, but with two, rather than the original one, free variables needing interpretation by τ_1, τ_2, respectively.

For $k = 0$, the formula ϕ is atomic, and the result holds by definition of the interpretation $R^{\mathcal{A}}$ of each relation symbol R (including $=$ if this is one of the symbols of the language).

Now suppose that the result holds for all formulas of length $\leq k$ and that ϕ has length $k + 1$. We shall leave the cases when ϕ is one of the forms $\neg\theta$ and $(\theta \rightarrow \psi)$ as an exercise for you, as these are essentially the same as in the proof of Theorem 3.13. The new case is when $\phi(x_1, x_2, \ldots, x_n)$ is of the form $\exists x\psi$, which we shall look at now.

Let's look first at the case when the variable x does not occur free in ψ. Then by the result of Exercise 4.24 in Section 4.2,

$$\mathcal{A} \vDash_{x_1/\tau_1, x_2/\tau_2, \ldots, x_n/\tau_n} \exists x\psi$$

if and only if

$$\mathcal{A} \vDash_{x_1/\tau_1, x_2/\tau_2, \ldots, x_n/\tau_n} \psi,$$

which by the induction hypothesis holds if and only if

$$\psi(\tau_1, \tau_2, \ldots, \tau_n) \in \Delta^*.$$

We shall leave it for you to show that this holds if and only if

$$\exists x\psi(\tau_1, \tau_2, \ldots, \tau_n) \in \Delta^*$$

as Exercise 5.35, which then deals with the situation of x not free in ψ.

The interesting case is of course when x is free in ψ. As we are taking the free variables in $\exists x\psi$ to be in the list $x_1, x_2, \ldots, x_n$, we shall assume for simplicity that x isn't one of these variables.

Suppose that

$$\mathcal{A} \vDash_{x_1/\tau_1, x_2/\tau_2, \ldots, x_n/\tau_n} \exists x\psi.$$

Then for some closed term τ in A,

$$\mathcal{A} \vDash_{x_1/\tau_1, x_2/\tau_2, \ldots, x_n/\tau_n, x/\tau} \psi,$$

There is an interesting difference between Theorem 3.13 and Theorem 5.18. In the former, given a complete set of propositional formulas Σ, there is a *unique* truth assignment v satisfying Σ. But in Theorem 5.18 there is no claim that $\mathcal{A}$ is the only structure satisfying Δ^*. In general there will be several different (non-isomorphic) structures satisfying this set, as you will see in the next chapter when we discuss the Löwenheim–Skolem theorems.

All we said was that the free variables of $\exists x\psi$ appeared in the list $x_1, x_2, \ldots, x_n$, which doesn't mean that all of these are free variables. So in principle one of the x_i isn't free in $\exists x\psi$ and the x could be this x_i! Our argument can easily be adjusted to cope with such a case.

so that, by the induction hypothesis,

$$\psi(\tau_1, \tau_2, \ldots, \tau_n, \tau) \in \Delta^*.$$

By the result of Exercise 5.10(a) in Section 5.2,

$$\vdash (\psi(\tau_1, \tau_2, \ldots, \tau_n, \tau) \rightarrow \exists x \psi(\tau_1, \tau_2, \ldots, \tau_n, x)),$$

so that

$$\Delta^* \vdash (\psi(\tau_1, \tau_2, \ldots, \tau_n, \tau) \rightarrow \exists x \psi(\tau_1, \tau_2, \ldots, \tau_n, x)),$$

and as

$$\Delta^* \vdash \psi(\tau_1, \tau_2, \ldots, \tau_n, \tau) \quad (\text{as } \psi(\tau_1, \tau_2, \ldots, \tau_n, \tau) \in \Delta^*)$$

a use of MP gives

$$\Delta^* \vdash \exists x \psi(\tau_1, \tau_2, \ldots, \tau_n, x).$$

As Δ^* is complete, the result of Exercise 5.33 gives

$$\exists x \psi(\tau_1, \tau_2, \ldots, \tau_n, x) \in \Delta^*,$$

as required.

For the converse, suppose that

$$\exists x \psi(\tau_1, \tau_2, \ldots, \tau_n, x) \in \Delta^*.$$

Then by the witness property of Δ^* (perhaps its crucial use in this proof!), there is a constant symbol $\mathbf{c}_\psi$ in L^*, which is of course also a closed term, such that

$$\psi(\tau_1, \tau_2, \ldots, \tau_n, \mathbf{c}_\psi) \in \Delta^*.$$

By the induction hypothesis,

$$\mathcal{A} \vDash_{x_1/\tau_1, x_2/\tau_2, \ldots, x_n/\tau_n, x/\mathbf{c}_\psi} \psi(x_1, x_2, \ldots, x_n, \mathbf{c}_\psi),$$

so that by the basic definition of $\mathcal{A} \vDash \exists x \psi$

$$\mathcal{A} \vDash_{x_1/\tau_1, x_2/\tau_2, \ldots, x_n/\tau_n} \exists x \psi.$$

■

The following exercises all relate to details of the proof of Theorem 5.18, so that Δ^* is a complete set of sentences in the language L^* with the witness property.

Exercise 5.34 ______________________________

Prove the inductive step of Theorem 5.18 in the case when ϕ of length $k + 1$ is one of the forms $\neg\theta$ and $(\theta \rightarrow \psi)$.

Exercise 5.35 ______________________________

Suppose that θ is a sentence of L^*. Show that for any variable x,

$$\theta \in \Delta^* \quad \text{if and only if} \quad \exists x \theta \in \Delta^*.$$

Solution

Suppose that $\exists x\theta \notin \Delta^*$ or, equivalently, that

$$\neg\forall x\neg\theta \notin \Delta^*.$$

Then as Δ^* is complete, we have

$$\forall x\neg\theta \in \Delta^*,$$

so that

$$\Delta^* \vdash \forall x\neg\theta.$$

But

$$\Delta^* \vdash (\forall x\neg\theta \to \neg\theta) \quad (\text{using } \text{Ax}\,4)$$

so that by MP

$$\Delta^* \vdash \neg\theta.$$

By Exercise 5.33 we then have $\neg\theta \in \Delta^*$, so that as Δ^* is complete and $\neg\theta$ is a sentence, we have $\theta \notin \Delta^*$. From this we can infer that if $\theta \in \Delta^*$, then $\exists x\theta \in \Delta^*$.

Conversely, suppose that $\theta \notin \Delta^*$, so that as Δ^* is complete we have $\neg\theta \in \Delta^*$ and thus

$$\Delta^* \vdash \neg\theta.$$

As Δ^* consists of sentences, so that x is not free in a member of Δ^*, use of the rule Gen gives

$$\Delta^* \vdash \forall x\neg\theta,$$

so that by Exercise 5.33

$$\forall x\neg\theta \in \Delta^*,$$

and as Δ^* is complete, we have

$$\neg\forall x\neg\theta \notin \Delta^*,$$

or, equivalently, $\exists x\theta \notin \Delta^*$. This shows that if $\exists x\theta \in \Delta^*$, then $\theta \in \Delta^*$.

The point of Theorem 5.18 is of course that as the set Δ^* includes the original consistent set Δ_0, the canonical structure $\mathcal{A}$ is a model for Δ_0. If L_0 is not a language with equality, we have proved version (D) of the completeness theorem. What happens for a language with equality?

Let L_0 be a language with equality. We shall show that the canonical structure $\mathcal{A}$, which is a model of Δ_0 but not in general normal, satisfies the equality axioms $\text{Ax}\,6$ and $\text{Ax}\,7$. It will then follow from Theorem 5.16 of Section 5.4 that Δ_0 has a model, namely the normal contraction $[\![\mathcal{A}]\!]$ of $\mathcal{A}$.

Consider axiom Ax 6. A simple proof using no assumptions,

$$(1) \quad x = x \qquad \text{Ax 6}$$
$$(2) \quad \forall x \, x = x \quad \text{Gen, 1,}$$

shows that $\vdash \forall x \, x = x$, so that

$$\Delta^* \vdash \forall x \, x = x,$$

which, as $\forall x \, x = x$ is a sentence, Δ^* is complete and using Exercise 5.33, gives

$$\forall x \, x = x \in \Delta^*,$$

so that

$$\mathcal{A} \vDash \forall x \, x = x.$$

But this means that $\mathcal{A}$ satisfies $x = x$. Similarly, by placing sufficient universal quantifiers in front of any instance of axiom Ax 7 to obtain a sentence which must be derivable using no assumptions and hence from the complete set Δ^*, we can show that $\mathcal{A}$ satisfies each instance of this axiom. Thus $\mathcal{A}$ satisfies the equality axioms and its normal contraction $[\![\mathcal{A}]\!]$ is then a model of Δ_0.

Exercise 5.36 ———————————————————————————————

Suppose that L_0 and thus L^* are languages with equality. The normal contraction on $\mathcal{A}$ has as domain the set of equivalence classes of the set of closed terms of L^* under the equivalence relation $=^{\mathcal{A}}$. Show that each equivalence class contains at least one constant symbol.

Solution

Take any closed term τ of L^*. Then by Exercise 5.11(c) of Section 5.2,

$$\vdash \exists x \, x = \tau,$$

so that by the now familiar argument using that Δ^* is complete,

$$\exists x \, x = \tau \in \Delta^*.$$

Then by the witness property, there is a constant $\mathbf{c}_{x=\tau}$ in L^* such that

$$\mathbf{c}_{x=\tau} = \tau \in \Delta^*,$$

which by definition of $=^{\mathcal{A}}$ gives

$$(\mathbf{c}_{x=\tau}, \tau) \in =^{\mathcal{A}}.$$

Thus the equivalence class $[\![\tau]\!]$ in the domain of the normal contraction of $\mathcal{A}$ contains the constant symbol $\mathbf{c}_{x=\tau}$.

———

This last exercise shows that the domain of the normal contraction $[\![\mathcal{A}]\!]$ of $\mathcal{A}$, for a language with equality, consists of the equivalence classes $[\![\mathbf{c}_{x=\tau}]\!]$ determined by the new constant symbols added to the original language L_0 to produce the language L^*. There is an alternative normal model of Δ^* which could have been constructed directly from these new constant symbols, rather than via equivalence classes of closed terms; but we suspect it is messier proving that this structure is a model of Δ^*.

If the original language L_0 contains no function or constant symbols, then the domain of $\mathcal{A}$ is actually the set of the new constant symbols $\mathbf{c}_\phi$.

We have yet to prove that the special set of sentences Δ^* and language L^* exist, so we turn to this now. We shall need to make repeated use of two results, each explaining that a consistent set of sentences can be extended to another consistent set of a special nature. The first is as follows.

Theorem 5.19

Let Σ be a consistent set of sentences in a countable language L. Then there is a complete set Σ' of sentences in L such that $\Sigma \subseteq \Sigma'$.

Here the language L is a general (countable) language.

Proof

As L is countable we can enumerate the sentences of L as

$$\phi_0, \phi_1, \phi_2, \ldots, \phi_n, \ldots.$$

Define a sequence of sets of sentences $\{\Sigma_n\}_{n \in \mathbb{N}}$ recursively by

$$\Sigma_0 = \Sigma,$$
$$\Sigma_{n+1} = \Sigma_n \cup \begin{cases} \{\phi_n\}, & \text{if } \Sigma_n \vdash \phi_n, \\ \{\neg\phi_n\}, & \text{if } \Sigma_n \nvdash \phi_n, \end{cases} \quad \text{for } n \geq 0.$$

and put

$$\Sigma' = \bigcup_{n \in \mathbb{N}} \Sigma_n.$$

This is the same construction as done in the proof of Theorem 3.9 for the formal system S for propositional calculus in Chapter 3. The proof is essentially identical, but we have to check that details which depend on what's derivable in S still hold for our system of predicate calculus.

First we can show that each Σ_n is consistent, using mathematical induction, which we leave as an exercise for you.

It follows that Σ' is consistent, as if Σ' is inconsistent, there are derivations of $\Sigma' \vdash \theta$ and $\Sigma' \vdash \neg\theta$ for some formula θ. These derivations are finitely long, so exploit just finitely many assumptions from Σ', each of which appears in a Σ_n for some n. Taking N to be the largest of these finitely many ns, we have $\Sigma_N \vdash \theta$ and $\Sigma_N \vdash \neg\theta$, so that Σ_N is inconsistent, contradicting that each Σ_n is consistent.

Lastly we show that Σ' is complete. Take any sentence ϕ in the language L. Then ϕ must be a ϕ_n for some n, so that one of ϕ and $\neg\phi$ must have been put into Σ_{n+1} and thus in Σ'. As Σ' is consistent, we cannot have both ϕ and $\neg\phi$ in Σ', so that exactly one of them is in Σ', as is needed to show that Σ' is complete. $\blacksquare$

Exercise 5.37 ____

Show that each set Σ_n in the proof of Theorem 5.19 above is consistent. [*Hints*: The same proof as in that of Theorem 3.9 does work, but you will need to check that details depending on facts about derivations for propositional calculus also hold for our system of predicate calculus. As we are handling sentences of the language L, rather than formulas which might have free variables, we avoid potential complications over the use of the rule Gen.]

The second result which we shall use repeatedly is as follows.

Theorem 5.20

Suppose that Γ is a consistent set of sentences in a countable language L. For each sentence $\exists x\phi(x)$ in Γ, introduce a new distinct constant symbol $\mathbf{c}_\phi$. Let L^+ be the union of L with the set of all such $\mathbf{c}_\phi$. Now define a set Γ^+ of sentences of L^+ by

$$\Gamma^+ = \Gamma \cup \{\phi(\mathbf{c}_\phi) : \text{the sentence } \exists x\phi(x) \in \Gamma\}.$$

Then Γ^+ is consistent.

> The constant $\mathbf{c}_\phi$ 'witnesses' the sentence $\exists x\phi(x)$.

Proof

The proof makes use of the following variant of Theorem 5.5.

> Suppose that the constant symbol $\mathbf{c}$ occurs neither in any sentence in the set Γ nor in the sentence σ. Then
>
> $$\text{if } \Gamma, \phi(\mathbf{c}) \vdash \sigma \text{ then } \Gamma, \exists x\phi(x) \vdash \sigma.$$

> To obtain this result from Theorem 5.5, treat the symbol $\mathbf{c}$ as a new variable symbol. As $\mathbf{c}$ appears nowhere in Γ of σ, a very innocuous proof by induction on the derivation of $\Gamma, \phi(\mathbf{c}) \vdash \sigma$ treating $\mathbf{c}$ as a constant shows that $\Gamma, \phi(\mathbf{c}) \vdash \sigma$ treating $\mathbf{c}$ as a variable.

Suppose that Γ^+ is inconsistent. Then some formula θ and its negation $\neg\theta$ are both derivable from Γ^+. These derivations use between them only finitely many assumptions from Γ^+, so only finitely many formulas of the form $\phi(\mathbf{c}_\phi)$ where $\exists x\phi(x) \in \Gamma$, let's say $\phi_1(\mathbf{c}_{\phi_1}), \phi_2(\mathbf{c}_{\phi_2}), \ldots, \phi_n(\mathbf{c}_{\phi_n})$. So both θ and $\neg\theta$ are derivable from $\Gamma \cup \{\phi_1(\mathbf{c}_{\phi_1}), \phi_2(\mathbf{c}_{\phi_2}), \ldots, \phi_n(\mathbf{c}_{\phi_n})\}$. By Theorem 5.4(c) we can then derive

$$\Gamma, \phi_1(\mathbf{c}_{\phi_1}), \phi_2(\mathbf{c}_{\phi_2}), \ldots, \phi_n(\mathbf{c}_{\phi_n}) \vdash \sigma,$$

for any sentence σ, and in particular we can construct σ to be a contradiction chosen so that the constants $\mathbf{c}_{\phi_i}$ for $i = 1, 2, \ldots, n$, which by construction don't appear in Γ or in $\phi_j(\mathbf{c}_{\phi_j})$ for any $j \neq i$, also do not appear in σ. Then repeated use of the variant of Theorem 5.5 above gives that

$$\Gamma, \exists x\phi_1(x), \exists x\phi_2(x), \ldots, \exists x\phi_n(x) \vdash \sigma.$$

But as $\Gamma \vdash \exists x\phi_i(x)$ for each i, Theorem 5.4(b) gives us that $\Gamma \vdash \sigma$, contradicting that Γ is consistent. ∎

Before exploiting these results to construct L^* and Δ^*, we need to look at the sizes of the sets involved. The key result is that for a language with countably many symbols, the set of finitely long strings of these symbols is also countable, and in fact is countably infinite. As all sentences are amongst these strings, the set Σ' constructed in the proof of Theorem 5.19 is countable. Likewise there are countably many sentences $\exists x\phi(x)$ in the set Γ in Theorem 5.20, so that countably many constant symbols $\mathbf{c}_\phi$ are added to the language L to obtain L^+, which is thus also a countable language.

> Recall that countable means finite or countably infinite.

We can now show how to construct the language L^* and the complete set Δ^* with the witness property. Define a sequence of sets of formulas $\{\Delta_n\}_{n \in \mathbb{N}}$ and one of languages $\{L_n\}_{n \in \mathbb{N}}$ as follows:

Δ_0 is the original consistent set Δ and L_0 is the original countable language.

Given the set of sentences Δ_n in the countable language L_n, let Δ'_n be a complete extension of Δ_n in this language, as exists using Theorem 5.19. Then taking the set Γ in Theorem 5.20 to be this Δ'_n and L to be L_n, let L_{n+1} be the language L_n^+ constructed by adding constants to L_n and Δ_{n+1} the set $(\Delta'_n)^+$ given by the proof of Theorem 5.20.

Now put

$$\Delta^* = \bigcup_{n \in \mathbb{N}} \Delta_n \text{ and } L^* = \bigcup_{n \in \mathbb{N}} L_n.$$

Of course you still have to check that with this language L^*, Δ^* is complete and has the witness property!

Exercise 5.38

Show that with Δ^* and L^* as constructed above, Δ^* is complete and has the witness property. [*Hint*: You'll need to check that each Δ_n and Δ'_n is consistent and that the relevant languages are countable along the way.]

Note that each language L_n above is countable and L^* is defined as the countable union of countable sets, so that the set of closed terms of L^* is countable. That means that the domain of the canonical model $\mathcal{A}$ is countable; and in the case when L_0 is a language with equality, the normal contraction $[\![\mathcal{A}]\!]$ also has a countable domain. Note that in the latter case, the domain of $[\![\mathcal{A}]\!]$ could be finite rather than countably infinite. This is because the complete set Δ^* could include sentences of the form $\mathbf{c}_\phi = \mathbf{c}_\psi$, in which case the equivalence classes $[\![\mathbf{c}_\phi]\!]$ and $[\![\mathbf{c}_\psi]\!]$ are equal. As the domain of $[\![\mathcal{A}]\!]$ is, by Exercise 5.36, essentially the set of $[\![\mathbf{c}_\phi]\!]$ for the extra constant symbols $\mathbf{c}_\phi$ in L^+, it could thus be that be that distinct constant symbols don't give rise to distinct elements of the domain.

Putting together all the results of thus section, we have essentially proved version (D) of the completeness theorem in Exercise 5.31 for a countable language.

Theorem 5.21 Completeness theorem

If Δ is a consistent set of sentences in a countable language, then Δ has a countable model.

Now that we have both a soundness and completeness theorem, we can answer the informal question of whether a sentence ϕ follows from a set of sentences Γ either by showing that $\Gamma \vDash \phi$ or that $\Gamma \vdash \phi$. Once one holds, the other must also hold. It is not entirely clear to the author that if one knew that one of these held, one would then try to show the other held from first principles.

For instance, there might be non-trivial results well-known from everyday maths and expressible using a first-order language for which the 'normal' mathematical proof seems a long way from a derivation within our formal system; and yet by the completeness theorem, such a derivation must exist. An example is a result about groups which we gave in Chapter 4. If you know some group theory, the result in question is normally seen as an easy consequence of Lagrange's theorem. This theorem cannot be expressed in a first-order language, but some of its consequences can, in particular Result 2 which we gave on page 201 in Section 4.4 of Chapter 4:

> if G is a finite group with N elements, then each element of G has finite order dividing N.

For any given positive integer N, we can write down a first-order sentence ϕ which expresses that G has exactly N elements and that for all x, the product of x with itself N times equals the identity element $\mathbf{e}$ of the group. Letting GPs be the set of axioms for groups given on page 198 and using our knowledge of everyday group theory, we thus have GPs $\vDash \phi$. If you know some group theory and are very keen, you might like to see whether you can construct a formal derivation of GPs $\vdash \phi$!

A perhaps rather loose thought on the author's part is that the formal system seems prima facie a likelier environment within which to provide an algorithmic procedure to decide whether some sentence is derivable than checking for its validity by seeing whether it holds in every possible structure for the language – there will surely be infinitely many of the latter structures, which will be hard to catalogue, whereas it might be feasible to catalogue all derivations in a systematic way. It turns out that there is no such algorithmic decision procedure, but we won't prove this major result in this book.

The completeness theorem is a great result in its own right and would make a worthy result with which to culminate this book. However in the next, final, chapter, we shall explore some of the consequences of the completeness theorem, in particular the compactness theorem which says that a set of sentences has a model if every finite subset has a model. This innocent sounding result has all sorts of interesting ramifications, as you will soon see, and is a stepping stone to one of the most important modern areas of study in logic, called *model theory*.

Lagrange's theorem says that for any subgroup H of a finite group G, the number of elements in H divides the number of elements of G. This involves both quantifying over subsets of G and counting elements, so would be tricky, if not actually impossible, to represent using a first-order language.

This is the content of what is called Church's theorem, after Alonzo Church (1903–1995).

Further exercises

Exercise 5.39 ————————————————————————————————

Suppose that L is the language with just the $=$ symbol and Σ be the set consisting of the single formula $\forall x_1 \exists x_2 \neg x_1 = x_2$.

(a) Suppose further that the sentences of L are enumerated as $\phi_0, \phi_1, \phi_2, \ldots,$ $\phi_n, \ldots,$ where ϕ_0 is the sentence

$$\exists x_1 \exists x_2 \exists x_3 \exists x_4 \bigwedge_{1 \leq i < j \leq 4} \neg x_i = x_j$$

and ϕ_1 is the sentence

$$\forall x_1 \forall x_2 \forall x_3 \forall x_4 \forall x_5 \bigwedge_{1 \leq i < j \leq 5} x_i = x_j.$$

(i) What are the corresponding sets Σ_0, Σ_1 and Σ_2 given by the construction in the proof of Theorem 5.19?

(ii) Suppose that L and Σ are taken as the L_0 and Δ_0, respectively, with the same enumeration of the sentences of L_0, and that the language L^* and complete set Δ^* with the witness property are then constructed as on page 262 in our proof of the completeness theorem. What can you say about the possible sizes of the domain of the canonical model $\mathcal{A}$ of Δ^* and of its normal contraction $[\![\mathcal{A}]\!]$?

(b) Now suppose that L and Σ are again taken as the L_0 and Δ_0, respectively, but with some different enumeration of the sentences of L_0, and that L^* and Δ^* are then constructed as on page 262. What can you say about the possible sizes of the domains of the canonical model $\mathcal{A}$ of Δ^* and its normal contraction $[\![\mathcal{A}]\!]$ in general?

6 SOME USES OF COMPACTNESS

6.1 Introduction: the compactness theorem

In this chapter we shall have a further look at the expressive power of first-order languages. We have already looked at the use of such languages to provide axioms for various mathematical theories, in Section 4.4 of Chapter 4. Our formal proof system for predicate calculus of course gives us a framework for establishing the deductive power of such axioms. However, thanks to the completeness and soundness theorems, we have a connection between the deductive theory and the structures which satisfy the axioms, usually called the *models* of the theory. These models provide a rich source of exploration into the power of axioms and are the basis of *model theory*, a modern area of mathematical research which has grown from (and perhaps outgrown) its historical roots in the foundations of mathematics. The jumping off point for model theory is the *compactness* theorem, which is a consequence of the completeness and soundness theorems, just as for propositional calculus.

Theorem 3.11 of Chapter 3.

Theorem 6.1 Compactness theorem

Let Γ be a set of sentences in a first-order language L. If every finite subset of Γ has a model, then so does Γ.

Proof

Suppose that Γ doesn't have a model. Then by the completeness theorem Γ is inconsistent. Thus there are proofs of $\Gamma \vdash \theta$ and $\Gamma \vdash \neg\theta$ for some sentence θ. These proofs are finitely long, so each involves at most finitely many assumptions from Γ. Putting these assumptions together gives a finite subset Δ such that the two proofs above are also proofs of $\Delta \vdash \theta$ and $\Delta \vdash \neg\theta$. Thus Δ is inconsistent. By the soundness theorem we then have that Δ has no models.

This shows that if Γ has no models then some finite subset of Γ has no models, which is equivalent to what we are trying to prove. ∎

So far, we have only proved the completeness theorem for a countable language. Our proof of the compactness theorem thus only works for a countable language. We shall discuss uncountable languages in Section 6.4. Until then, all our applications will use countable languages.

Using version (B) of the soundness theorem in Exercise 5.12 in Section 5.3 of Chapter 5.

Exercise 6.1 ⎯⎯⎯⎯⎯⎯⎯⎯⎯⎯⎯⎯⎯⎯⎯⎯⎯⎯⎯⎯⎯⎯⎯⎯⎯⎯⎯⎯

Is the converse of the compactness theorem true?

Exercise 6.2 ⎯⎯⎯⎯⎯⎯⎯⎯⎯⎯⎯⎯⎯⎯⎯⎯⎯⎯⎯⎯⎯⎯⎯⎯⎯⎯⎯⎯

Show that for all sets of sentences Σ and sentences ϕ, if $\Sigma \vDash \phi$ then $\Delta \vDash \phi$, for some finite subset Δ of Σ.

This chapter will consist mainly of applications of the compactness theorem. We shall illustrate a variety of ways of using the theorem, sometimes proving the same result in different ways – please bear with this seeming redundancy, but it's all in aid of gaining experience in using the theorem.

Several of the results in this chapter will concern theories and it is time to revisit the ways in which the word 'theory' is used. We began Section 4.4 of Chapter 4 by mentioning that in mathematics 'theory' tends to be used to mean either all the mathematical consequences of a set of axioms or all the properties shared by some class of structures. We then put our money on the first of these meanings and defined a first-order theory as the set of all sentences which are logical consequences of a given set of sentences (its axioms). But in this chapter we shall also look at the second meaning, as so many important mathematical theories are described in this way, for instance the theory of the real numbers and the theory of finite groups. The question arises of whether each theory with this second meaning can be axiomatized in a first-order language, so that it is also a theory with the first meaning of the word. In Sections 6.3 and 6.4 we shall give some important negative answers to this question. Before then, in Section 6.2, we shall gain experience of using the compactness theorem by asking the question about first-order theories for which we have an *infinite* set of axioms of whether they can also be axiomatized by a *finite* set of axioms, again obtaining some negative answers. More positively, we shall use the compactness theorem in Sections 6.4 and 6.5 to obtain models of first-order theories with interesting properties, giving an idea of what model theory is really all about. We conclude the chapter and book in Section 6.6 with a brief look at the decidability of some first-order theories.

We also then used the word 'theory' as shorthand for a first-order theory, but will make more use of the fuller description to aid clarity.

6.2 Finite axiomatizability

A first-order theory is *finitely axiomatizable* if there is some finite set of axioms for it. Most of the examples of theories in Section 4.4 of Chapter 4 were given by a finite set of axioms, but some were not. When we are given a theory with infinitely many axioms, the question arises of whether there is some alternative and finite set of axioms for it. In this section we shall look at a use of the compactness theorem to show that for certain theories there can be no such finite axiomatization.

Of course, if a theory is finitely axiomatizable, then by taking the conjunction of these axioms, it can be axiomatized by a single sentence!

Let's look at the example of the theory of infinite sets, which in Section 4.4 we axiomatized by the set Σ, where

$$\Sigma = \{\exists_{\geq n} : n \in \mathbb{N}\},$$

with the sentence $\exists_{\geq n}$ expressing that 'there are at least n elements'. Is there an alternative finite set of axioms for this theory?

One suitable version of the sentence $\exists_{\geq n}$ is
$$\exists x_1 \exists x_2 \ldots \exists x_n \bigwedge_{1 \leq i < j \leq n} \neg\, x_i = x_j.$$

First note that no finite subset of Σ will do. Why not?

Exercise 6.3 ―――――――――――――――――――――――――

Let Δ be a finite subset of the set $\Sigma = \{\exists_{\geq n} : n \in \mathbb{N}\}$. Show that Δ has a finite model.

Thus Δ does not axiomatize the theory of infinite sets.

Solution

As Δ is a finite set, there is a largest n for which $\exists_{\geq n}$ appears in Δ. As any other sentence in Δ is a $\exists_{\geq m}$ for $m < n$, any set with at least this number n of elements would be a model of Δ.

But could there be some different, cunningly constructed, finite set of sentences axiomatizing the theory? If so, we could take the conjunction of these finitely many sentences and obtain a single sentence, σ say, which axiomatizes the theory. It turns out that no such σ can exist, as follows, in an argument exploiting the compactness theorem

Actually, if there was such a finite set, then there would also be a finite subset of Σ itself which would axiomatize the theory. This is left for you as Exercise 6.12.

Theorem 6.2

There is no finite set of axioms for the theory of infinite sets.

Proof

Suppose that there is a sentence σ which has the same effect as Σ, namely axiomatizing the theory of infinite sets. Consider the set of sentences Γ, where

$$\Gamma = \Sigma \cup \{\neg\sigma\}$$
$$= \{\exists_{\geq n} : n \in \mathbb{N}\} \cup \{\neg\sigma\}.$$

The construction of Γ ensures that it has no models! Any model of Σ, and thus also of σ, is an infinite set, while a model of $\neg\sigma$ must be something that *isn't* an infinite set – so Γ has no models. But we shall show that any finite subset of Γ has a model, contradicting the compactness theorem. The key assumption behind this argument, that the theory can be axiomatized by finitely many sentences and thus their conjunction σ, must then be false.

Let Δ be a finite subset of Γ. Γ includes infinitely many sentences of the form $\exists_{\geq n}$, so that as Δ is finite there must be a largest n for which $\exists_{\geq n} \in \Delta$. We want a model for Δ and need to allow for the possibility that Δ includes the sentence $\neg\sigma$ of Γ. This gives two requirements for a model of Δ: it has at least this largest n elements (which caters for any other sentences $\exists_{\geq m}$ for $m < n$ which might appear in Δ); and it isn't an infinite set. These are perfectly compatible constraints. Take any n element set: this will be a model of Δ.

A finite subset of Γ might or might not include some of the $\exists_{\geq n}$s and the $\neg\sigma$ of Γ. Our argument copes with a worst-case scenario when Δ includes all of these.

As every finite subset of Γ has a model but Γ has no models, this contradicts the compactness theorem. We conclude that the theory of finite sets cannot be axiomatized by finitely many axioms. ∎

Exercise 6.4 ⎯⎯⎯⎯⎯⎯⎯⎯⎯⎯⎯⎯⎯⎯⎯⎯⎯⎯⎯⎯⎯

Suppose that, in the proof above, the finite subset Δ of Γ contains none of the $\exists_{\geq n}$s. Show that Δ has a model.

Solution

If Δ contains no $\exists_{\geq n}$s, it is either empty (so vacuously has models) or just consists of $\neg\sigma$, so any finite non-empty set is a model of Δ.

There are other first-order theories which can be axiomatized using infinitely many axioms, but are not finitely axiomatizable. Two of these are explored in the following exercises. Inevitably, they rely on you knowing various bits of mathematics! In each case, you must first show that there is some infinite set of axioms for the theory – it's not so much the case that you're asked to find an infinite set of axioms: rather it's that the only set of axioms it's easy to

We shall be building on work done in Section 4.4 of Chapter 4.

think of happens to be infinite! – and then show that there cannot be a finite set of axioms. This essentially means there's no single sentence (a boring long conjunction!) which on its own axiomatizes the theory, by a similar argument to Theorem 6.2 above. The first example is the theory of torsion-free groups.

Definition   *Torsion-free groups*

A group is *torsion-free* if it contains no element of finite order besides its identity element.

Every group must contain an identity element, which has order 1.

Simple examples of such groups are $\mathbb{Z}$, $\mathbb{Q}$ and $\mathbb{R}$ under addition. The group with only one element (which must be the identity element of the group) is trivially a torsion-free group. All other examples have an infinite number of elements, with each non-identity element having infinite order.

Exercise 6.5 ⎯⎯⎯⎯⎯⎯⎯⎯⎯⎯⎯⎯⎯⎯⎯⎯⎯⎯⎯⎯⎯⎯⎯⎯⎯⎯⎯

Let L be a language (chosen to be suitable for axiomatizing the theory of groups) with equality, a binary function symbol $\cdot$, intended to be used for the group operation using infix notation, and a constant symbol $\mathbf{e}$, intended to represent the identity element of the group.

(a) (i) Write down axioms for group theory using this language.

(ii) Write down a sentence to represent the statement that, for a given $n \in \mathbb{N}$ with $n > 1$, there is no element of order n. [*Hints*: You are welcome to use x^n as a shorthand for $\underbrace{(x \cdot (x \cdot (x \cdot \ldots)))}_{n \text{ times}}$. You may find it helpful to exploit Result 1 on page 201 that if in a group $x^k = e$, where e is the identity element of the group, then the order of x divides k (but might not be k itself). And don't forget that as $e^2 = e$, $e^n = e$ for all $n \geq 1$.]

This result and other relevant facts about groups can be found in Section 4.4 of Chapter 4.

(iii) Hence write down an infinite set of axioms for the theory of torsion-free groups.

(b) Show that the theory of torsion-free groups is not finitely axiomatizable. [*Hints*: Structure your argument like the proof of Theorem 6.2. There are suitable examples of groups in Section 4.4 of Chapter 4 which have at least one non-identity element of finite order, so are not torsion-free, but with all such elements of at least some appropriately chosen large order. These will be needed to provide suitable models of the finite subsets Δ.]

Exploit Exercise 4.87.

The next example is the theory of fields of characteristic 0. We gave axioms for the theory of fields in Section 4.4 of Chapter 4 using a language with equality, binary function symbols $+$ and $\cdot$, unary function symbols $-$ and $^{-1}$ and constant symbols $\mathbf{0}$ and $\mathbf{1}$. We shall first explain what is meant by 'characteristic' and get to the point about finite axiomatizability in Exercise 6.10.

The symbols $+$ and $\cdot$ are intended to represent the addition and multiplication of a field, $-$ and $^{-1}$ the additive and multiplicative inverses, and the constant symbols $\mathbf{0}$ and $\mathbf{1}$ the additive and multiplicative identities.

We have seen some examples of fields in Section 4.4. For some of these examples, there is a finite sum of the multiplicative identity 1 of the field which equals its additive identity 0; that is, there is some positive integer n for which

$$\underbrace{1 + 1 + 1 + \ldots + 1}_{n} = 0;$$

and if there is such an n, there is a smallest such $n > 0$ with this property for that field. For instance, in the field $\mathbb{Z}_5$, consisting of the set $\{0, 1, 2, 3, 4\}$ with addition and multiplication modulo 5, the multiplicative identity is 1, the additive identity is 0 and $1 + 1 + 1 + 1 + 1 = 0$, so adding 1 to itself 5 times gives 0, while for no smaller positive n does adding 1 to itself n times give 0. Of course, there are fields for which there is no such $n > 0$, for instance $\mathbb{Q}$, $\mathbb{R}$ and $\mathbb{C}$: in each of these fields, adding 1 to itself finitely many times never gives 0. The following definition distinguishes between fields on this basis.

Recall that for any integer $n \geq 2$, the ring $\mathbb{Z}_n$ is the set $\{0, 1, 2, \ldots, n - 1\}$ with addition and multiplication modulo n. If n is prime, this ring is a field.

Definition   Characteristic

Let F be a field with additive and multiplicative identities 0 and 1 respectively. If there is some positive integer n such that

$$\underbrace{1 + 1 + 1 + \ldots + 1}_{n} = 0,$$

then F is said to have *finite characteristic* and the least such positive n is called the *characteristic* of F. If there is no such n, then F is said to have *characteristic* 0.

Some books use *characteristic infinity*, ∞, rather than characteristic 0. We use the 0, despite the number 0 sounding like a finite characteristic!

We ask you to explore the idea of characteristic in the next few exercises, which we hope are straightforward.

Exercise 6.6 ————————————————————

Suppose that F is a field with characteristic $n > 0$. Show that for all elements $x \in F$:

(a) $\underbrace{x + x + x + \ldots + x}_{n} = 0$;

(b) $\underbrace{x + x + x + \ldots + x}_{kn} = 0$, for all $k \in \mathbb{Z}$.

Exercise 6.7 ————————————————————

Suppose that F is a field and that there are positive integers m, n with $m < n$ such that

$$\underbrace{1 + 1 + \ldots + 1}_{m} = \underbrace{1 + 1 + 1 + \ldots + 1}_{n},$$

where 1 is the multiplicative identity of F. Show that F has finite characteristic.

As a consequence of the last exercise, if a field has characteristic 0, then the elements

$$0, \; 1, \; 1 + 1, \; 1 + 1 + 1, \; \ldots, \; \underbrace{1 + 1 + 1 + \ldots + 1}_{n}, \; \ldots$$

of the field are distinct, so that the field is infinite.

Exercise 6.8

Show that if F is a finite field, then it has finite characteristic.

There are infinite fields with finite characteristic. See Exercise 6.22.

It turns out that only certain positive integers can arise as the characteristic of a field.

Exercise 6.9

(a) Show that the field $\mathbb{Z}_p$, where p is a prime, has characteristic p.

(b) Let F be a field.

(i) Suppose that n is a composite positive integer with $n = ab$ for integers a, b with $a \geq 2$, $b \geq 2$. Explain why

$$\underbrace{1 + 1 + 1 + \ldots + 1}_{n} = \underbrace{(1 + 1 + \ldots + 1)}_{a}\underbrace{(1 + \ldots + 1)}_{b},$$

where 1 is the multiplicative identity of F.

(ii) Suppose that $n \geq 2$ is an integer such that

$$\underbrace{1 + 1 + 1 + \ldots + 1}_{n} = 0.$$

Show that the characteristic of F is one of the primes dividing n. [*Hints*: If n is composite, use part (b)(i) and the result of Exercise 4.91(a) in Chapter 4.]

The result of the last exercise can be summarized as the following theorem.

Theorem 6.3

Let F be a field with finite characteristic n. Then n is a prime.

We have also shown that for each prime p, there is at least one field with characteristic p, namely $\mathbb{Z}_p$. An observation stemming from this which will be useful later is that there are fields of arbitrarily large finite characteristic.

We can now get back to the business of finite axiomatizability, armed with these various facts. Our goal is to show that the theory of fields of characteristic 0 is axiomatizable, but not finitely axiomatizable.

Exercise 6.10

(a) Given a positive integer n, write down a set of axioms for a field whose characteristic is *not* n. Hence write down a set of axioms (rather likely to be an infinite set!) for the theory of fields of characteristic 0.

(b) Show that the theory of fields of characteristic 0 is not finitely axiomatizable. [*Hints*: Structure your argument like the proof of Theorem 6.2. You'll need to exploit the fact established above that there are fields of characteristic p for all primes p.]

Here's an interesting general result about finite axiomatizability.

> **Theorem 6.4**
>
> Suppose that a theory T can be axiomatized using a first-order language L by a set of sentences Σ and that Λ is a set of sentences in L such that for all structures $\mathcal{A}$ for L, $\mathcal{A}$ is a model for Σ if and only if it is not a model of Λ. Then the theory T is finitely axiomatizable.

We could regard Λ as axiomatizing the theory 'not T'.

Thus if θ is the conjunction of these finitely many axioms of T, then $\neg\theta$ axiomatizes 'not T'.

The proof of this theorem requires a reasonably similar compactness argument to those earlier in this section and the next exercise leads you through it.

Exercise 6.11

Suppose that there are sets of sentences Σ and Λ in the language L such that for all structures $\mathcal{A}$ for L, $\mathcal{A}$ is a model for Σ if and only if it is not a model of Λ, where the set Σ axiomatizes the theory T. We shall assume that T is not finitely axiomatizable and obtain a contradiction.

(a) Under the assumption that T is not finitely axiomatizable, Σ must of course be an infinite set. Show in addition that any finite subset of Σ must have a model which is *not* a model of T.

(b) Still assuming that T is not finitely axiomatizable, apply the compactness theorem to the set Γ, where $\Gamma = \Sigma \cup \Lambda$ to obtain the desired contradiction. [*Hints*: What models does Γ have? What does part (a) tell us about finite subsets of Γ?]

We can infer from Theorem 6.4 that at least one reasonable-sounding mathematical theory *cannot* be axiomatized in a first-order language, namely the theory of finite sets. The complementary theory, the theory of infinite sets, can be axiomatized, but not finitely axiomatized, in a language with equality. Theorem 6.4 then prevents the theory of finite sets from being axiomatizable.

The question arises of whether other mathematical theories of interest cannot be axiomatized in a first-order language. The answer is 'yes', but we'll need something more subtle than Theorem 6.4. Knowing that, say, the theory of fields of characteristic 0 is axiomatizable but not finitely axiomatizable, leads, via this theorem, to the result that the complementary theory is not axiomatizable. But this complementary theory is that of all structures which aren't fields of characteristic 0, including ones that aren't fields at all, as well as fields that are of finite characteristic – this theory isn't so interesting! Of more interest, were it true, would be a result that one cannot axiomatize fields of finite characteristic. This is what we'll investigate in the next section.

Further exercises

Exercise 6.12

Suppose that a theory T can be axiomatized by a set of sentences Σ in a language L. Show that T is finitely axiomatizable if and only if there is some finite subset of Σ which axiomatizes T.

6.3 Some non-axiomatizable theories

We saw at the end of the last section that the theory of finite sets cannot be axiomatized in a first-order language with equality. In this section we shall use the compactness theorem to show that there are further examples of interesting mathematical theories which aren't first-order axiomatizable.

One method of showing that a theory cannot be axiomatized in a language L is as follows. Assume that it can be axiomatized by a set of sentences Σ and bolt onto Σ a cunningly chosen infinite set of sentences to get a set Γ with the property that Γ has no models but every finite subset Δ of Γ does have a model. This contradicts the compactness theorem and we can conclude that no set of axioms Σ exists. We shall illustrate this method with the theory of finite sets – yes, we know we've already shown that this cannot be axiomatized, doubtless using a method which is in some sense equivalent to the one we are about to use! – and give you a more interesting example to sort out by this method.

Suppose that the theory of finite sets can be axiomatized by a set of sentences Σ in some first-order language L with equality. Guided by the principle of this method that we want to end up with a set Γ which has no models, the cunning sentences to bolt on are the very useful $\exists_{\geq n}$s saying, for each natural number n, 'there are at least n elements'. So define

A nice feature of this method is that it covers languages of arbitrary complexity.

$$\Gamma = \Sigma \cup \{\exists_{\geq n} : n \in \mathbb{N}\}.$$

Any model of Γ would be a model of Σ, hence a finite set, and thereby couldn't make all the $\exists_{\geq n}$s true: so no models of Γ exist. But a, we hope by now familiar, argument shows that every finite subset of Γ has a model, contradicting the compactness theorem. Thus no set Σ exists axiomatizing the theory.

Exercise 6.13 ─────────────────────────────────

Show that if Δ is a finite subset of Γ above, then Δ has a model.

Solution

If Δ is finite, then there is a largest n for which $\exists_{\geq n}$ appears in Δ. As Δ is then a subset of $\Sigma \cup \{\exists_{\geq m} : m \leq n\}$, it is enough to show that this latter set has a model, as the model will also be a model for Δ. Such a model would need to be finite, to be a model for Σ, but also contain at least n elements, to cope with the $\exists_{\geq m}$s for $m \leq n$ – this is no problem! Just take any n-element set!

As in our earlier worked examples, we consider only a worst-case scenario when Δ includes some $\exists_{\geq n}$s.

Let's try out this method on something a bit more interesting, namely the theory of fields of finite characteristic. It's easy to write down axioms for the theory of fields of characteristic p for a fixed prime p. It's easy to adapt this to axiomatize fields of one of a finite number of finite characteristics. But can we axiomatize at one go all fields of all possible finite characteristics?

Exercise 6.14

Let L be a language with equality including constant symbols $\mathbf{0}, \mathbf{1}$, binary function symbols $+$ and $\cdot$ and anything else you'd expect for fields.

(a) Let p be a prime number. Write down a sentence which, when added to the axioms for the theory of fields, gives a set of axioms for the theory of fields of characteristic p.

(b) Let $p_1, p_2, \ldots, p_n$ be finitely many prime numbers. Explain how to adapt your solution to part (a) to axiomatize the theory of fields of characteristic one of $p_1, p_2, \ldots, p_n$.

(c) Adapt the method used for the non-axiomatizability of the theory of finite sets to show that the theory of fields of finite characteristic cannot be axiomatized. [*Hints*: Our solution bolts on certain well-chosen sentences which use $\mathbf{0}$, $\mathbf{1}$, $+$ and $=$. Also we know that for each prime p there is at least one field of characteristic p, namely $\mathbb{Z}_p$.]

We don't think that this method quite works for the next new example of a non-axiomatizable theory we'd like you to look at, namely the theory of groups in which every element has finite order. An extra ingredient is required, namely introducing extra constant symbols to the language and exploiting them in a clever way – this technique turns out to be a vital tool in many other applications of compactness, as we'll see later. Inevitably, our first illustration of this refinement is with the good old theory of finite sets, for which you've already seen two different arguments showing it's not axiomatizable!

As before, we shall assume that the theory of finite sets is axiomatizable by a set of sentences Σ in a language L with equality and go for a contradiction. This time, we add some new constant symbols to the language: our argument will use the countably many extra symbols $\mathbf{c}_n$, for $n \in \mathbb{N}$, giving an enriched language L'. We'll now add extra sentences to Σ using these new symbols to get a set Γ with no models, but such that every finite subset of Γ has a model, giving us the familiar contradiction of the compactness theorem. In particular, we'll take Γ defined by

If L includes some constants $\mathbf{c}_n$, we just choose some other symbols!

$$\Gamma = \Sigma \cup \{\neg\, \mathbf{c}_i = \mathbf{c}_j : i \neq j, \; i, j \in \mathbb{N}\}.$$

Any potential model $\mathcal{A}$ of Γ has to be a structure for the enriched language L', which means that $\mathcal{A}$ has to specify which element of its domain interprets the constant symbol $\mathbf{c}_i$ for each $i \in \mathbb{N}$. The sentences $\neg\, \mathbf{c}_i = \mathbf{c}_j$ for $i \neq j$ force such a model to interpret the $\mathbf{c}_i$s by distinct elements of its domain, so that its domain is forced to be infinite. But this is incompatible with being a model for Σ, which has only finite models. So Γ has no models.

Whether or not $\mathcal{A}$ satisfies Σ in L doesn't depend on how it interprets the extra symbols in L', by Exercise 4.26 in Section 4.2 of Chapter 4.

Exercise 6.15

Show that if Δ is a finite subset of the set Γ above, then Δ has a model. Conclude that the theory of finite sets is not axiomatizable.

Let's use this refined method to show that the theory of groups where every element has finite order is not first-order axiomatizable. The models of this theory include all finite groups, as by Result 2 on page 201 in Section 4.4 of Chapter 4, the order of an element of a finite group divides the number of elements in the group. But there are also infinite groups in which every element has finite order. Take for instance the group G with domain all rationals in the interval $[0, 1)$ with operation $\oplus$ defined in terms of the usual $+$ on $\mathbb{Q}$ by

$$a \oplus b = \begin{cases} a + b, & \text{if } a + b < 1, \\ a + b - 1, & \text{if } a + b \geq 1, \end{cases}$$

in which a typical element expressed in lowest terms as $\dfrac{m}{n}$ has order n. (Those with a suitable background in group theory might recognize this as the quotient group $\mathbb{Q}/\mathbb{Z}$, where $\mathbb{Q}$ is the group of rationals under addition and $\mathbb{Z}$ is the subgroup of all integers.) This group contains elements of all possible finite orders.

> More fully, m and n are integers with $0 \leq m < n$ and $\gcd(m, n) = 1$.

Assume that the language L includes the binary function symbol $\cdot$ for the group operation and the constant $\mathbf{1}$ for the identity. We'll allow ourselves to use the notation x^n as a shorthand for $\underbrace{(x \cdot (x \cdot (x \cdot \ldots)))}_{n \text{ times}}$. Let's suppose that the theory of these groups can be axiomatized in L by a set of sentences Σ and go for a contradiction by the sort of methods we've used so far.

Exercise 6.16

(a) Show that Γ defined by

$$\Gamma = \Sigma \cup \{\exists x \, \neg \, x^n = 1 : n \in \mathbb{N}\}$$

has a model. (This means that Γ fails to give us a set of sentences with no models, so is no use for our method!)

(b) Add a new constant symbol $\mathbf{c}$ to the language. Find a set of sentences Γ in the enriched language extending the set Σ such that Γ has no models, but every finite subset of Γ does have a model. [*Hints*: Find sentences which ensure that in any group $\mathcal{G}$ which is a model of these, the interpretation $\mathbf{c}^{\mathcal{G}}$ of $\mathbf{c}$ has infinite order. You'll also need to recall suitable examples of groups to show that any finite subset of a well-chosen Γ has a model.]

The result of Exercise 6.16 contradicts the compactness theorem, so that the theory of groups in which every element has finite order cannot be axiomatized in a first-order language.

Our next example is the theory of well-ordered sets. As we mentioned in Section 4.4 of Chapter 4, a well-order on a set A is a linear order on A for which every non-empty subset B of A contains a least element. Somewhat trivially, every finite linear order is a well-order, as any non-empty subset contains finitely many elements, so must have a minimum element. Any non-trivial example thus has to be infinite and the classic example is the set $\mathbb{N}$ of natural numbers with the usual order. The standard axiomatization of the

> Another example of an infinite well-order is the subset
> $$\{m - \tfrac{1}{n} : m, n = 1, 2, 3, \ldots\}$$
> of the set of rationals $\mathbb{Q}$ with its usual order.

theory of well-order can be done in a second-order language, quantifying over subsets B as well as elements b of the domain, using an axiom like

$$\forall B(B \neq \varnothing \rightarrow \exists b_0(b_0 \in B \;\wedge\; \forall b(b \in B \rightarrow b_0 \leq b)))$$

alongside the usual first-order axioms for a linear order. We will use the same method as above for groups of finite order to show that the theory of well-order is not axiomatizable in any first-order language.

Exercise 6.17 ____

Let L be a language with equality and including a binary relation symbol $<$ suitable for axiomatizing strictly linearly ordered sets.

(a) Suppose that Σ is a set of sentences in L which axiomatizes the theory of well-order. Add infinitely many new constant symbols $\mathbf{c}_n$ for all $n \in \mathbb{N}$ to L to get the language L'. Define

$$\Gamma = \Sigma \cup \{\mathbf{c}_{n+1} < \mathbf{c}_n : n \in \mathbb{N}\}.$$

    As earlier, if the language already uses some of the symbols $\mathbf{c}_n$, just use other new symbols.

    (i) Explain why Γ has no models.

    (ii) Show that every finite subset Δ of Γ has a model.

(b) Explain why the theory of well-order is not axiomatizable in L.

____

We shall make further use of this method of adding new constant symbols and sentences involving them in the next two sections.

Actually, the intended method for obtaining the result about fields of finite characteristic in Exercise 6.14 could be regarded as an illustration of this refined method based on adding constant symbols, in the following sense. Suppose that for some reason, the given language for representing the theory of fields didn't include the constant symbols $\mathbf{0}$ and $\mathbf{1}$ intended to represent the additive and multiplicative identities of the field. It's still perfectly possible (but perhaps a bit more tedious) to write down axioms for fields: for instance, the sentence

$$\exists x \forall y (x + y = y \;\wedge\; y + x = y)$$

    You could try writing down the axiom that every non-zero element has a multiplicative inverse using the language with just the binary function symbols $+$ and $\cdot$, if you'd like the practice.

captures that there is an additive identity. One can still capture the property that a field has characteristic p, where p is prime, by the sentence

$$\exists x(\forall y(x + y = y \;\wedge\; y + x = y) \;\wedge\; \forall z(\underbrace{z + z + z + \ldots + z}_{p} = x)),$$

which will imply that the multiplicative identity (as a particular z here) when added to itself p times gives the additive identity. If we'd used such a language for fields, then to show that the theory of fields of finite characteristic is not axiomatizable we could have added 'new' constant symbols $\mathbf{0}$ and $\mathbf{1}$ to the language and then bumped up any presumed set of axioms Σ for this theory with suitable sentences involving these new constants to get a Γ with no models etc., pretty well as we hope you did in Exercise 6.14.

Let's now look at a different method exploiting a result which is a foretaste of a very important consequence of compactness, which we'll look at in the next section. The method involves straightforward application of the following result.

Theorem 6.5

Let Σ be a set of sentences in a first-order language L with equality. If Σ has models of arbitrarily large finite domain, then it has an infinite model.

Proof

The proof requires use of the compactness theorem and for a change we'll use it in a positive way, not to obtain a contradiction as in all our applications of it so far. From now on, most of our uses of compactness will be similarly positive.

The proof also uses a very familiar looking set of sentences Γ defined by

$$\Gamma = \Sigma \cup \{\exists_{\geq n} : n \in \mathbb{N}\}.$$

For a change we actually want to show that Γ does have a model, as such a model would have to be infinite – and that's what we are looking for. So let Δ be any finite subset of Γ. Then as Δ is finite there is a largest n such that $\exists_{\geq n}$ is in Δ. We are given that Σ has arbitrarily large finite models, so in particular it has a model with at least n elements – this model must then also be a model of Δ.

As every finite subset of Γ has a model, the compactness theorem tells us that Γ also has a model, which in turn means that Σ has an infinite model. ∎

We can now provide yet another argument that the theory of finite sets cannot be axiomatized in a first-order language! This theory has arbitrarily large finite models, so that if it was axiomatizable, then by Theorem 6.5 its axioms would have an infinite model, contradicting that the axioms are for the theory of finite sets. Theorem 6.5 can be used in a similar way to show that neither of the theories of finite groups and of finite fields can be axiomatized in a first-order language – because both theories have arbitrarily large finite models.

The business of cardinalities of models, if any exist, is taken a lot further in the next section.

We shall end this section with a theorem that supersedes much of the work we have asked you to do earlier to show that certain theories were not axiomatizable.

We don't apologize for making you trawl through results with specific theories, even though this theorem gives a shortcut. The extra bit of experience of these theories is probably good for you!

Theorem 6.6

Suppose that T is a theory in a language L with finitely many axioms. Suppose that Σ and Λ are sets of sentences with the property that for all models $\mathcal{A}$ of T,

$$\mathcal{A} \vDash \Sigma \text{ if and only if } \mathcal{A} \nvDash \Lambda.$$

Then both the theories $T \cup \Sigma$ and $T \cup \Lambda$ are finitely axiomatizable.

Exercise 6.18

Prove the above theorem. [*Hint*: The set of sentences $T \cup \Sigma \cup \Lambda$ has no models, so there are finite subsets Σ' of Σ and Λ' of Λ such that $T \cup \Sigma' \cup \Lambda'$ has no models. What can one say about models of $T \cup \Sigma'$ and $T \cup \Lambda'$?]

As a consequence of this theorem, when one has a theory which can be axiomatized but not finitely axiomatized, what we might describe as its complementary theory cannot be axiomatized at all. For instance, take T to be the theory of fields, which can be finitely axiomatized, and Σ to be the theory of fields of characteristic 0. We showed in Exercise 6.10 that this theory can be axiomatized but not finitely axiomatized. Suppose that the theory of fields with finite characteristic could be axiomatized by a set Λ. As all fields have characteristic 0 or a finite number (namely a prime), we then have that for all fields $\mathcal{A}$, i.e. all models $\mathcal{A}$ of T,

$$\mathcal{A} \vDash \Sigma \text{ if and only if } \mathcal{A} \nvDash \Lambda.$$

Then by Theorem 6.6 both Σ and Λ could be finitely axiomatized, contradicting that Σ cannot be. The resulting contradiction leads to the conclusion that there is no such Λ axiomatizing the theory of fields of finite characteristic.

Exercise 6.19

Use Theorem 6.6 and relevant results from Section 6.2 to show each of the following theories cannot be axiomatized.

(a) The theory of finite sets.

(b) The theory of groups with at least one non-identity element of finite order.

6.4 The Löwenheim–Skolem theorems

In this section we look at two major results about the possible sizes of the infinite models of a first-order theory.

Inevitably we shall require some knowledge of the theory of infinite 'size', properly called the theory of cardinal numbers, which would normally be acquired from the study of set theory. Likewise we shall need some knowledge of the theory of ordinal numbers. We shall summarize most of what we need to know without giving very much background. Some of the marginal notes will give extra detail or give caveats. We hope that the main results of the section can be appreciated without having previously studied cardinal and ordinal numbers. If you do happen to have the necessary background, do see whether you can flesh out our sketchy details!

There are plenty of books on set theory, e.g. those by Enderton [12], Halmos [17] and Goldrei [16]. There are some useful potted accounts, e.g. in Cameron [4].

Set theory background

The most commonly used framework for the theory of cardinal numbers is Zermelo–Fraenkel set theory (abbreviated as ZF) which is based on various first-order axioms expressed in a language with equality and the 2-place relation symbol $\in$, intended to represent 'is an element of'. To give you a flavour of the simpler axioms and the use of this language, one axiom is

$$\forall x \forall y (x = y \leftrightarrow \forall z(z \in x \ \leftrightarrow \ z \in y)),$$

with the intended interpretation that two sets are equal if and only if they contain the same elements; and another is

$$\exists x \forall y \, \neg \, y \in x,$$

with the intended interpretation that there is an empty set. Overall the intended interpretation of the theory is the universe of sets, with $\in$ interpreted as 'is an element of'. We are not even going to attempt to explain what 'sets' are! They are an undefined concept of which all that can be said is that their properties are the logical consequences of the Zermelo-Fraenkel axioms. This approach – explaining important mathematical objects in terms of the properties they have and basing all mathematics on the logical consequences of these properties – is central to the development of much of what is in this book. It is perhaps unsurprising, with the benefit of hindsight, that this approach, previously adopted for the likes of $\mathbb{N}$ and $\mathbb{R}$, should then be used for set theory. The basis of modern set theory is the work of Cantor in the second half of the 19th century on infinite sets and two sorts of infinite number, cardinals and ordinals. This work produced exciting results, which we shall be using, but was and remains something of a minefield in terms of its potential for generating contradictions and paradoxes, not to mention a spot of controversy. Hence the adoption of the axiomatic approach to attempt to place it on a secure foundation.

> We use the term 'universe' of sets as the existence of a 'set of all sets' would result in contradictions, most famously *Russell's paradox*.

> The German mathematician Georg Cantor (1845–1918) in addition initiated the theory of point-set topology.

Some of the behaviour of sets that we have been educated to expect and exploit is uncontroversial, for instance pretty well anything to do with finite unions and intersections of finite sets. But the existence of any infinite set at all was very controversial and our casual acceptance and manipulation of sets like $\mathbb{N}$, $\mathbb{Q}$ and $\mathbb{R}$ perhaps testifies to the influence of our education as much as to the power of the underlying ideas! The modern theory of infinite sets is usually based on the Zermelo-Fraenkel axioms and their underpinning of Cantor's theory. Much of this theory is based on a principle of Zermelo called the *axiom of choice* (usually abbreviated as AC), given as follows.

Axiom of choice

Suppose that $\mathcal{F}$ is a set of non-empty sets. Then there is a function $h \colon \mathcal{F} \longrightarrow \bigcup \mathcal{F}$, called a *choice function* on $\mathcal{F}$, such that for each $A \in \mathcal{F}$, $h(A) \in A$.

> This can all be expressed in the first-order language using $\in$ and $=$.

The function h 'chooses' an element, namely $h(A)$, for each member of the set $\mathcal{F}$. The axiom doesn't say *how* the choice is made, only that such a function exists. Broadly speaking, AC isn't needed to justify the existence of a choice function when the set $\mathcal{F}$ is finite or when the sets $A \in \mathcal{F}$ all have enough

> We used AC in our solution to Exercise 4.50 in Section 4.3 to construct a Skolem function.

'useful' structure to define the function, rather than just say it exists. But AC is needed when $\mathcal{F}$ is infinite and its elements A aren't equipped with enough structure. As many of the most important results in this book make use of infinite sets (e.g. infinite lists of formulas, as well as structures with infinite domains), these results are crucially underpinned by the theory of the Zermelo-Fraenkel axioms along with the axiom of choice, usually abbreviated as *ZFC*, with AC playing a critical role.

One sort of useful structure on the sets A is that they are all well-ordered, so h could choose the least element of each of them.

The main results we need are as follows. First of all, we can compare the sizes of sets using the following relations involving sets X and Y.

Definitions   Comparing sizes of sets

We say that $X \approx Y$ if there is a bijection from X to Y.

We say that $X \preceq Y$ if there is a one–one function from X to Y; and that $X \prec Y$ if $X \preceq Y$ and it is not the case that $X \approx Y$.

The relation $\approx$ is easily shown to be an equivalence relation on sets. Important examples involving infinite sets include

$$\mathbb{N} \approx \mathbb{Z} \approx \mathbb{Q} \approx \mathbb{N} \times \mathbb{N},$$

which are all *countable* sets, and

$$\mathscr{P}(\mathbb{N}) \approx \mathbf{2}^{\mathbb{N}} \approx \mathbb{R} \approx \mathbb{R}^2,$$

'Countable' includes finite as well as sets of the same size as $\mathbb{N}$.

where for any set X, $\mathscr{P}(X)$ is the set of all subsets of X and $\mathbf{2}^X$ is the set of all functions from X to the two element set $\mathbf{2} = \{\mathbf{0}, \mathbf{1}\}$.

In general, for any sets X, Y, the set Y^X is the set of all functions from X to Y.

The relation $\preceq$ on sets is easily shown to be reflexive and transitive and is almost anti-symmetric – the *Schröder–Bernstein theorem* shows that if $X \preceq Y$ and $Y \preceq X$ then $X \approx Y$ (rather than the $X = Y$ required by anti-symmetry). This can all be proved without using AC. Without AC, we cannot prove what looks like the linearity axiom, namely that for all sets X, Y, one of $X \preceq Y$ and $Y \preceq X$ holds. But with AC, we can prove that this holds.

The principle $X \preceq Y$ and $Y \preceq X$ for all sets X, Y is called *dichotomy*. It is equivalent to AC.

A very important result, *Cantor's theorem*, is that

$$X \prec \mathbf{2}^X,$$

for all sets X. One consequence of this is the existence of infinite sets that aren't countable, for instance $\mathbb{R}$. Another consequence is that there is no 'largest' set – if there was such a set X, then

Another example of an uncountable set consists of all truth assignments on a countable set of propositional variables $\{p_i : i \in \mathbb{N}\}$, which are essentially functions from $\{p_i : i \in \mathbb{N}\}$ to the two element set $\{T, F\}$.

$$X \prec \mathbf{2}^X \quad \text{(by Cantor's theorem)}$$
$$\preceq X \quad \text{(as } X \text{ is the largest set)},$$

so that $X \prec X$, which is a contradiction (thanks to the definition of $\prec$). Important examples include $X \prec \mathbb{N}$, for all finite sets X and

$$\mathbb{N} \prec \mathbb{R} \prec \mathbf{2}^{\mathbb{R}} \prec \mathbf{2}^{2^{\mathbb{R}}}.$$

With AC we can define a special class of sets, called the *cardinal numbers*, or cardinals for short, which measure all possible sizes of sets. We can define for each set X its cardinal number $\mathrm{Card}(X)$ in such a way that

- $X \approx \mathrm{Card}(X)$;
- if $\mathrm{Card}(X) \approx \mathrm{Card}(Y)$ then $\mathrm{Card}(X) = \mathrm{Card}(Y)$.

So there's a cardinal number for every set and there's only one cardinal number of each size.

Using the Schröder–Bernstein theorem for cardinals, these results mean that if $\mathrm{Card}(X) \preceq \mathrm{Card}(Y)$ and $\mathrm{Card}(Y) \preceq \mathrm{Card}(X)$ then $\mathrm{Card}(X)$ actually equals $\mathrm{Card}(Y)$. Putting this all together, the relation $\preceq$ defines a linear order on the class of all cardinals – actually it defines a well-order on these too. We'll often use Greek letters like κ, λ, μ to stand for cardinal numbers and write $\leq$ ($<$) instead of $\preceq$ ($\prec$).

For finite sets X, $\mathrm{Card}(X)$ is simply defined as the (natural) number of elements in the set. For infinite sets X, $\mathrm{Card}(X)$ is usually defined to be a special well-ordered set called an *initial ordinal*. The smallest infinite cardinal is the cardinal of $\mathbb{N}$ and thus also of all countably infinite sets. It is usually written as $\aleph_0$. ($\aleph$ is the first letter of the Hebrew alphabet and is pronounced 'alef'.) All larger cardinals must be uncountable. The next larger cardinal after $\aleph_0$ is written as $\aleph_1$, the next larger as $\aleph_2$ and so on. As $\mathbb{N} \approx \aleph_0$ and $\mathbb{R} \approx 2^{\mathbb{N}}$, we can use the $\aleph$ notation to describe the cardinality of the set of real numbers by $\mathrm{Card}(\mathbb{R}) = 2^{\aleph_0}$. An important mathematical problem which is not resolved within the theory *ZFC* is which of the $\aleph$s is the cardinal of $\mathbb{R}$. As $\mathbb{R}$ is also known as the *continuum*, the hypothesis that $2^{\aleph_0} = \aleph_1$, the first uncountable cardinal, is called the *continuum hypothesis*.

The ordinal numbers are another of Cantor's ground-breaking creations. An *ordinal* α is a set which is well-ordered by the membership relation (the usual $\in$) such that for each $\beta \in \alpha$, $\beta \subseteq \alpha$. What makes an ordinal α *initial* is that for each $\beta \in \alpha$ we have $\beta \prec \alpha$. The set $\mathbb{N}$, regarded as an ordinal, is the smallest initial ordinal.

There is an arithmetic on cardinal numbers defined as follows.

Both the continuum hypothesis and its negation can be shown to be consistent with *ZFC*, always assuming that *ZFC* is itself consistent. In this book, we shall avoid results requiring a decision about the continuum hypothesis.

Definitions   Cardinal arithmetic

Let κ, λ be cardinals (so they are also sets). Then operations of $+$, $\cdot$ and exponentiation are defined by:

$$\kappa + \lambda = \mathrm{Card}((\kappa \times \{0\}) \cup (\lambda \times \{1\})), \text{ i.e. the cardinal of the union}$$
$$\text{of disjoint sets of cardinal } \kappa \text{ and } \lambda;$$
$$\kappa \cdot \lambda = \mathrm{Card}(\kappa \times \lambda), \text{ i.e. the cardinal of the set of pairs } \kappa \times \lambda;$$
$$\kappa^{\lambda} = \mathrm{Card}\left(\kappa^{\lambda}\right), \text{ i.e. the set of all functions from } \lambda \text{ to } \kappa.$$

With AC, simple additions and multiplications of infinite cardinals don't generate larger cardinals. We have, provided that at least one of κ, λ is infinite, that

$$\kappa + \lambda = \kappa \cdot \lambda = \max\{\kappa, \lambda\}.$$

However, thanks to Cantor's theorem, exponentiation often gives larger cardinals: provided $2 \leq \kappa < \lambda$, we have $\kappa < \kappa^{\lambda}$.

The intertwining of the axiom of choice with the theory of cardinals is quite considerable. For instance, AC is equivalent to the statement that $X \times X \approx X$ for all infinite sets X, that is, $\kappa \cdot \kappa = \kappa$ for all infinite cardinals κ.

This equivalence can be demonstrated within the framework of the Zermelo-Fraenkel axioms.

Cardinalities of infinite models of a theory

We can now turn to the major business of this section, which is to look at two major results about the possible infinite cardinalities (within the framework of ZFC) of models of a theory, should it have any infinite models at all.

Exercise 6.20 ____

(a) Give an example of a first-order theory with only finite models.

(b) Give an example of a first-order theory with only infinite models.

____

Although the main application of these results that we'll discuss here will be yet more negative results about whether certain theories can be axiomatized, the results really open up a door in the study of models of theories. The results are usually called the Löwenheim–Skolem theorems: one theorem is described as 'upward' and the other as 'downward', for reasons which will become obvious. Perhaps the most important of the two in modern model theory is the upward theorem, as it holds out the prospect of an unlimited number of models for most theories; and it's the one whose proof exploits the compactness theorem. We shall state and prove it before we deal with the downward result.

The original Löwenheim–Skolem theorem is what we call the downward result. It was an important milestone in the development of the subject, overtaken subsequently by Gödel's completeness theorem, which is a stronger result. Leopold Löwenheim (1878–1957) was a German mathematician.

Theorem 6.7 *Upward Löwenheim–Skolem theorem*

Let T be a theory in a first-order language L with equality. If T has an infinite model, then T has models of arbitrarily large infinite cardinality, i.e. for any infinite cardinal κ, there is a model of T whose domain is a set of cardinality at least κ.

As ever, when L is a language with equality, 'model' means a normal model, unless we say otherwise.

Proof

Suppose that T has an infinite model – call this model $\mathcal{B}$ – and let κ be an infinite cardinal.

As we mentioned earlier, many – maybe most – applications of compactness involve adding new constant symbols to the language. That's what we'll do here. Add to the language L a set $\{\mathbf{c}_\alpha : \alpha \in \kappa\}$ of new distinct constant symbols not already in L. Define a set of sentences Γ in the enlarged language by

Recall that the cardinal κ is also a set.

$$\Gamma = T \cup \{\neg\, \mathbf{c}_\alpha = \mathbf{c}_\beta : \alpha \neq \beta,\ \alpha, \beta \in \kappa\}.$$

Then any model $\mathcal{A}$ for Γ is not just a model of T, but also has to have enough distinct elements in its domain to interpret the constants $\mathbf{c}_\alpha$, $\alpha \in \kappa$ in a way that satisfies all the sentences $\neg\, \mathbf{c}_\alpha = \mathbf{c}_\beta$ whenever $\alpha \neq \beta$. Thus the cardinality of $\mathcal{A}$ would be at least κ as required.

We shall show that Γ indeed has a model by using the compactness theorem. Let Δ be any finite subset of Γ. As Δ is finite, it can involve at most finitely many of the sentences $\neg\, \mathbf{c}_\alpha = \mathbf{c}_\beta$. Any model of Δ would need a domain with enough distinct elements with which to interpret the $\mathbf{c}_\gamma$s involved in these finitely many sentences to make these latter true; but that only requires the model to have at least some large finite domain. Of course, any model of Δ

Slightly more poshly, if $\mathcal{A}$ has domain A and interprets each $\mathbf{c}_\alpha$ by $\mathbf{c}_\alpha^{\mathcal{A}}$, then the map

$$f : \kappa \longrightarrow A$$
$$\alpha \longmapsto \mathbf{c}_\alpha^{\mathcal{A}}$$

is a one–one from κ to A, so that $\kappa \preceq A$.

must also be a model for whichever axioms of T are in Δ. Luckily we are assuming that T has an infinite model $\mathcal{B}$, making these sentences in Δ true. So as the domain of $\mathcal{B}$ is infinite, we can expand the structure $\mathcal{B}$ to interpret those $\mathbf{c}_\gamma$s appearing in Δ by different elements of its domain (as there are only finitely many of these $\mathbf{c}_\gamma$s), thereby obtaining a model for Δ.

The required result now follows by the compactness theorem. $\blacksquare$

As we can axiomatize theories like those for groups and fields which have known infinite models, we can deduce from the upward Löwenheim–Skolem theorem that there arbitrarily large groups and fields – without having to construct them!

There's a somewhat important detail required for the proof above which we neither mentioned nor justified, taking the view that mere details shouldn't get in the way of such a good story! However, now is the place to remind that while we have stated and used the compactness theorem for an arbitrary first-order language, we have effectively only proved it for languages with at most countably many symbols – the proof of the compactness theorem (in Section 6.1) exploits the completeness theorem and we have proved the latter only for countable languages. The completeness theorem does however hold for uncountable languages. We shall state the theorem for languages with equality and give a sketch proof which you can fill out if you have enough background in set theory.

In the proof of Theorem 6.7, by introducing the new constant symbols $\{\mathbf{c}_\alpha : \alpha \in \kappa\}$ for an arbitrarily large cardinal κ, we have made the language uncountable.

> ### Theorem 6.8 *Completeness theorem for uncountable languages*
>
> If Δ is a consistent set of sentences in an uncountable language L, then Δ has a model. If L is a language with equality, then Δ has a normal model.

Proof

We'll give only a brief idea of how this might be proved on the basis of our proof of the completeness theorem in Section 5.5 of Chapter 5. This proof starts with a consistent set of sentences Δ_0 in a language L_0. It then interleaves processes of

> extending a consistent set Σ of sentences in some language L to a maximal consistent set Σ' in that language

This was done in Theorem 5.19 in Section 5.5.

and

> taking a consistent set Γ of sentences, adding to the language L extra constant symbols for each sentence in Γ beginning with a $\exists$ to obtain a larger language L^+ and adding sentences exploiting these constants to 'witness' these particular $\exists$s to obtain a larger consistent set Γ^+.

This was done in Theorem 5.20.

This interleaving is done often enough (countably many times) to build up the original Δ into a complete set Δ^* in an enlarged language L^+ with lots of new constant symbols. We then construct the canonical structure $\mathcal{A}$ with domain the set of of closed terms of L^+ which is a model of Δ^* and thus of

Δ_0. If the original language L includes $=$, the normal contraction $[\![\mathcal{A}]\!]$ of $\mathcal{A}$ is a normal model of Δ^* and Δ_0.

Our proof of Theorem 5.19 was given for a countable language. The proof for an uncountable language of cardinality λ would be along essentially the same lines. A finite or countable language has countably infinitely many sentences, and correspondingly a language with λ symbols has the same number λ of sentences. As the cardinal λ is an initial ordinal, we can list the sentences of such a language in order as $\{\phi_\alpha : \alpha \in \lambda\}$. We then construct sets Σ_α for $\alpha < \lambda$ corresponding to this listing taking Σ_0 to be Σ, defining $\Sigma_{\alpha+}$ from Σ_α and ϕ_α in the same way that Σ_{n+1} was defined from Σ_n and ϕ_n in the proof of Theorem 5.19, and the extra definition to cope with limit ordinals $\mu < \lambda$ of Σ_μ as $\bigcup_{\alpha<\mu} \Sigma_\alpha$. The set Σ' is defined essentially as in the proof of Theorem 5.19 to be $\bigcup_{\alpha<\lambda} \Sigma_\alpha$. Proving that each Σ_α and Σ' are consistent is virtually the same as in our earlier proof.

The proof of Theorem 5.20 stays the same for an uncountable language, but the statement of the theorem changes to saying that if the original language L has uncountable cardinality λ, adding the new constant symbols gives a language L^+ of the same cardinality λ. This is essentially because at most λ sentences of L begin with a $\exists$, so that the new language has cardinality $\lambda + \lambda$, which equals λ.

The interleaving of these results to expand Δ_0 into the complete set Δ^* in the enlarged language L^+ still only has to be done countably many times. This is because sentences are only finitely long, so that the process of 'witnessing' all existential quantifiers is completed after countably many passes through the cycle. Just as in the proof of the completeness theorem, a canonical structure $\mathcal{A}$ is constructed out of the set of closed terms of the enlarged language L^+, treating the $=$ symbol, if included in the language L, as any old 2-place relation symbol at this stage, so that the domain of $\mathcal{A}$ has cardinality λ. Theorem 5.18 can be used as it stands to show that $\mathcal{A}$ is a model for Δ^* and thus of the original consistent set Δ. Finally if L is a language with equality, then the normal contraction $[\![\mathcal{A}]\!]$ of $\mathcal{A}$ is a normal model of Δ, just as in Section 5.5 of Chapter 5. ∎

The members of an infinite cardinal λ fall into three sorts. There is the natural number 0 which is the least element of λ. Next are ordinals which are *successors*: for $\alpha, \beta \in \lambda$, β is called the successor of α if β is the least element of λ greater than α, in which case β is written as α^+. The remaining elements μ of λ are called *limit* ordinals, which can be characterised as not equal to 0 and such that for any $\alpha < \mu$, one also has $\alpha^+ < \mu$.

Combining this with the completeness theorem for countable languages, we obtain the full version of the completeness theorem.

Theorem 6.9   Completeness theorem for an arbitrary language

For all sentences ϕ and sets Γ of sentences in a first-order language L, possibly uncountable, if $\Gamma \vDash \phi$ then $\Gamma \vdash \phi$.

Armed with this version of the completeness theorem, we can then prove the compactness theorem for an arbitrary language. We state it below. Its proof is identical to the one given in Section 6.1 except that the reference made

Note, just to tie these results with our earlier comments on the set theoretic background, that both the completeness and compactness theorems for arbitrary languages are equivalent to AC. In the proof of Theorem 6.8, AC is needed at the stage of listing the sentences of the language as $\{\phi_\alpha : \alpha \in \lambda\}$.

to the completeness theorem should now be to the one above, which covers uncountable as well as countable languages.

Theorem 6.10 Compactness theorem for an arbitrary language

Let Γ be a set of sentences in a first-order language L, possibly uncountable. Then if every finite subset of Γ has a model, so does Γ.

Our proof of the next theorem, the downward version of the Löwenheim–Skolem theorem, relies on details within the proof of the completeness theorem for uncountable languages above.

We give a sketch of Skolem's original proof of the theorem, done (in 1920) some years before the completeness theorem was proved, at the end of this section.

Theorem 6.11 Downward Löwenheim–Skolem theorem

Let T be a theory in a first-order language L with equality of cardinality λ (meaning that the set of symbols of L has this cardinality, which could be finite). If T has a model, then T has a model of cardinality at most $\max\{\lambda, \aleph_0\}$.

Proof

We shall need the following detail which can be extracted from the end of our proof of Theorem 6.8 for an uncountable language of cardinality λ. The domain of the canonical structure $\mathcal{A}$ consists of the closed terms of the language L^* and has cardinality λ. As the complete set Δ^* in the language L^* may include sentences saying that some of these closed terms are equal, the cardinality of the normal contraction $[\![A]\!]$ of $\mathcal{A}$ could be smaller than λ, so that this model has cardinality at most λ.

Now to prove the downward Löwenheim–Skolem theorem! Suppose that T has a model. Then by the soundness theorem T is consistent. If the language L has only finitely or countably many symbols, Theorem 5.21 tells us that T has a countable model, that is, one of cardinality at most $\aleph_0$. If the language L is uncountable, our initial remark in this proof shows that T has a model of cardinality at most λ. Thus T has a model of cardinality at most $\max\{\lambda, \aleph_0\}$. ∎

The soundness theorem doesn't depend on the cardinality of the language.

The upward and downward Löwenheim–Skolem theorems can be combined to give something that we can call the full Löwenheim–Skolem theorem:

Theorem 6.12 Full Löwenheim–Skolem theorem

Let T be a theory in a language L with equality of cardinality λ. If T has an infinite model and κ is any cardinal with $\max\{\lambda, \aleph_0\} \leq \kappa$, then T has a model of cardinality precisely κ.

Proof

Suppose that T has an infinite model and $\max\{\lambda, \aleph_0\} \leq \kappa$. Add to the language a set of κ new constant symbols $\{\mathbf{c}_\alpha : \alpha \in \kappa\}$ and define Γ by

$$\Gamma = T \cup \{\neg\, \mathbf{c}_\alpha = \mathbf{c}_\beta : \alpha \neq \beta,\ \alpha, \beta \in \kappa\}.$$

As T has an infinite model, the upward Löwenheim–Skolem theorem says that it also has a model of cardinality at least κ, so that Γ has a model. Then by the downward Löwenheim–Skolem theorem Γ also has a normal model of cardinality at most $\max\{\kappa, \aleph_0\}$; but the inequalities between the $\mathbf{c}_\gamma$s in Γ guarantees that this model has size at least κ, so it must have cardinality exactly κ. $\blacksquare$

So there are groups and fields of all infinite cardinalities – both theories can be axiomatized in finite languages, i.e. λ is finite, so the $\max\{\lambda, \aleph_0\}$ in the statement of the full Löwenheim–Skolem theorem equals $\aleph_0$, and the κ can thus be any infinite cardinal.

We can now explain a remark we made in Section 5.5 of Chapter 5 alongside the proof of Theorem 5.18. This theorem essentially showed that a special structure $\mathcal{A}$ satisfied a complete set of sentences Δ^* and the remark was that this structure was in general not unique. You can now see why this is so. If Δ^* has an infinite model, then it has models of all infinite cardinalities, rather than a unique model.

As a by-way, the next exercise gets you to prove that the full Löwenheim–Skolem theorem implies a statement about infinite cardinals equivalent to the axiom of choice, giving one of the links in a proof of the equivalence of AC to all sorts of things via the route

$$\mathrm{AC} \ \Rightarrow\ \text{completeness} \ \Rightarrow\ \text{compactness}$$
$$\Rightarrow\ \text{Löwenheim–Skolem} \ \Rightarrow\ \ldots\ \Rightarrow\ \mathrm{AC}.$$

Exercise 6.21 ⸺

One equivalent of AC is the statement that $\kappa \cdot \kappa = \kappa$ for all infinite cardinals κ. As κ is a set, this really means that there is a bijection between $\kappa \times \kappa$ and κ for each of these infinite sets κ. Let L be the language with equality and a 2-place function symbol f. By considering models of the sentence

$$(\forall x \forall y \forall x' \forall y'\,(f(x, y) = f(x', y') \to (x = x' \land y = y')) \land \forall z \exists x \exists y\, f(x, y) = z),$$

show that the full Löwenheim–Skolem theorem implies this equivalent of AC. [*Hint*: Think about whether the sentence has a countably infinite model.]

See e.g. Goldrei [16] for a proof that this is equivalent to AC.

The Löwenheim–Skolem theorems have a variety of positive applications, showing for many theories the existence of infinite models of given cardinalities, as in the following exercise.

Exercise 6.22

Let L be a language with equality and binary function symbols $+$ and $\cdot$ and constant symbols $\mathbf{0}$ and $\mathbf{1}$. Let p be a prime number.

(a) Write down axioms in L for the theory of fields of characteristic p.

(b) It can be shown that for each positive integer n there is exactly one field (up to isomorphism) of characteristic p with p^n elements. Use this information to show that there is a countably infinite field of characteristic p. [*Hints*: First show that there is an infinite model of the theory and then use the downward Löwenheim–Skolem theorem to get a countable model.]

Actually, there is a straightforward direct construction of a field as required in (b). Let $\mathbb{Z}_p[t]$ be the ring of polynomials in t with coefficients in $\mathbb{Z}_p$. Then the set of quotients

$$\{f/g : f, g \in \mathbb{Z}_p[t],\ g \neq 0\}$$

with the natural addition and multiplication of rational functions is a countably infinite field of characteristic p. (The $g \neq 0$ above means g is not the zero polynomial.)

The Löwenheim–Skolem theorems also have a variety of negative applications, in the sense of showing that certain mathematical theories cannot be axiomatized in a first-order language. For instance, the theory of countably infinite groups cannot be axiomatized in this way – any such axioms would have infinite models, for instance the group $\langle \mathbb{Z}, +, = \rangle$ of integers under addition, so by the upward Löwenheim–Skolem theorem would also have uncountable models, which cannot be models of the desired theory, simply because we insisted that the groups should be countably infinite. Such theories are perhaps just a tad artificial, but the Löwenheim–Skolem theorems can be used to show that some very important theories are not first-order axiomatizable. These are well-known theories which have the properties that all models are isomorphic and the one model (up to isomorphism) is infinite. This would appear to contradict the upward Löwenheim–Skolem theorem, as follows: if the cardinality of this one infinite model is κ, the upward Löwenheim–Skolem theorem says that there is a model of cardinality at least 2^κ; but this model cannot be isomorphic to the unique model as they are of different cardinalities – an isomorphism of models is, amongst other things, a bijection between their domains. From this contradiction one concludes that these theories cannot be axiomatized in a first-order language. What sort of theories come into this category? The most interesting ones are the well-known axiomatizations of the natural numbers and of the real numbers.

A theory for which all its models are isomorphic is called *categorical*.

The natural numbers can be axiomatized as follows by *Peano's axioms*.

The Italian mathematician Giuseppe Peano (1858–1932) introduced much of the notation used in modern mathematical logic. Peano's axioms, along with a proof that any two models of them are isomorphic, can be found in many books, e.g. Goldrei [16].

Peano's axioms for the natural numbers

X is a set with a special element $0_X \in X$ and a function $s \colon X \longrightarrow X$ such that the following also hold:

1. the function S is one–one, i.e. for all $x, y \in X$, if $S(x) = S(y)$ then $x = y$;

2. for all $x \in X$, $0_X \neq S(x)$;

3. for all subsets $A \subseteq X$, if A contains 0_X and contains $S(x)$ whenever $x \in A$, then A is all of X.

This axiom is the principle of mathematical induction.

This set of axioms can be shown to be categorical and has the infinite model $\mathbb{N}$ with the usual 0 and successor function. Our argument above using the upward Löwenheim–Skolem theorem shows that there is no such axiomatization of the natural numbers in a first-order language. So where do Peano's axioms fail to be first-order? The problem lies with the final axiom, the induction principle, which might be written symbolically as

$$\forall A((\mathbf{0} \in A \ \wedge \ \forall x(x \in A \ \rightarrow \ s(x) \in A) \ \rightarrow \ \forall x \ x \in A).$$

There's nothing wrong with $\in$ as a binary relation symbol, s as a unary function symbol and $\mathbf{0}$ as a constant symbol. But this sentence involves quantification not only over elements x of the domain of any interpretation (the $\forall x$), but also over subsets A of the domain (the $\forall A$). This sort of sentence belongs in a second-order language, which allows quantification over subsets, rather than a first-order language.

The reason that the usual axioms for well-ordered sets are not all first-order, which we asked you to explain in Exercise 6.17 in the last section, is the same: the well-ordering property conceals quantification over (non-empty) subsets.

Exercise 6.23

Axioms for the real numbers which are categorical are given in Chapter 1, the introduction to this book. Which ones fail to be first-order?

Solution

The only axiom which is not first-order is axiom 16, the completeness axiom: any non-empty subset A of S which is bounded above has a least upper bound in S. The 'bounded above' and 'has a least upper bound' can be dealt with in a suitable first-order language, but the 'any non-empty subset' requires quantification over subsets as well as elements.

Exercise 6.24

Show that it is impossible to give axioms in a first-order language which describe $\mathbb{R}$ in the way as the standard axioms given in Chapter 1.

A consequence of the downward Löwenheim–Skolem theorem, called *Skolem's paradox*, concerns the Zermelo–Fraenkel axiomatization of set theory, ZF. This axiomatization is done using a first-order language with equality and the 2-place relation symbol $\in$ and has an infinite model (or so we believe!). So by the downward Löwenheim–Skolem theorem ZF has a countable model. But this seems to contradict results like Cantor's theorem which, given that there are some infinite sets, entails the existence of uncountable sets, e.g. once $\mathbb{N}$ is a set, $\mathscr{P}(\mathbb{N})$ is an uncountable set. Surely in any model of the ZF axioms, there should be some uncountable sets: so how can a model only be countable? The resolution of the paradox is rather subtle. The statement $\mathbb{N} \prec \mathscr{P}(\mathbb{N})$ says that there is a one–one function from $\mathbb{N}$ into $\mathscr{P}(\mathbb{N})$, but that there is no bijection between the sets: the one–one function and any possible bijection would themselves be sets, namely sets of ordered pairs. All of these statements could be true in a countable model, provided the domain contains elements corresponding to $\mathbb{N}$ and $\mathscr{P}(\mathbb{N})$, an element which the model thinks is a set of pairs representing a one–one function from $\mathbb{N}$ to $\mathscr{P}(\mathbb{N})$ and no element which it thinks codes a bijection between the sets.

This is a considerable simplification, of course! The model also has to recognize certain elements of the domain as subsets of its representation of $\mathbb{N}$ and then ensure these elements are in its representation of $\mathscr{P}(\mathbb{N})$, and so on.

To end the section, it is interesting to see how Skolem proved the downward Löwenheim–Skolem theorem well before the completeness theorem for predicate calculus, which is what our proof above exploits, had been proved. We shall sketch the proof for a theory T in a countable language L. We suppose that T has an infinite model $\mathcal{A}$ and want to show T has a countable model.

For each sentence $\psi \in T$ we construct a Skolem form ψ^{Sk} for ψ, as discussed at the end of Section 4.3 of Chapter 4, making sure that the function and constant symbols added to the language to produce each ψ^{Sk} are distinct from each other. By Theorem 4.6 of Section 4.3, for each $\psi \in T$ there is an expansion $\mathcal{A}^*$ of $\mathcal{A}$, adding interpretations of the extra function and constant symbols, which satisfies ψ^{Sk}. As the extra symbols for each ψ^{Sk} are distinct from each other, these expansions of $\mathcal{A}$ don't conflict with each other over the interpretations of these symbols. This means that these expansions can be bundled together to give a single expansion $\mathcal{A}^{\mathrm{Sk}}$ of $\mathcal{A}$ for the language L^{Sk} obtained by adding all the new function and constant symbols to L such that $\mathcal{A}^{\mathrm{Sk}}$ satisfies ψ^{Sk} for all $\psi \in T$.

> As we mentioned in Section 4.3, this requires use of AC.

Now let $\mathcal{B}$ be any substructure of $\mathcal{A}^{\mathrm{Sk}}$, so $\mathcal{B}$ is a structure for the language L^{Sk}. As each ψ^{Sk} is a universal sentence, Exercise 4.7(b) of Section 4.5 of Chapter 4 tells us that $\mathcal{B}$ satisfies ψ^{Sk}. Then as $(\psi^{\mathrm{Sk}} \to \psi)$ is universally valid (also by Theorem 4.6), this means that $\mathcal{B}$ satisfies ψ for each $\psi \in T$, so that $\mathcal{B}$ is a model of T.

All we need to do now is investigate possible cardinalities of the substructures $\mathcal{B}$. First note that as the original language L is countable, and any sentence of L is a finite string of symbols in L, there are at most countably many sentences ψ in T. As production of the prenex normal form and then the Skolem form of each ψ adds only finitely many extra function and constant symbols, there are at most countably many extra function and constant symbols added to L to give the language L^{Sk}. Now let C be any countable subset of the infinite domain A of both the structures $\mathcal{A}^{\mathrm{Sk}}$ and $\mathcal{A}$. As there are at most countably many terms in any countable language (as each term is just a finite string of symbols), the substructure $\mathcal{B}$ of $\mathcal{A}^{\mathrm{Sk}}$ generated by C is countable.

Putting it together, if the theory T has an infinite model $\mathcal{A}$, it also has a countable model.

Further exercises

Exercise 6.25

Show that none of the following theories can be axiomatized in a countable first-order language.

(a) The theory of all uncountable sets.

(b) The theory of countable linearly ordered sets.

(c) The theory of vector spaces over the field $\mathbb{R}$ of real numbers.

(d) The theory of finite-dimensional vector spaces over the finite field $\mathbb{Z}_p$, where p is a prime number.

> To do parts (c) and (d) of this exercise, one needs some elementary knowledge of vector spaces.

6.5 New models from old ones

The compactness theorem has many positive consequences. One is to show
the existence of models of a theory T with some special property, using the
method of adding new constant symbols and new sentences involving them.
We have already seen several examples of this method, for instance in the
proof of the upward Löwenheim–Skolem theorem where the special property
is having a certain minimum size, and we shall look at further applications of
it in this section.

An outline of the method of adding new constant symbols is as follows. We
have a theory T in a first-order language L with some known models and
we want to show that T has some sort of special model. This can often be
achieved by adding new constant symbols to L and (usually infinitely many)
sentences involving them to T, so that any model of this extension of T is
what's required. If any finite subset of the extended theory can be shown to
have a model, probably using one of the known models, then the compactness
theorem tells us that the special model exists.

In this section we're interested in
theories which *are* first-order
axiomatizable. In Section 6.3 we
were essentially using this
technique to show that some
mathematical theories were *not*
axiomatizable: we assumed the
theory could be axiomatized,
added new constants to the
language and sentences involving
them in a clever way so that any
model of them could not also be a
model of the theory, and arrived at
a contradiction via the
compactness theorem.

In the proof of the upward Löwenheim–Skolem theorem, Theorem 6.7, enough
new constant symbols, namely κ of them, are added to L and sentences saying
that they are unequal are added to T. As T has an infinite model, it is always
possible to find a model for any finite subset of the extended theory. So by
the compactness theorem the extended theory has a model, which is thus a
model for T with at least κ distinct elements.

See if you can apply this technique in the next few exercises.

Exercise 6.26 ————————————————————————

Let L be a language with equality including a binary function symbol $\cdot$ and
constant symbol $\mathbf{c}$.

(a) Let T be a theory in this language with axioms including those for groups
such that the models of T include all the finite cyclic groups. Show
that T has a model which is a group with an element of infinite order.
[*Hint*: Use a new constant symbol $\mathbf{c}$ and introduce sentences to force any
interpretation of $\mathbf{c}$ to have infinite order.]

(b) Use the result above to show that the theory of groups where every element has finite order is not axiomatizable.

(c) This part is a digression from the theme of this section, but useful practice!
Take the same theory T as in part (a). Devise suitable sentences involving
no new constant symbols which when added to T have an infinite model
in which every non-identity element has infinite order. (As ever, you'll
probably need the compactness theorem to show that this model exists.)

The set-up here has a great deal in
common with Exercise 6.16 in
Section 6.3. The main point of the
current exercise is in part (a), using
the compactness theorem in a
positive way to show the existence
of a special model of the theory.

This model will be a torsion-free
group.

Let's move to an application of the technique of some mathematical significance. This is cast in the form of an exercise.

Exercise 6.27

Let OF be the theory of ordered fields expressed in the language L with equality containing 2-place function symbols $+$ and $\cdot$, 1-place function symbols $-$ and $^{-1}$, constant symbols $\mathbf{0}$ and $\mathbf{1}$ and the 2-place relation symbol $<$.

(a) Prove that any model of OF has characteristic 0. [*Hint*: You may find the logical consequence of T that $\mathbf{0} < \mathbf{1}$ useful. This was the result of Exercise 4.93(a)(i) in Section 4.4.]

(b) Use the method of adding new constant symbol(s) to show that there exists a model $\mathcal{A}$ of OF in which there is an element c with $0 <_{\mathcal{A}} c$ such that for no natural number n does $c <_{\mathcal{A}} \mathbf{n}^{\mathcal{A}}$ hold in $\mathcal{A}$. (We are using $\mathbf{n}$ as a shorthand for $\underbrace{(\mathbf{1} + \mathbf{1} + \ldots \mathbf{1})}_{n}$, so that $\mathbf{n}^{\mathcal{A}}$ is the interpretation in $\mathcal{A}$ of this term.)

(c) Deduce that in the model $\mathcal{A}$ of part (b) there is an element d such that $\mathbf{0} <_{\mathcal{A}} d <_{\mathcal{A}} (\mathbf{n}^{\mathcal{A}})^{-1}$ holds in $\mathcal{A}$ for all natural numbers n.

Axioms for this theory were given on page 205 in Section 4.4 of Chapter 4.

Although the theory of fields of characteristic 0 cannot be finitely axiomatized, it is of interest to see how an extension of the field axioms by finitely many axioms can entail the characteristic 0 property.

If you have studied real analysis, you should have met the *Archimedean property* of the fields $\mathbb{Q}$ and $\mathbb{R}$:

> if $a > 0$ and b is any member of the field, then there is some $n \in \mathbb{N}$ such that $n \cdot a > b$.

Exercise 6.27 shows that there is an ordered field $\mathcal{A}$ which does not have this property (by taking $\mathbf{1}^{\mathcal{A}}$ as a and c as b). We call $\mathcal{A}$ a *non-Archimedean ordered field*. If we identify each element $\mathbf{n}^{\mathcal{A}}$ with the natural number n, we could then regard the element c of $\mathcal{A}$ as an infinite element, regarding an element a as finite if there is some $n \in \mathbb{N}$ for which $-\mathbf{n}^{\mathcal{A}} <_{\mathcal{A}} a <_{\mathcal{A}} \mathbf{n}^{\mathcal{A}}$. It is easy and fun to see that $\mathcal{A}$ contains many more infinite elements, e.g. $c - 1$ and c^2. An element like d in part (c) of Exercise 6.27 is called an *infinitesimal* element.

The result of Exercise 6.27 is of considerable interest. In the original formulation of the differential and integral calculus, vital use was made of infinitesimals, meaning infinitely small quantities. Although many powerful and now familiar results were obtained thereby, this use of the infinitely small led to confusion and, arguably, to contradictions. The process of making calculus more rigorous in the 19th century resulted in the banishment of infinitesimals – definitions of various forms of limit in real analysis are all couched in terms of ordinary real numbers, albeit arbitrarily small positive ones. The existence of infinitesimals shown in Exercise 6.27 within a modern mathematical framework enabled the renewal of the study of the calculus using infinitesimals in a subject called *non-standard analysis*.

See for instance Robinson [26] or [27], works by the founder of the theory of non-standard analysis, Abraham Robinson (1918–1974).

The basis of a further rich vein of modern model theory in the study of the arithmetic of the natural numbers is given in the following exercise.

Exercise 6.28

Let L be a language with equality, binary function symbols $+$ and $\cdot$ and constant symbols $\mathbf{0}$ and $\mathbf{1}$, suitable for expressing statements about the arithmetic of $\mathbb{N}$, the natural numbers. Let $\mathcal{N}$ be the structure with domain $\mathbb{N}$ and the usual interpretations in $\mathbb{N}$ of these symbols: $\mathcal{N}$ is called the *standard model* of this arithmetic. Let T be the set of all sentences in L true in $\mathcal{N}$, giving a theory we shall call *complete arithmetic*. Show that there are countable models (described as *non-standard*) of T which are not isomorphic to $\mathcal{N}$. [*Hints*: Every element of $\mathbb{N}$ in the standard model is of the form $\underbrace{1 + 1 + 1 + \ldots + 1}_{n}$

for $n \in \mathbb{N}$. How would you set about creating a model in which there is some element not of this form? Once you have such a model, don't forget the detail that a countable model is required!]

The set of sentences true in a structure is often of great interest, so merits a definition and some notation.

Definition   *Theory of a structure*

Let $\mathcal{A}$ be a structure for a language L. The *theory of* $\mathcal{A}$, written as $\mathrm{Th}(\mathcal{A})$, is the set of sentences of L true in $\mathcal{A}$.

So taking L to be the language with equality and $+, \cdot, \mathbf{0}, \mathbf{1}$ as in Exercise 6.28 and $\mathcal{N}$ the standard model of arithmetic, $\mathrm{Th}(\mathcal{N})$ is the theory complete arithmetic.

Exercise 6.29

Let L be the language with equality, symbols $+, \cdot, \mathbf{0}, \mathbf{1}$ as above and the binary relation symbol $<$ and let $\mathcal{R}$ be the structure with domain $\mathbb{R}$, the set of real numbers, and the usual interpretations in $\mathbb{R}$ of these symbols. Show that $\mathrm{Th}(\mathcal{R})$ has a model of the same cardinality as the continuum $\mathbb{R}$ which is not isomorphic to $\mathcal{R}$.

A different sort of application of the compactness theorem is as follows. We shall show that a strict partial order $<$ on a set A can be extended to a strict linear order $<'$ on the same set; that is, $<'$ is a strict linear order on A and for all $a, b \in A$, if $a < b$ then $a <' b$.

We first proved this result in Section 3.3 of Chapter 3 using the compactness theorem for propositional calculus. But the result is more naturally proved within the framework of predicate calculus.

It can be shown that this result holds for any finite partially ordered set. We shall use this fact to show that it holds for any (in particular infinite) partially ordered set. The argument will exploit the compactness theorem.

Let $\mathcal{A} = \langle A, <_{\mathcal{A}}, = \rangle$ be a strict partial order. We add to the language the set $\{\mathbf{a} : a \in A\}$ of new distinct constant symbols corresponding to the elements of A. Let Σ be the set of sentences

$$\Sigma = \mathrm{SLO} \cup \{\mathbf{a} < \mathbf{b} : a <_{\mathcal{A}} b\} \cup \{\neg\, \mathbf{a} = \mathbf{b} : a \neq b\},$$

where SLO is the set of axioms for a strict linear order.

With this notational convention using boldface symbols, the constant symbols corresponding to distinct elements a, a_2, b of A are, respectively, $\mathbf{a}, \mathbf{a_2}, \mathbf{b}$.

We shall first show that any finite subset of Σ has a model. Any such subset Δ involves at most finitely many of the new constant symbols, let's say $\mathbf{a_1}, \mathbf{a_2}, \ldots, \mathbf{a_n}$, corresponding to the elements $a_1, a_2, \ldots, a_n$ in a finite subset A' of A. The sentences involving these constant symbols in Δ must be a subset of

$$\{\mathbf{a} < \mathbf{b} : a <_{\mathcal{A}} b, \text{ where } a, b \in A'\} \cup \{\neg\,\mathbf{a} = \mathbf{b} : a \neq b, \text{ where } a, b \in A'\}.$$

These latter sentences effectively describe the substructure $\mathcal{A}'$ of $\mathcal{A}$ generated by the subset A'. As the axioms of a strict partial order are universal, the result of Theorem 4.7(b) of Chapter 4 tells us that $\mathcal{A}'$ is also a strict partial order (ordered by the restriction of $<_{\mathcal{A}}$ to A'). We are given that any finite partial order can be extended to a linear order on the same set, so that taking this extension for the finite set A' we obtain a model for the given finite subset Δ of Σ.

It follows by the compactness theorem that Σ has a model

$$\mathcal{C} = \langle C, <_{\mathcal{C}}, =, \{\mathbf{a}^{\mathcal{C}} : a \in A\}\rangle,$$

which is thus a strict linear order. We shall exploit this model to show that the order $<_{\mathcal{A}}$ on A can be extended to a linear order on A. We define a relation $<^*$ on A by

$$a <^* b \quad \text{if and only if} \quad \mathbf{a}^{\mathcal{C}} <_{\mathcal{C}} \mathbf{b}^{\mathcal{C}},$$

for all $a, b \in A$. Note that $<^*$ extends the original partial order $<_{\mathcal{A}}$, as if for any $a, b \in A$, $a <_{\mathcal{A}} b$, then

$$\mathbf{a} < \mathbf{b} \in \Sigma,$$

so that as $\mathcal{C}$ is a model of Σ

$$\mathbf{a}^{\mathcal{C}} <_{\mathcal{C}} \mathbf{b}^{\mathcal{C}},$$

so that by definition

$$a <^* b.$$

> Note that $\mathcal{C}$ is a structure for the bigger language including the extra constant symbols.

But the definition also guarantees that the structure $\langle A, <^*, =\rangle$ is isomorphic to the substructure of $\mathcal{C}$ generated by the subset $\{\mathbf{a}^{\mathcal{C}} : a \in A\}$ consisting of the interpretations in $\mathcal{C}$ of the new constant symbols. As the language has no function symbols, the domain of this substructure is simply $\{\mathbf{a}^{\mathcal{C}} : a \in A\}$ itself. As the axioms of the theory of strict linear order are universal, the substructure is a strict linear order, again using Theorem 4.7(b) of Chapter 4. Thus $<^*$ extends the original partial order on A to a linear order, as required.

> The sentences $\{\neg\,\mathbf{a} = \mathbf{b} : a \neq b\}$ were included in Σ to ensure that $a \longmapsto a^{\mathcal{C}}$ is a bijection (needed for the isomorphism), by preventing the interpretation of $\mathbf{a}, \mathbf{b}$ in $\mathcal{C}$ for distinct a, b in A being the same element of C.

There are other examples of this sort, where we are given a structure $\mathcal{A}$ for a language L which is a model for a theory T and want to show that $\mathcal{A}$ has a particular property. If this property holds for all finitely generated substructures of $\mathcal{A}$, then by adding constants and clever sentences involving them, using compactness to find a model and then taking the right substructure of it, one can often show that the original structure also has this property. We shall look at one example, which is called a *colouring of a graph*.

For our purposes here, a *graph* is a non-empty set with a symmetric binary relation on it.

Exercise 6.30

Let L be a language with equality and a binary relation symbol E. Using the above description of a graph, give axioms for the theory of graphs.

Solution

We just need the one axiom to express the symmetry property:

$$\forall x \forall y (E(x, y) \leftrightarrow E(y, x)).$$

If $\mathcal{G} = \langle G, E^{\mathcal{G}}, = \rangle$ is a graph, it is customary in graph theory to call an element of G a *vertex* of the graph, and if $(a, b) \in E^{\mathcal{G}}$ (so that also $(b, a) \in E^{\mathcal{G}}$) to say that the vertices a, b are *connected by an edge*. If you have never met any graph theory before, this might seem rather dull so far. But imagine that the elements of G are countries on a map, for instance of Europe, and that $(a, b) \in E^{\mathcal{G}}$ whenever the countries a and b share a common boundary. This is the background for the Four Colour Conjecture, now proved, which says that each country on the map can be coloured with one of four colours in such a way that countries with a common boundary have a different colour. As an abbreviation, we shall say that the corresponding graph is *4-colourable*.

See Exercise 6.36 for more about graphs which do represent maps, as not every graph does represent a map. But every map gives a graph! For more about graph theory, see the excellent introduction by Wilson [30]. For a history of the Four Colour Conjecture and its proof, see Wilson [31].

Exercise 6.31

Suppose that the language L with equality and the binary relation symbol E now also includes the 1-place relation symbols C_1, C_2, C_3, C_4 and that the intended interpretation of $C_i(x)$ is that the vertex (or country) x has colour i, for $i = 1, 2, 3, 4$. Write down sentences to express that a graph is 4-colourable.

Solution

One of the sentences needed might be $\forall x \forall y ((E(x, y) \wedge C_2(x)) \rightarrow \neg C_2(y))$. We leave the rest to you.

Now for the example in the spirit of this section, given as an exercise! Note that if $\mathcal{G} = \langle G, E^{\mathcal{G}}, = \rangle$ is a graph, then as the axiom for the theory of graphs is universal, every substructure of $\mathcal{G}$ is also a graph, normally called a *subgraph* of $\mathcal{G}$.

For any subset H of G, the subgraph generated by H has domain H.

Exercise 6.32

Let $\mathcal{G} = \langle G, E^{\mathcal{G}}, = \rangle$ be a graph with the property that every finite subgraph is 4-colourable. Show that $\mathcal{G}$ is 4-colourable. [*Hints*: Take the language with equality and the relation symbols E, C_1, C_2, C_3, C_4 and add new distinct constant symbols for each element of G. Now construct a set Σ of sentences including axioms for the theory of 4-colourable graphs, sentences $E(\mathbf{a}, \mathbf{b})$ whenever $(a, b) \in E^{\mathcal{G}}$ and sentences $\neg \mathbf{a} = \mathbf{b}$ for each $a \neq b$ in G. Show that Σ has a model and use this model to show that the original graph $\mathcal{G}$ is 4-colourable.]

The idea of adding new constant symbols for each element of a structure $\mathcal{A}$ and using these symbols to write sentences which express some of the properties of the corresponding elements of $\mathcal{A}$ is very fruitful in model theory. One way of doing this is given in the following definition.

Definition   Diagram of a structure

Let $\mathcal{A} = \langle A, \ldots \rangle$ be a structure for a language L with equality. Let $L(\mathcal{A})$ be the language obtained by adding to L new distinct constant symbols for each element of the domain A. The *diagram* of $\mathcal{A}$, written as $\Delta(\mathcal{A})$, is the set consisting of all sentences in the language $L(\mathcal{A})$ of the form

$$\phi(\mathbf{a_1}, \mathbf{a_2}, \ldots, \mathbf{a_n})$$

where ϕ is atomic and $\mathcal{A} \vDash_{x_1/a_1, x_2/a_2, \ldots, x_n/a_n} \phi(x_1, x_2, \ldots, x_n)$ and also all sentences of the form

$$\neg\phi(\mathbf{a_1}, \mathbf{a_2}, \ldots, \mathbf{a_n})$$

where ϕ is atomic and $\mathcal{A} \not\vDash_{x_1/a_1, x_2/a_2, \ldots, x_n/a_n} \phi(x_1, x_2, \ldots, x_n)$.

We shall use the same convention with boldface symbols as earlier, e.g. $\mathbf{a}, \mathbf{a_2}, \mathbf{b}$ for distinct elements a, a_2, b of A.

An alternative way of describing the diagram of $\mathcal{A}$, $\Delta(\mathcal{A})$, is as the set of all atomic sentences and negated atomic sentences of the language $L(\mathcal{A})$ true in the expansion of $\mathcal{A}$ obtained by interpreting the extra constant symbols by the corresponding elements of A. So, for instance, if L is the language with equality with the binary symbol $<$ and $\mathcal{A} = \langle A, <_{\mathcal{A}}, = \rangle$ is a structure for L, the diagram of $\mathcal{A}$ would be the set

$$\{\mathbf{a} < \mathbf{b} : (a, b) \in \,<_{\mathcal{A}}\} \cup \{\mathbf{a} = \mathbf{a} : a \in A\}$$
$$\cup \{\neg\, \mathbf{a} < \mathbf{b} : (a, b) \notin \,<_{\mathcal{A}}\} \cup \{\neg\, \mathbf{a} = \mathbf{b} : a \neq b\}.$$

This set includes the sentences which we used in our earlier example about extending a partial order into a linear order.

The main point of the definition is given by the following theorem, which says that any structure satisfying the diagram of $\mathcal{A}$ must contain a substructure which is isomorphic to $\mathcal{A}$.

Theorem 6.13

Let $\mathcal{A}, \mathcal{B}$ be structures for the language L with equality with domains A, B respectively. Let $\{b_a : a \in A\}$ be elements of B such that the expansion $\mathcal{B}^*$ of $\mathcal{B}$ to the language $L(\mathcal{A})$ obtained by interpreting each new constant symbol $\mathbf{a}$ by b_a (for the $a \in A$ corresponding to $\mathbf{a}$) is a model for $\Delta(\mathcal{A})$, the diagram of $\mathcal{A}$. Then $\mathcal{A}$ is isomorphic to the substructure $\mathcal{C}$ of $\mathcal{B}$ generated by $\{b_a : a \in A\}$.

Proof

We shall sketch a proof.

First note that the domain of the substructure $\mathcal{C}$ is the set $\{b_a : a \in A\}$ itself, as we can show that $\{b_a : a \in A\}$ is closed under the interpretation of any function and constant symbols of the original language L. For instance, if f is a 2-place function symbol in L and $a_1, a_2, a_3 \in A$ with $a_3 = f^{\mathcal{A}}(a_1, a_2)$, then the set includes elements $b_{a_1}, b_{a_2}, b_{a_3}$. As one of the atomic sentences in the diagram $\Delta(\mathcal{A})$ is $\mathbf{a_3} = f(\mathbf{a_1}, \mathbf{a_2})$ and $\mathcal{B}^*$ is a model of the diagram, we have

$b_{a_3} = f^{\mathcal{B}}(b_{a_1}, b_{a_2})$. So the set $\{b_a : a \in A\}$ is closed under the function $f^{\mathcal{B}}$. Likewise, if $\mathbf{c}$ is a constant symbol of the original language L interpreted in $\mathcal{A}$ by a, then one of the atomic sentences in the diagram $\Delta(\mathcal{A})$ is $\mathbf{a} = \mathbf{c}$ for some $a \in A$, and the corresponding element b_a interprets the constant symbol $\mathbf{c}$ in $\mathcal{B}$.

Now we know the domain of the substructure $\mathcal{C}$, the obvious candidate for the isomorphism is the map

$$\theta \colon A \longrightarrow \{b_a : a \in A\}$$
$$a \longmapsto b_a$$

We need θ to be a bijection (property (i) in the definition of an isomorphism). It is plainly onto. As for the one–one property, if $a_1, a_2 \in A$ with $a_1 \neq a_2$, then $\Delta(\mathcal{A})$ includes $\neg\, \mathbf{a_1} = \mathbf{a_2}$ (which is the negation of an atomic sentence), so that as $\mathcal{B}^*$ is a model of $\Delta(\mathcal{A})$, $b_{a_1} \neq b_{a_2}$, so that $\theta(a) \neq \theta(a_2)$. Thus θ is one–one.

From our earlier discussion of the interpretation of the 2-place function symbol f we have

$$\theta(f^{\mathcal{A}}(a_1, a_2)) = b_{a_3}, \quad \text{where } a_3 = f^{\mathcal{A}}(a_1, a_2)$$
$$= f^{\mathcal{B}}(b_{a_1}, b_{a_2})$$
$$= f^{\mathcal{C}}(b_{a_1}, b_{a_2})$$
$$= f^{\mathcal{C}}(\theta(a_1), \theta(a_2)),$$

which gives an example to show that property (iii) of an isomorphism is satisfied. Likewise, it follows from our discussion of any constant symbol $\mathbf{c}$ of L that $\theta(\mathbf{c}^{\mathcal{A}}) = \mathbf{c}^{\mathcal{B}} = \mathbf{c}^{\mathcal{C}}$, so that property (iv) of an isomorphism is satisfied.

You will note that the mathematically very useful method of proof by example is not a feature of our formal proof system!

We leave checking an example of the remaining property of an isomorphism, for relations symbols, as an exercise for you. ∎

Exercise 6.33 ——————————————————————

Complete the sketch proof of Theorem 6.13 by showing that the function θ obeys property (iv) of an isomorphism in the case of a 3-place relation symbol R of the language L, that is, show that

$$(a_1, a_2, a_3) \in R^{\mathcal{A}} \quad \text{if and only if} \quad (\theta(a_1), \theta(a_2), \theta(a_3)) \in R^{\mathcal{C}}.$$

Exercise 6.34 ——————————————————————

Suppose that the theory T in a countable language L has a countably infinite model $\mathcal{A}$. Show that for any infinite cardinal κ, T has a model of cardinality κ containing a substructure isomorphic to $\mathcal{A}$.

———————————————————————————————

In Chapter 4 we proved that if a theory T can be axiomatized by universal sentences, then for any model of T, all of its substructures are also models of T. To conclude this section, we shall use the idea of a diagram to prove the converse of this theorem.

In Theorem 4.5(c).

Theorem 6.14

Let T be a theory in a language L with equality whose models are *closed under substructures*, that is, if $\mathcal{B}$ is a model of T and $\mathcal{A}$ is a substructure of $\mathcal{B}$, then $\mathcal{A}$ is a model of T. Then T can be axiomatized by a set of universal sentences in L.

Proof

We define a set $T_\forall$ of universal sentences by

$$T_\forall = \{\psi : \psi \text{ is a universal sentence and } T \vDash \psi\}.$$

We shall show that $T_\forall$ axiomatizes the theory T, that is, $\mathcal{A}$ is a model of T if and only if $\mathcal{A}$ is a model of $T_\forall$.

Plainly the definition of $T_\forall$ ensures that any model of T is a model of $T_\forall$. It's the other direction which needs the work! So suppose that $\mathcal{A}$ is a model of $T_\forall$. We must show that $\mathcal{A}$ is a model of T. Consider the set of sentences

$$\Sigma = T \cup \Delta(\mathcal{A})$$

in the language $L(\mathcal{A})$. If Σ has a model, then this model is the expansion of a structure $\mathcal{B}$ for the language L to the language $L(\mathcal{A})$ obtained by giving an interpretation b_a for each new constant symbol $\mathbf{a}$. Then by Theorem 6.13 the substructure of $\mathcal{B}$ generated by $\{b_a : a \in A\}$ is isomorphic to $\mathcal{A}$. Also as $T \subseteq \Sigma$ and the only symbols used in T are in the language L, $\mathcal{B}$ is a model of T. By supposition, any substructure of $\mathcal{B}$ is a model of T, so that as $\mathcal{A}$ is isomorphic to such a substructure, $\mathcal{A}$ is a model of T (by Theorem 4.8 of Chapter 4). Thus so long as Σ does have a model, we have proved that every model of $T_\forall$ is a model of T, proving the theorem.

We shall show that Σ does have a model by supposing it doesn't and obtaining a contradiction. If Σ has no models, then by the compactness theorem, there is a finite subset Σ' of Σ with no models. For this Σ'

$$T \cup \Sigma'$$

has no models. As Σ' is finite, it involves only finitely many sentences from $\Delta(\mathcal{A})$ which we can list as

$$\phi_1(\mathbf{a_1}, \mathbf{a_2}, \ldots, \mathbf{a_n}), \quad \phi_2(\mathbf{a_1}, \mathbf{a_2}, \ldots, \mathbf{a_n}), \quad \ldots, \quad \phi_r(\mathbf{a_1}, \mathbf{a_2}, \ldots, \mathbf{a_n}),$$

where the (finitely many) new constants involved in Σ' are included in the list $\mathbf{a_1}, \mathbf{a_2}, \ldots, \mathbf{a_n}$ and each $\phi_j(x_1, x_2, \ldots, x_n)$ is an atomic formula of L or its negation. Then

$$T \cup \{\phi_1(\mathbf{a_1}, \mathbf{a_2}, \ldots, \mathbf{a_n}), \quad \phi_2(\mathbf{a_1}, \mathbf{a_2}, \ldots, \mathbf{a_n}), \quad \ldots, \quad \phi_r(\mathbf{a_1}, \mathbf{a_2}, \ldots, \mathbf{a_n})\}$$

has no models. As none of the constant symbols $\mathbf{a_i}$ are in L, so none appear in a sentence in T, we can infer that

$$T \cup \left\{\exists x_1 \exists x_2 \ldots \exists x_n \bigwedge_{j=1}^{r} \phi_j(x_1, x_2, \ldots, x_n)\right\}$$

has no models, so that

$$T \vDash \neg \exists x_1 \exists x_2 \dots \exists x_n \bigwedge_{j=1}^{r} \phi_j(x_1, x_2, \dots, x_n),$$

or equivalently,

$$T \vDash \forall x_1 \forall x_2 \dots \forall x_n \neg \bigwedge_{j=1}^{r} \phi_j(x_1, x_2, \dots, x_n).$$

But each of the formulas ϕ_j is atomic or the negation of an atomic formula, so that the formula

$$\neg \bigwedge_{j=1}^{r} \phi_j(x_1, x_2, \dots, x_n)$$

is quantifier-free. Therefore

$$\forall x_1 \forall x_2 \dots \forall x_n \neg \bigwedge_{j=1}^{r} \phi_j(x_1, x_2, \dots, x_n)$$

is a universal sentence, and as it is a logical consequence of T, it is in the set $T_\forall$ and thus is true in $\mathcal{A}$. But as each $\phi_j(\mathbf{a_1}, \mathbf{a_2}, \dots, \mathbf{a_n})$ for $j = 1, 2, \dots, r$ is in the diagram of $\mathcal{A}$,   Recall that $\mathcal{A}$ is a model of T!

$$\mathcal{A} \vDash_{x_1/a_1, x_2/a_2, \dots, x_n/a_n} \phi_j(x_1, x_2, \dots, x_n)$$

for each j, so that

$$\mathcal{A} \vDash_{x_1/a_1, x_2/a_2, \dots, x_n/a_n} \bigwedge_{j=1}^{r} \phi_j(x_1, x_2, \dots, x_n),$$

contradicting that

$$\mathcal{A} \vDash \forall x_1 \forall x_2 \dots \forall x_n \neg \bigwedge_{j=1}^{r} \phi_j(x_1, x_2, \dots, x_n).$$

We can conclude that for each model $\mathcal{A}$ of $T_\forall$, Σ does have a model. ∎

The method of adding constants, the concept of diagram and, above all, the use of the compactness theorem are major jumping off points for the modern subject of model theory.

To read more about model theory, see e.g. Bridge [3], Cori and Lascar [8], Hodges [20], Marker [24] and Chang and Keisler [6].

Further exercises

Exercise 6.35 ____

Let $\mathcal{A} = \langle A, \dots \rangle$ be a structure for a countable language L with equality and let $L(\mathcal{A})$ be the language obtained by adding to L new distinct constant symbols for each element of the domain A. Let $\mathrm{Th}_{L(\mathcal{A})}(\mathcal{A})$ be the set of all sentences $\phi(\mathbf{a_1}, \mathbf{a_2}, \dots, \mathbf{a_n})$, where $\phi(x_1, x_2, \dots, x_n)$ is a formula of L with free variables in the list $x_1, x_2, \dots, x_n$ and

$$\mathcal{A} \vDash_{x_1/a_1, x_2/a_2, \dots, x_n/a_n} \phi(x_1, x_2, \dots, x_n).$$

Each model of $\mathrm{Th}_{L(\mathcal{A})}(\mathcal{A})$ has a substructure isomorphic to $\mathcal{A}$ and is called an *elementary extension* of $\mathcal{A}$.

Show that if A is infinite, then $\mathrm{Th}_{L(\mathcal{A})}(\mathcal{A})$ has models of all infinite cardinalities.

Exercise 6.36

The generalization of the idea of a graph which can represent a map of countries on the globe is called a *planar graph*. Crudely speaking, this means that
one can draw a diagram representing the vertices and edges of the graph in
a plane with any two edges meeting each other only at vertices common to
them both – an idea which makes sense at least for finite graphs! It can be
shown that a graph is planar if and only if it contains no subgraphs of one of
the following two types.

See e.g. Wilson [30].

1.   Five vertices with each pair of vertices connected by an edge;

This is called the *complete* graph K_5.

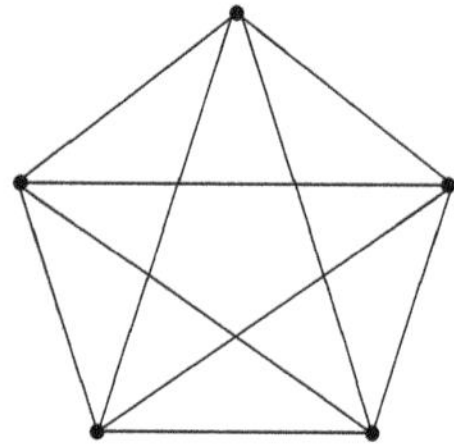

2.   Six vertices grouped into two sets A and B each with three vertices, such
     that each vertex in A and each vertex in B are connected by an edge, but
     the vertices in A are not connected to each other and the vertices in B
     are not connected to each other.

This is called the *complete bi-partite* graph $K_{3,3}$.

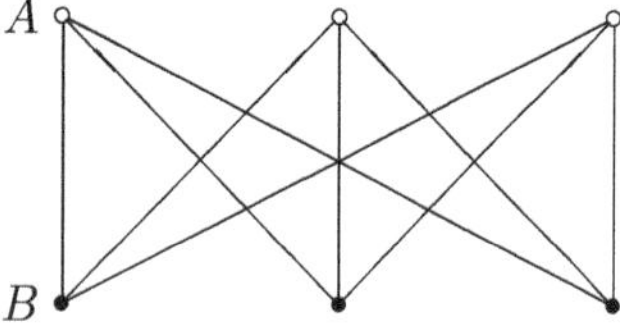

(a) Let L be a language with equality and a binary relation symbol E. Using
    the above result about planar graphs, give axioms for their theory.

(b) Let $\mathcal{G} = \langle G, E^{\mathcal{G}}, = \rangle$ be a graph with the property that every finite sub-
    graph is planar. Show that $\mathcal{G}$ is planar.

6.6   Decidable theories

We shall conclude the book by looking at the decidability of some first-order
theories. To say that a theory T in a language L is decidable means that
there is some algorithmic procedure which, given a sentence ϕ of L, decides
after a finite number of steps of this procedure whether or not ϕ is a logical
consequence of T. You can perhaps imagine how desirable it might seem that
interesting mathematical theories, for instance, the theory of the real numbers
or number theory, are decidable. Famously, these theories are not decidable,
but there are nevertheless some positive results which we shall discuss in this
section.

Our first positive results about decidability will stem from an important model-theoretic result, Vaught's test, for the statement of which we need some new definitions.

Definitions *κ-categorical, complete theory*

Let T be a consistent theory in a language L and κ an infinite cardinal. T is *κ-categorical* if all models of T of cardinality κ are isomorphic.

T is *complete* if for each sentence ϕ of L, exactly one of ϕ and $\neg\phi$ is a logical consequence (or equivalently a theorem) of T.

This use of the word 'complete' is very similar to its use in describing a complete set of sentences. The connection is that the logical consequences of a complete theory form a complete set of sentences.

Theorem 4.9 of Chapter 4 tells us that the theory of unbounded dense linear orders is $\aleph_0$-categorical. As we ask you to show as an exercise, a simple but important source of examples of a complete theory is obtained by taking, for any structure $\mathcal{A}$, the theory of $\mathcal{A}$, $\mathrm{Th}(\mathcal{A})$.

Exercise 6.37

Let $\mathcal{A}$ be a structure for a language L. Show that $\mathrm{Th}(\mathcal{A})$, the set of sentences of L true in $\mathcal{A}$, is complete.

This is why we can describe $\mathrm{Th}(\mathcal{N})$ as complete arithmetic.

Exercise 6.38

The theory of infinite sets is axiomatized by the set of sentences $\{\exists_{\geq n} : n \in \mathbb{N}\}$. For which infinite cardinals κ, if any, is this theory κ-categorical?

Theorem 6.15 *Vaught's test*

Let T be a consistent theory in a countable language L with no finite models. If all models of T of cardinality κ are isomorphic, for some infinite κ, then T is complete.

The result, first proved by the American logician Robert Vaught, applies more generally when the language has cardinality λ and κ is an infinite cardinal with $\kappa \geq \lambda$.

Proof

Let ϕ be any sentence of L. We need to show that either $T \vDash \phi$ or $T \vDash \neg\phi$.

As T is consistent, it has a model, so that we cannot have both $T \vDash \phi$ and $T \vDash \neg\phi$.

Suppose that neither $T \vDash \phi$ nor $T \vDash \neg\phi$. We shall show that this gives a contradiction. As $T \nvDash \phi$, $T \cup \{\neg\phi\}$ has a model. Similarly as $T \nvDash \neg\phi$, $T \cup \{\phi\}$ has a model. As these models are both infinite (by the given property of T), the full Löwenheim–Skolem theorem (Theorem 6.12 of Section 6.4) ensures that each set has a model of cardinality κ. These models are isomorphic, so by Theorem 4.8 of Section 4.5 they satisfy the same sentences. But one model satisfies the sentence $\neg\phi$ while the other satisfies ϕ, giving a contradiction as required. ∎

Thus, using Vaught's test, the theory of dense linear orders (in the language $\{<, =\}$) and the theory of infinite sets (in the language $\{=\}$) are both complete.

Knowledge that a theory T is complete can sometimes be exploited to show that it is decidable. The key extra ingredients are that the theory T is in a *countable* language L and, of crucial importance, that it has a *decidable* set of axioms, that is to say, there is an algorithmic procedure for deciding whether a sentence of L is one of the axioms for T. The countability of the language means that there is an algorithmic procedure for listing all finite sequences of formulas and then deciding which are formal derivations from assumptions – this would plainly be impossible for an uncountable language. The decidability of the set of axioms then means there is an algorithmic procedure for deciding which of these formal derivations uses these axioms as assumptions and is thus a theorem of T. It can thus be shown that there is an algorithmic procedure which lists all the theorems of T in such a way that theorem $T \vdash \phi$ appears in the list after finitely many steps. Given the extra information that T is complete, to decide whether ϕ or $\neg\phi$ is a theorem, we generate this list of theorems and as one of these is a theorem (thanks to T being complete), one of these will materialize in the list after finitely many steps. Hence if T is a complete theory in a countable language with a decidable set of axioms, then T is decidable.

If T wasn't complete, and neither of ϕ and $\neg\phi$ was a theorem of T, this listing procedure wouldn't generate either sentence, but we'd only be aware of this at the end of an infinite process. This doesn't in itself mean that T isn't decidable, as there might be a completely different procedure which shows up after finitely many steps that neither ϕ nor $\neg\phi$ is a theorem.

The theory of dense linear orders has finitely many axioms and we hope that it is plausible that there is an algorithmic procedure which will decide whether a formula is one of these axioms. Although there are countably infinitely many axioms for the theory of infinite sets, the axioms $\exists_{\geq n}$ are of a very regular shape and again we hope it is plausible that there is an algorithmic procedure which will decide whether a formula is one of these axioms. So both theories are also decidable, as well as complete.

As a further example, we shall look at a theory which attempts to axiomatize the theory of the successor function on the natural numbers $\mathbb{N}$.

Let T be the theory in the first-order language with equality with a 1-place function symbol s and a constant symbol $\mathbf{0}$ with the following axioms:

1. $\forall x \neg s(x) = \mathbf{0}$

2. $\forall x \forall y (s(x) = s(y) \rightarrow x = y)$

3. $\forall y(\neg y = \mathbf{0} \rightarrow \exists x\, y = s(x))$

4. $\forall x \neg x = s^n(x)$, for all positive integers n, where $s^n(x)$ denotes s applied to x successively n times.

$s^2(x)$ means $s(s(x))$, $s^3(x)$ means $s(s(s(x)))$ and so on.

Exercise 6.39 ───────────────────────────────────

Explain why T has no finite models.

───

One countably infinite model is $\langle \mathbb{N}, S, 0, = \rangle$, where $S \colon \mathbb{N} \longrightarrow \mathbb{N}$ is the successor function defined by $S(n) = n + 1$, for all $n \in \mathbb{N}$.

Furthermore, any model $\mathcal{A} = \langle A, S, a_0, = \rangle$ must contain the subset of distinct elements

Why are the elements $S^n(a_0)$ for $n \geq 0$ distinct?

$$\{a_0, S(a_0), S(S(a_0)), \ldots, S^n(a_0), \ldots\},$$

so contains a substructure isomorphic to $\langle \mathbb{N}, S, 0, = \rangle$. Now suppose that such a model contains an element a which isn't one of the elements $S^n(a_0)$ for some $n \geq 0$ (where $S^0(a_0) = a_0$). Then the model contains not only $S^n(a)$ for each $n \geq 1$, but (as $a \neq a_0$) an element $b \in A$ such that $S(b) = a$. Could b equal

$S^k(a_0)$ for some $k \geq 0$? No, as if $b = S^k(a_0)$, then $a = S(b) = S^{k+1}(a_0)$, contradicting our original assumption about a. Write b as $S^{-1}(a)$ – a reasonable notation because $S(S^{-1}(a)) = S(b) = a = S^0(a)$. Similarly b must equal $S(c)$ for some $c \in A$ which we shall write as $S^{-2}(a)$. We leave you to convince yourself that A contains distinct elements in the set $\{S^k(a) : k \in \mathbb{Z}\}$, which is a set disjoint from $\{S^n(a_0) : n \in \mathbb{N}\}$, so that the model A also contains a subset isomorphic to a copy of the set $\mathbb{Z}$ of integers with the '+1' function. If the model contains an element c which isn't this copy of $\mathbb{Z}$ or in the copy of $\langle \mathbb{N}, S, 0, = \rangle$, then c lives in a further copy of $\mathbb{Z}$, and so on.

Exercise 6.40

(a) Classify the countably infinite models of T (i.e. ones with a countably infinite domain). That is, give an example of each possible countably infinite model, avoiding examples which are isomorphic to earlier models in your list.

(b) Find a countably infinite model A of T such that any countable model of Σ is isomorphic to a substructure of A.

We hope that you discovered in the last exercise that any model of T consists of a substructure isomorphic to $\langle \mathbb{N}, S, 0, = \rangle$ along with disjoint copies of $\mathbb{Z}$ with the $+1$ function. The arithmetic of cardinal numbers means that the only way to construct a model of T with cardinality $2^{\aleph_0}$, the cardinality of the continuum $\mathbb{R}$, is to take the union of a copy of $\langle \mathbb{N}, S, 0, = \rangle$ with $2^{\aleph_0}$ copies of $\mathbb{Z}$ with the $+1$ function. It follows that all models of cardinality $2^{\aleph_0}$ are isomorphic. One concrete representation of the model is the structure

$$\langle \mathbb{R} \setminus \{k \in \mathbb{Z} : k < 0\}, S, 0 \rangle,$$

where S is the function defined on the domain by $s(x) = x + 1$. So by Vaught's test, T is complete. As we hope that it is plausible that there is an algorithmic procedure which will test whether a formula is one of the axioms of T, T is decidable.

The set $\mathbb{R} \setminus \{k \in \mathbb{Z} : k < 0\}$ is the set of real numbers, excluding the negative integers.

We prefaced this last example by saying that the axioms of T were an attempt to axiomatize the theory of the successor function on the natural numbers $\mathbb{N}$, by which we mean the theory $\mathrm{Th}(\langle \mathbb{N}, S, 0, = \rangle)$. Now that we know T is complete, we can conclude that our attempt was successful, as we ask you to show in the following exercise.

Exercise 6.41

Show that the set S of logical consequences of T coincides with the theory $\mathrm{Th}(\langle \mathbb{N}, S, 0, = \rangle)$.

Solution

First note that $\langle \mathbb{N}, S, 0, = \rangle$ is a model for T and hence for all its logical consequences, so that $S \subseteq \mathrm{Th}(\langle \mathbb{N}, S, 0, = \rangle)$. But T and $\mathrm{Th}(\langle \mathbb{N}, S, 0, = \rangle)$ are complete theories for the same language, so that S must equal $\mathrm{Th}(\langle \mathbb{N}, S, 0, = \rangle)$.

6 Some uses of compactness

The theory T we have just considered can be regarded as a first step towards trying to axiomatize the theory of the natural numbers. Of course, we know from Section 6.4 that the theory of the natural numbers cannot be axiomatized using a first-order language in such a way that the only model is the standard model $\mathcal{N}$. But we can investigate just how much of the theory one can axiomatize in such a language. For instance, we do have the first-order theory $\mathrm{Th}(\mathcal{N})$, complete arithmetic. As the underlying language is finite, so countable, if there was a decidable set of axioms for this complete theory, then $\mathrm{Th}(\mathcal{N})$ would be decidable. However, it can be shown that there are no such axioms. One way to establish this involves looking at the following fragment of the theory of natural numbers, the theory Q.

Definition The theory Q

The system Q has the following seven axioms expressed in a language with equality and two 2-place functions $+$ and $\cdot$, a 1-place function s and constant symbol $\mathbf{0}$.

$$\forall x \forall y (s(x) = s(y) \to x = y)$$
$$\forall x \; \neg \; \mathbf{0} = s(x)$$
$$\forall x (\neg \; x = \mathbf{0} \to \exists y \; x = s(y))$$
$$\forall x \; (x + \mathbf{0}) = x$$
$$\forall x \forall y \; (x + s(y)) = s((x + y))$$
$$\forall x \; (x \cdot \mathbf{0}) = \mathbf{0}$$
$$\forall x \forall y \; (x \cdot s(y)) = ((x \cdot y) + x)$$

This system was devised by the American mathematician Raphael Robinson (1911-1995).

All the axioms of Q are easily seen to be true in the standard model of complete arithmetic $\mathcal{N} = \langle \mathbb{N}, +, \times, 0, S \rangle$, where S is the *successor* function defined by $S(n) = n + 1$. So the theory Q is consistent. The theory is on the surface very weak, in the sense that one cannot prove a very rich selection of theorems from these axioms. Its theorems include

$$Q \vdash (s(\mathbf{0}) + s(s(\mathbf{0}))) = s(s(s(\mathbf{0}))),$$
$$Q \vdash (s(s(\mathbf{0})) + s(\mathbf{0})) = s(s(s(\mathbf{0}))),$$

and so it includes

$$Q \vdash (s(\mathbf{0}) + s(s(\mathbf{0}))) = (s(s(\mathbf{0})) + s(\mathbf{0})).$$

More generally, using the abbreviation $s^n(\mathbf{0})$ for $\underbrace{s(s(\ldots s(\mathbf{0}) \ldots))}_{n}(\mathbf{0})$ for any natural number n (with the convention that $s^0(\mathbf{0})$ stands for $\mathbf{0}$),

$$Q \vdash (s^n(\mathbf{0}) + s^m(\mathbf{0})) = (s^m(\mathbf{0}) + s^n(\mathbf{0})),$$

for any specific positive integers m, n. But it can be shown that

$$Q \nvdash \forall x \forall y (x + y) = (y + x),$$

that is, we cannot prove in Q that addition is in general commutative. But Q does have the very significant property that any consistent theory with axioms including those of Q is undecidable. So, in particular the theory $\mathrm{Th}(\mathcal{N})$, complete arithmetic, is undecidable. Another consequence is that

When we defined the standard model $\mathcal{N}$ on page 291, we did so relative to a slightly different language, but with essentially the same expressive power.

Likewise it can be shown that for any natural numbers m, n,
$$Q \vdash (s^n(\mathbf{0}) \cdot s^m(\mathbf{0})) = s^{n+m}(\mathbf{0}),$$
$$Q \vdash (s^m(\mathbf{0}) \cdot s^n(\mathbf{0})) = s^{n+m}(\mathbf{0}),$$
$$Q \vdash (s^n(\mathbf{0}) \cdot s^m(\mathbf{0}))$$
$$= (s^m(\mathbf{0}) \cdot s^n(\mathbf{0})),$$
but, despite this,
$$Q \nvdash \forall x \forall y (x \cdot y) = (y \cdot x).$$
See Exercises 6.45 and 6.46 for practice on theorems and non-theorems of Q.

See, for instance, Enderton [12] or Epstein and Carnielli [14] for details.

the first-order predicate calculus is undecidable, as follows. Let ϕ to be the conjunction of the finitely many axioms of Q. Then for any sentence θ in the same language as Q, the sentence $(\phi \to \theta)$ is decidable if and only if $\phi \vdash \theta$ is decidable, i.e. one can decide whether θ is a theorem of Q – but the latter isn't decidable, so the former isn't either.

Despite the impossibility of deciding $\mathrm{Th}(\mathcal{N})$, there is considerable value in studying fragments of the theory and we would like to mention the most important of these, which is the first-order language version of the Peano axioms. As we saw on page 286 in Section 6.4, Peano's axioms for the natural numbers give a categorical set of axioms for $\mathbb{N}$ in a language which is not first-order, and by the upward Löwenheim–Skolem theorem there is no possible set of axioms in a first-order language. It is not the first two of Peano's axioms which are a problem. If we take the first-order language L with equality, a 1-place function symbol s, the two 2-place function symbols $+$ and $\cdot$, and constant symbol $\mathbf{0}$, these axioms can be represented by

1. $\forall x \forall y(s(x) = s(y) \to x = y)$

2. $\forall x \neg \mathbf{0} = s(x)$

As remarked in Section 6.4, it is the third of Peano's axioms, the principle of mathematical induction, which presents a problem. This says

> for all subsets $A \subseteq X$, if A contains 0_X and contains $S(x)$ whenever $x \in A$, then A is all of X.

This can be rephrased in terms of the property $P(x)$ possessed by the elements x in a set A as

$P(x)$ holds exactly when $x \in A$.

> for all properties P of elements $x \in X$, if $P(0_X)$ holds and whenever $P(x)$ holds then $P(S(x))$ also holds, then $P(x)$ holds for all $x \in X$.

Quantifying over properties P is no more first-order than quantifying over subsets, as well as elements, of the domain of any interpretation. However, this second formulation does lead us to some useful first-order axioms for a fragment of the theory of the natural numbers, as follows. For each first-order formula $\phi(x)$ with one free variable x, consider the sentence

$$((\phi(\mathbf{0}) \wedge \forall x(\phi(x) \to \phi(s(x)))) \to \forall x \phi(x)).$$

This asserts the induction principle for the property represented by the first-order formula $\phi(x)$. The first-order theory with axioms consisting of all sentences of this form, i.e. corresponding to all formulas $\phi(x)$ in L with one free variable, along with the first two for the function symbol s, is given the special name *Peano Arithmetic*. Despite inevitably failing to axiomatize the natural numbers and nothing but, this theory is powerful enough to derive a considerable part of the standard mathematical theory of the natural numbers.

We shall end the section with a proof that the monadic first-order calculus is decidable. The proof has nothing to do with compactness, and illustrates that the solution of one genuinely interesting sort of problem, in this case about decidability, isn't *always* linked to some other very clever and important piece of mathematics (here, compactness). The description *monadic* just means that the language consists only of 1-place relation symbols, so there's nothing like equality in the language. The issue we consider is whether there

This is Church's theorem. The American mathematician Alonzo Church (1903–1995) was one of the key founders of the modern theory of computability and decidability.

For a very accessible account of some fragments of $\mathrm{Th}(\mathcal{N})$ which are decidable, see Enderton [12].

The failure is thanks to the Löwenheim–Skolem theorems.

is a decision procedure that establishes whether or not, for a given formula $\phi(x_1, x_2, \ldots, x_m)$ with free variables amongst $x_1, x_2, \ldots, x_m$,

$$\vDash \phi(x_1, x_2, \ldots, x_m).$$

Our procedure will be essentially one of brute force! We shall see whether for all structures $\mathcal{A}$ for the language involved in ϕ (so just finitely many relations are involved) and interpretations $\vec{a}$ of the free variables, it is the case that $\mathcal{A} \vDash_{\vec{x}/\vec{a}} \phi$. If this is always the case then ϕ is universally valid. We hope that you are highly sceptical that this will work! There are surely structures with domains of all possible sizes, finite and infinite, and on any given domain there are surely many different ways of interpreting the relation symbols and interpreting the free variables. Also testing truth for infinitely many structures and interpretations cannot give an algorithmic procedure, which must come to a conclusion in a finite number of steps. However, it turns out that there are essentially only finitely many structures, each with a finite domain and thus finitely many different ways of interpreting the free variables, for which one has to test the truth of ϕ.

Let's look at the case where ϕ involves just three 1-place relation symbols R_1, R_2, R_3. Take any structure for this language, $\mathcal{A} = \langle A, A_1, A_2, A_3 \rangle$, where the domain A might be any set, perhaps infinite. In the following sense, the language cannot tell the difference very well between distinct elements of A. All that one can say about an element a of A is whether or not it is in the subset A_i for each of $i = 1, 2, 3$. So if two distinct elements a and b have the same pattern of behaviour in regard to being in or out of each A_i, there's no formula in the language which can tell them apart – remember that the language doesn't contain any useful symbol like equality, which would at once allow a and b to be told apart by the formula $\neg\, x = y$, in the sense that we then have $\mathcal{A} \vDash_{x/a, y/b} \neg\, x = y$. More precisely we define a relation $\sim$ on A by

> Of course, as the R_i are 1-place relation symbols, the A_i are subsets of A.

> There are similarities between what follows and the discussion about non-normal structures in the Section 5.4.

$$a \sim b \text{ if and only if } \quad \text{for each } i = 1, 2, 3, \text{ both of } a \text{ and } b$$
$$\text{are in } A_i, \text{ or neither are.}$$

> So if a, b are in A_2 and A_3, while neither is in A_1, then $a \sim b$.

This is easily shown to be an equivalence relation. We then define a structure $\mathcal{B} = \langle B, B_1, B_2, B_3 \rangle$ with domain B consisting of the equivalence classes $[\![a]\!]$ of $\sim$ and with $B_i = \{[\![a]\!] : a \in A_i\}$ for each $i = 1, 2, 3$. Note that there are at most $2^3 = 8$ equivalence classes of $\sim$, as for each element a of A, there are two possibilities, a in A_i or not in it, for each of the three subsets A_1, A_2, A_3. So the set B has at most 8 elements. One can then show that for all $\phi(x_1, x_2, \ldots, x_m)$ in this language and all interpretations $a_1, a_2, \ldots, a_m$ of the free variables $x_1, x_2, \ldots, x_m$,

$$\mathcal{A} \vDash_{x_1/a_1, x_2/a_2, \ldots, x_m/a_m} \phi(x_1, x_2, \ldots, x_m)$$

if and only if

$$\mathcal{B} \vDash_{x_1/[\![a_1]\!], x_2/[\![a_2]\!], \ldots, x_m/[\![a_m]\!]} \phi(x_1, x_2, \ldots, x_m).$$

Thus rather than test whether $\phi(x_1, x_2, \ldots, x_m)$ is satisfied in the perhaps infinite set A for all possible interpretations of the variables $x_1, x_2, \ldots, x_m$, all we need to test is whether the formula is satisfied in the set B with at most 8 elements for all possible interpretations of these variables – and if B has k elements, with $k \leq 8$, there are a mere k^m such interpretations, doubtless possibly a large number, but finite!

To test for $\vDash \phi(x_1, x_2, \ldots, x_m)$, we need to look at all possible structures $\mathcal{A}$, but this now means looking only at all possible structures for the language with up to 8 elements. There are only finitely many of these, up to isomorphism. For each there are just finitely many possible interpretations of the free variables. And voila! an algorithmic decision procedure.

Plainly there are infinitely many possible domains with up to 8 elements. But there are only finitely many structures with such domains up to isomorphism; and isomorphic structures satisfy precisely the same formulas, by Theorem 4.8 of Section 4.5.

Exercise 6.42 ______________

For the language above, how many essentially different structures, i.e. not double-counting ones which are isomorphic, does it have with a two-element domain?

Solution

Take the domain to be $\{a, b\}$. The interpretation of R_1 can be any of the $2^2 = 4$ subsets of the domain. Similarly the interpretation of each of R_2 and R_3 can be any of the 4 subsets of the domain. So the number of distinct structures with this domain is $4 \times 4 \times 4 = 64$. But some of these are isomorphic! For instance the structures

$$\langle \{a, b\}, \{a\}, \varnothing, \{a, b\} \rangle$$

and

$$\langle \{a, b\}, \{b\}, \varnothing, \{a, b\} \rangle$$

are isomorphic, via the isomorphism θ, where $\theta(a) = b$ and $\theta(b) = a$. We think that there are 36 distinct structures up to isomorphism.

An example of a structure not isomorphic to this one is $\langle \{a, b\}, \varnothing, \{a\}, \{a, b\} \rangle$.

____

Perhaps you would like to check the details in a slightly more general setting, with n 1-place relation symbols, in the following exercise.

Exercise 6.43 ______________

Let L be a language (without equality) consisting only of finitely many 1-place relation symbols $P_1, P_2, \ldots, P_n$.

(a) Let $\mathcal{A} = \langle A, A_1, A_2, \ldots, A_n \rangle$ be a structure for L. Define a binary relation $\sim$ on the domain A by

$$a \sim b \text{ if and only if} \quad \text{for each } i \in \{1, 2, \ldots, n\}, \text{ both of } a \text{ and } b$$
$$\text{are in } A_i, \text{ or neither are.}$$

Show that $\sim$ is an equivalence relation on A.

(b) Let B be the set of equivalence classes of $\sim$ on A, i.e. $B = \{[\![a]\!] : a \in A\}$. For each $i \in \{1, 2, \ldots, n\}$, define the subset B_i of B by $B_i = \{[\![a]\!] : a \in A_i\}$. Now let $\mathcal{B}$ be the structure $\langle B, B_1, B_2, \ldots, B_n \rangle$.

   (i) Explain why B is a finite set and give an upper bound for its size.

   (ii) You have shown that the domain B of $\mathcal{B}$ is finite. Explain briefly why for any formula $\phi(x_1, x_2, \ldots, x_m)$ with free variables amongst $x_1, x_2, \ldots, x_m$ and $b_1, b_2, \ldots, b_m \in B$, there is a procedure involving finitely many steps that establishes whether or not

$$\mathcal{B} \vDash_{x_1/b_1, x_2/b_2, \ldots, x_m/b_m} \phi(x_1, x_2, \ldots, x_m).$$

This is a result that holds for all languages, not just the special sort which we are considering here.

(iii) Show that for each $i \in \{1, 2, \ldots, n\}$ and all $a \in A$,

$$a \in A_i \text{ if and only if } [\![a]\!] \in B_i.$$

(iv) Show that for any formula $\phi(x_1, x_2, \ldots, x_m)$ with free variables amongst $x_1, x_2, \ldots, x_m$ and all $a_1, a_2, \ldots, a_m \in A$,

$$\mathcal{A} \vDash_{x_1/a_1, x_2/a_2, \ldots, x_m/a_m} \phi(x_1, x_2, \ldots, x_m)$$

if and only if

$$\mathcal{B} \vDash_{x_1/[\![a_1]\!], x_2/[\![a_2]\!], \ldots, x_m/[\![a_m]\!]} \phi(x_1, x_2, \ldots, x_m).$$

[*Hint*: What is the basic technique for proving something for all formulas ϕ?]

(c) By exploiting relevant previous parts of this exercise, describe an algorithmic procedure which decides whether or not, for a given formula $\phi(x_1, x_2, \ldots, x_m)$,

$$\vDash \phi(x_1, x_2, \ldots, x_m),$$

i.e. $\phi(x_1, x_2, \ldots, x_m)$ holds under all possible interpretations.

(d) Is the following sentence using the 1-place relation symbols R_1, R_2 universally valid?

$$\forall x_3 \big((\forall x_1 (R_1(x_1) \wedge R_2(x_1)) \vee \neg R_2(x_3))$$
$$\to (\exists x_2 (R_1(x_2) \leftrightarrow \neg R_2(x_2)) \vee \exists x_4 (R_2(x_4) \wedge \neg R_1(x_3)))\big)$$

A subtle point! Plainly, by definition, if $a \in A_i$ then $[\![a]\!] \in B_i$. But a particular property of $\sim$ is needed to guarantee that the converse holds.

You might like to think briefly why this technique for deciding formulas in a language with only 1-place relation symbols is likely to be of no use for a language with as little as just one binary relation symbol. In fact, the predicate calculus with just one binary relation symbol is undecidable. But that shouldn't deter you from further study into first-order theories and model theory. Good hunting!

Further exercises

Exercise 6.44

(a) Devise axioms for each of the following theories:

(i) dense linear order with a minimum element and unbounded above;

(ii) dense linear order with a maximum element and unbounded below;

(iii) dense linear order with distinct minimum and maximum elements.

(b) Show that each of the above theories has a countable model.

(c) Explain briefly how to adapt the argument of Exercise 4.112 in Section 4.5 of Chapter 4 to show that for each of the above theories, all countable models are isomorphic.

(d) By Theorem 6.15, Vaught's test, each of the above theories is complete. You should have been able to find just finitely many axioms for each of the theories, so each theory is decidable, as is the theory of unbounded dense linear orders (by Theorem 4.9).

(i) Devise axioms for the theory of dense linear orders with at least two elements.

Treating a one-element set as a dense linear order seems a bit silly!

(ii) Is it the case that any two countable models of this theory are isomorphic?

(iii) Show that this theory is decidable by outlining an algorithmic procedure for deciding whether a sentence is one of its theorems.

Exercise 6.45

Give formal proofs of the following theorems of Q.

(a) $Q \vdash \neg s(\mathbf{0}) = s(s(s(\mathbf{0})))$

(b) $Q \vdash (s(\mathbf{0}) + s(\mathbf{0})) = s(s(\mathbf{0}))$

(c) $Q \vdash (s(\mathbf{0}) + s(s(\mathbf{0}))) = (s(s(\mathbf{0})) + s(\mathbf{0}))$

(d) $Q \vdash (s(s(s(\mathbf{0}))) \cdot s(s(\mathbf{0}))) = s(s(s(s(s(s(\mathbf{0}))))))$

Exercise 6.46

A non-standard structure $\mathcal{N}^*$ for the language underlying Q is as follows. Its domain is the set $\mathbb{N} \cup \{\alpha, \beta\}$, where α, β are distinct elements not in $\mathbb{N}$. The symbol $\mathbf{0}$ is interpreted as the number 0, and the functions which are the interpretations of the symbols $s + \cdot$ are given by the following tables.

x	0 1 2 $\ldots$ n $\ldots$ α β
$S(x)$	1 2 3 $\ldots$ $n+1$ $\ldots$ α β

$+$	0	1	2	$\ldots$	n	$\ldots$	α	β
				b				
0	0	1	2	$\ldots$	n	$\ldots$	β	α
1	1	2	3	$\ldots$	$1+n$	$\ldots$	β	α
2	2	3	4	$\ldots$	$2+n$	$\ldots$	β	α
$\vdots$	$\vdots$	$\vdots$	$\vdots$		$\vdots$		$\vdots$	$\vdots$
m	m	$m+1$	$m+2$		$m+n$		β	α
$\vdots$	$\vdots$	$\vdots$	$\vdots$		$\vdots$		$\vdots$	$\vdots$
α	α	α	α	$\ldots$	α	$\ldots$	β	α
β	β	β	β	$\ldots$	β	$\ldots$	β	α

(Row-label column is a.)

$\cdot$	0	1	2	$\ldots$	n	$\ldots$	α	β
				b				
0	0	0	0	$\ldots$	0	$\ldots$	α	β
1	0	1	2	$\ldots$	n	$\ldots$	α	β
2	0	2	4	$\ldots$	$2n$	$\ldots$	α	β
$\vdots$	$\vdots$	$\vdots$	$\vdots$		$\vdots$		$\vdots$	$\vdots$
m	0	m	$2m$		mn	$\ldots$	α	β
$\vdots$	$\vdots$	$\vdots$	$\vdots$		$\vdots$		$\vdots$	$\vdots$
α	0	β	β	$\ldots$	β	$\ldots$	β	β
β	0	α	α	$\ldots$	α	$\ldots$	α	α

(Row-label column is a.)

(a) Show that $\mathcal{N}^*$ is a model of Q.

(b) Show each of the following.

    (i) $Q \nvdash \forall x \, \neg x = s(x)$

    (ii) $Q \nvdash \forall x \forall y (x + y) = (y + x)$

    (iii) $Q \nvdash \forall x \forall y (x \cdot y) = (y \cdot x)$

BIBLIOGRAPHY

1. Reg Allenby *Rings, Fields and Groups*, Arnold, 1983.

2. George Boole *The Laws of Thought*, Dover, 1973.

3. Jane Bridge *Beginning Model Theory*, Oxford University Press, 1977.

4. Peter J. Cameron *Sets, Logic and categories*, Springer, 1999.

5. Lewis Carroll *Symbolic Logic*, Dover, 1958.

6. C.C.Chang and H.J.Keisler *Model Theory*, North-Holland, 1990.

7. René Cori and Daniel Lascar *Mathematical Logic: a Course with Exercises, Part I*, trans Donald Pelletier, Oxford University Press, 2000.

8. René Cori and Daniel Lascar *Mathematical Logic: a Course with Exercises, Part II*, trans Donald Pelletier, Oxford University Press, 2001.

9. Nigel Cutland *Computability: An Introduction to Recursive Function Theory*, Cambridge University Press, 1980.

10. Martin Davis *The Universal Computer*, Norton, 2000.

11. Michael Dummett *Elements of Intuitionism*, Oxford University Press, 1977.

12. Herbert B. Enderton *A Mathematical Introduction to Logic*, Academic Press, 1972.

13. Herbert B. Enderton *Elements of Set Theory*, Academic Press, 1977.

14. Richard Epstein, Walter Carnielli *Computability: Computable Functions, Logic and the Foundations of Mathematics*, Wadsworth, 1999.

15. Gottlob Frege *Grundlagen der Arithmetik (Foundations of Arithmetic)*, Translated by JL Austin, Blackwell Publishers, 1980.

16. Derek Goldrei *Classic Set Theory*, Chapman and Hall, 1996.

17. Paul R. Halmos *Naive Set Theory*, Van Nostrand, 1960.

18. A.G.Hamilton *Logic for Mathematicians*, Cambridge University Press, 1978.

19. Jean van Heijenoort *From Frege to Gödel, a Source Book in Mathematical Logic 1879–1931*, Harvard University Press, 1967.

20. Wilfrid Hodges *A Shorter Model Theory*, Cambridge University Press, 1997.

21. Camilla Jordan and David Jordan *Groups*, Arnold, 1994.

22. Stephen Cole Kleene, *Mathematical Logic*, Wiley, 1967.

23. Walter Ledermann, Alan Weir *Introduction to Group Theory*, Longman, 1996.

24. David Marker, *Model Theory: an Introduction*, Springer, 2002.

25. Elliott Mendelson *Introduction to Mathematical Logic*, Van Nostrand, 1964.

Bibliography

26. Abraham Robinson *Introduction to Model Theory and to the Metamathe-matics of Algebra* 2nd edition, North Holland, 1965.

27. Abraham Robinson *Non-standard Analysis*, North Holland, 1966.

28. Raymond Smullyan *What is the Name of this Book?*, Penguin, 1990.

29. Raymond Smullyan *To Mock a Mockingbird: And Other Logic Puzzles*, Oxford University Press, 2000.

30. Robin Wilson *Introduction to Graph Theory*, Longman 1996.

31. Robin Wilson *Four Colours Suffice*, Allen Lane (Penguin) Books, 2002.

INDEX

GPSR Compliance
The European Union's (EU) General Product Safety Regulation (GPSR) is a set
of rules that requires consumer products to be safe and our obligations to
ensure this.

If you have any concerns about our products, you can contact us on

ProductSafety@springernature.com

In case Publisher is established outside the EU, the EU authorized
representative is:

Springer Nature Customer Service Center GmbH
Europaplatz 3
69115 Heidelberg, Germany